Universitext

Universitext

Universitext is a series of textbooks that presents material from a wide variety of mathematical disciplines at master's level and beyond. The books, often well class-tested by their author, may have an informal, personal even experimental approach to their subject matter. Some of the most successful and established books in the series have evolved through several editions, always following the evolution of teaching curricula, into very polished texts.

Thus as research topics trickle down into graduate-level teaching, first textbooks written for new, cutting-edge courses may make their way into *Universitext.*

More information about this series at http://www.springer.com/series/223

David A. Craven

Representation Theory of Finite Groups: a Guidebook

David A. Craven
School of Mathematics
University of Birmingham
United Kingdom

ISSN 0172-5939 ISSN 2191-6675 (electronic)
Universitext
ISBN 978-3-030-21791-4 ISBN 978-3-030-21792-1 (eBook)
https://doi.org/10.1007/978-3-030-21792-1

Mathematics Subject Classification (2010): 20C20, 20C15, 20C30, 20C33, 20C08

This Springer imprint is published by the registered company Springer Nature Switzerland AG.
The registered company address is: Gewerbestrasse 11, 6330 Cham, Switzerland

Preface

The representation theory of finite groups has, at its core, a collection of open problems. Taken together, they are called 'local-global conjectures', although 'local-global rough guesses' might be more appropriate. The missing pieces of the representation-theoretic puzzle lie right in the centre, leaving us with corners and edges to fit together.

Many of the conjectures are numerical in nature, telling us that there is some exquisite underlying structural theory, but we cannot yet perceive it. The numerical results are attractive in their own right, but it is the potential hidden structure that makes the subject so captivating. One of the main purposes of this text is to make this field accessible to anyone with a basic understanding of representation theory, such as that provided by an undergraduate course in it, and some general knowledge of algebra.

The local-global conjectures might form the centrepiece of finite group representation theory, but there is much else on the table around them: the theory of blocks with a cyclic defect group, the representation theory of the symmetric groups, and of the groups of Lie type, the structure of tame blocks, and much more besides.

The aim of this book is to give the reader a guided tour of these areas. Like a guide book, we seek to uncover the highlights of the field; proofs are more or less completely excised from this text, although the intrepid reader, using this as a springboard for further study, will be delighted to find that there are literally hundreds of references to consult for more details.

Of course, like any guide book, it starts to become out of date once it is written. This will set in aspic the state of the theory at the early part of 2019, but I believe that much of this theory will remain relevant for decades to come.

The first chapter quickly summarizes what is needed about representation theory, and the second and third chapters go through the basic building blocks of the theory, namely the character theory and the module theory.

The remaining chapters are more or less independent of one another, though there will be cross referencing. This should allow readers to dip in and out of the book as and when they want to know about a specific aspect of the theory.

The local-global conjectures occupy Chap. 4, the well-developed theory of blocks with cyclic defect groups following it in Chap. 5. Chapter 6 attempts to give some rough ideas about the structure of blocks with non-cyclic defect groups, although very little is known in this case. The next chapter deals with Clifford theory, the relationship between modules for a group and a normal subgroup. The last two chapters give some of the representation theory of symmetric groups and of the groups of Lie type.

I also have to thank people for helping find some of the many errors that occurred in previous versions of this text. In particular, I thank David Benson, Norman McGregor, John Murray, Gabriel Navarro, Jeremy Rickard, Geoffrey Robinson, Adam Thomas, Mark Wildon, and especially Gunter Malle for their suggestions.

Birmingham, UK David A. Craven

Contents

Chapter 1
The Basics

This book aims to introduce the reader to the modern representation theory of finite groups. To do so we need some of the basics of representation theory, which we will rather skim through, mostly to fix notation. We are going to assume that the reader is more or less familiar with this, and might need a refresher only. We also include a rather rushed description of the group theory we will need for this book, at least to understand the chapters that are not devoted to particular groups, most notably Chaps. 4 and 9.

1.1 Representation Theory

General references for the basic representation theory of finite groups include [313, 328, 388], although any textbook on representation theory or set of lecture notes from an advanced undergraduate/beginning graduate course will do. In particular, I personally have always found [328] an unparalleled first introduction to representations of finite groups.

Let k denote a field, which will have characteristic p (p is allowed to be 0). We will almost always consider k to be algebraically closed. A *representation* of a group G is a homomorphism $G \to \mathrm{GL}_n(k)$. This is equivalent to the notion of a (finite-dimensional) kG-module, which we define now. The *group algebra* kG of a group G over a field k is the k-algebra consisting of all formal finite linear combinations of group elements with coefficients in k (so has a basis consisting of the group elements) and with addition and multiplication given by the obvious linear actions. A right *kG-module* is a right module for the group algebra kG. We can simply think of this as a vector space with an action of G on it. In this book, all of our modules will be right modules, and finite-dimensional unless we explicitly mention otherwise. Of course, it is just a convention to use right modules, and all our theorems will hold true for left modules, except perhaps with a change of ordering

D. A. Craven, *Representation Theory of Finite Groups: a Guidebook*, Universitext,
https://doi.org/10.1007/978-3-030-21792-1_1

in some formulae.[1] A generalization of this is a ‘bimodule’. If G and H are both groups, then a *(kG, kH)-bimodule* is a vector space M that is a left kG-module, a right kH-module, and such that the two actions are compatible, so $(gm)h = g(mh)$.

If M is a kG-module then a *submodule* is a G-invariant subspace of M. If M is a kG-module and N is a submodule of M, denoted by $N \leq M$, then we can form the *quotient module* M/N, which is the standard quotient vector space, with the obvious inherited group action. A *kG-module homomorphism* $\phi : M \to N$ for two kG-modules M and N is a linear transformation between the underlying vector spaces such that, for all $m \in M$ and $g \in G$,

$$(mg)\phi = (m\phi)g.$$

(Of course, a module isomorphism is a bijective module homomorphism.)

A non-zero kG-module M is *simple* or *irreducible* if the only G-invariant subspaces of M are 0 and M. If M_1 and M_2 are kG-modules, then the *direct sum* of M_1 and M_2, denoted by $M_1 \oplus M_2$, is the vector space $M_1 \oplus M_2$ with the obvious group action, and the M_i are called *summands* of $M_1 \oplus M_2$; we write $M \mid N$ if M is a summand of N. Write $M^{\oplus n}$ for the n-fold direct sum of M with itself. If a kG-module M cannot be written as $M = M_1 \oplus M_2$ with each M_i non-zero, then M is said to be *indecomposable*. By the Krull–Schmidt theorem, if M is any finite-dimensional kG-module then the decomposition

$$M = M_1 \oplus M_2 \oplus \cdots \oplus M_r$$

into indecomposable summands is unique, in the sense that if

$$M = N_1 \oplus N_2 \oplus \cdots \oplus N_s$$

is another such decomposition, then $r = s$ and up to reordering $M_i \cong N_i$. Note that in general we cannot say that $M_i = N_i$, merely that they are isomorphic.

Similar to this are composition factors. A *composition factor* of a kG-module M is a simple module V such that there exist submodules $M_2 \leq M_1 \leq M$ with $V \cong M_1/M_2$. If M is finite-dimensional then the *composition factors* of M are the simple modules M_i/M_{i-1}, where

$$0 = M_0 < M_1 < M_2 < \cdots < M_r = M$$

[1] The reason I use right modules is my background in group theory, where most practitioners act on the right. Most representation theorists use left modules, however, as they often don't start off as finite group theorists. As I almost never have to write explicit maps down, this should not be much of an issue, but if you violently prefer left modules, you can either pencil in the maps yourself or use a mirror.

is a series for which M_i/M_{i-1} is simple for all $1 \leq i \leq r$, called a *composition series* for M. By the Jordan–Hölder theorem, the composition factors of a module (including multiplicities) do not depend on the composition series chosen.

If U and V are kG-modules then define the *tensor product* $U \otimes V$ of U and V to be the module with basis the symbols $u_i \otimes v_j$ for $\{u_i\}$ a basis of U and $\{v_j\}$ a basis of V, with addition linear in both variables (i.e., $(u_1 + u_2) \otimes v = u_1 \otimes v + u_2 \otimes v$ and so on), and such that

$$(u_i \otimes v_j)g = u_i g \otimes v_j g.$$

Write $V^{\otimes n}$ for the n-fold tensor product of V with itself. If U and V are two kG-modules, and H is a subgroup of G, then

$$\mathrm{Hom}_{kH}(U, V)$$

is the set of all kH-module homomorphisms from U to V. Note that this is an abelian group, in fact a k-vector space, under addition: $u(\phi + \psi) = u\phi + u\psi$ for $\phi, \psi \in \mathrm{Hom}_{kH}(U, V)$. In addition, if H is a normal subgroup of G then $\mathrm{Hom}_{kH}(U, V)$ becomes a kG-module under the map

$$\phi^g : u \mapsto \left((ug^{-1})\phi\right) g.$$

In particular, if U is a kG-module then the *dual* of U, denoted U^*, is the kG-module $\mathrm{Hom}_k(U, k)$. It is often conceptually easier to consider the dual representation: if we fix a basis of U, hence giving a map $\phi : G \to \mathrm{GL}_{\dim(U)}(k)$, then the dual is given by taking the inverse transpose of the matrix $g\phi$ for $g \in G$. The set of kH-module *endomorphisms*, i.e., homomorphisms from U to itself, is denoted by $\mathrm{End}_{kH}(U)$.

If M is a kG-module then there is an important direct sum decomposition of $M \otimes M$, at least if the characteristic of k is not 2. Define the *symmetric square* $S^2(M)$ of M to be the quotient of $M \otimes M$ by the submodule generated by all expressions of the form

$$m \otimes n - n \otimes m$$

for $m, n \in M$. Similarly, the *exterior square* $\Lambda^2(M)$ of M is the quotient of $M \otimes M$ by the submodule generated by

$$m \otimes n + n \otimes m$$

for $m, n \in V$. If $\mathrm{char}\, k \neq 2$, then $m \otimes m$ lies in this submodule, whereas for $\mathrm{char}\, k = 2$ it does not; thus if $\mathrm{char}\, k = 2$ we include $m \otimes m$ in the submodule with quotient $\Lambda^2(M)$. We see that if k does not have characteristic 2 then

$$M \otimes M \cong S^2(M) \oplus \Lambda^2(M),$$

and in fact the kernels of the two quotients intersect trivially. If $\operatorname{char} k = 2$ however, the exterior square is a quotient of the symmetric square.

Generalizing this, the *nth symmetric power* $S^n(M)$ is the quotient of $M^{\otimes n}$ by the submodule generated by

$$(m_1 \otimes \cdots \otimes m_n) - (m'_1 \otimes \cdots \otimes m'_n), \tag{1.1}$$

where $m_i = m'_i$ except for two indices a and b, and for these $m_a = m'_b$ and $m_b = m'_a$ (so two entries in the tensor product are swapped). The *nth exterior power* has the same definition except with a '+' sign in (1.1). (Again, if the characteristic is 2, we have to include all elements

$$m_1 \otimes m_2 \otimes \cdots \otimes m_n$$

where $m_i = m_j$ for some $1 \leq i < j \leq n$, as these are not in the submodule in characteristic 2.) If $p > n$ then $S^n(M)$ and $\Lambda^n(M)$ are summands of $M^{\otimes n}$.

If M is a simple module over k, then M is *absolutely simple* if M remains simple over any extension field of k, and in particular an algebraic closure. For example, if $G = C_4$ is cyclic of order 4 and $k = \mathbb{Q}$ then there are three simple modules, two of dimension 1 and one of dimension 2. The 2-dimensional module is simple but not absolutely simple, as over the extension field $\mathbb{Q}(\sqrt{-1})$ it splits as the sum of two simple modules. If M is an absolutely simple kG-module then *Schur's lemma* states that $\operatorname{Hom}_{kG}(M, M)$ is 1-dimensional, and consists solely of scalar linear transformations. As k is assumed to be algebraically closed we will not need to worry about this, but the fact that k is algebraically closed is sometimes absolutely necessary.

The concepts of tensor products and homomorphisms are related by the statement

$$\operatorname{Hom}_k(A \otimes B, C) \cong \operatorname{Hom}_k(A, B^* \otimes C) \cong \operatorname{Hom}_k(A, \operatorname{Hom}_k(B, C))$$

for kG-modules A, B, C, called *tensor-hom adjunction*, which hides a deeper category-theoretic statement about functors being adjoint. In particular, we see that $M \cong \operatorname{Hom}_k(k, M)$, which it might be useful to check directly.

If H is a subgroup of G and M is a kG-module then the *restriction* of M to H is simply the kH-module M, i.e., the same vector space with the inherited action of H on it. In this book we write $M{\downarrow_H}$ for the restriction, but some authors use $\operatorname{Res}_H^G(M)$ or $M|_H$, and a few use M_H. Along with the concept of restriction comes the concept of induction, which is much more difficult to define. The construction is formally a tensor product: if M is a kH-module for $H \leq G$, then the *induction* of M to G is the module $M \otimes_{kH} kG$. This is perhaps easier to see with a direct construction. We consider the vector space

$$N = \bigoplus_{i=1}^{n} M g_i,$$

where $|G : H| = n$, $\{g_1, \ldots, g_n\}$ is a right transversal to H in G, and Mg_i is an isomorphic copy of M. Since the g_i form a right transversal to H in G, there exists a permutation $\sigma \in S_n$ such that $g_i g = h_i g_{i\sigma}$ for some $h_i \in H$. (If you have ever read about the transfer in group theory then this idea should look very familiar.) We then define the action of $g \in G$ on N by

$$\sum_{i=1}^{n}(m_i g_i) \cdot g = \sum_{i=1}^{n}(m_i h_i) g_{i\sigma}.$$

To emphasize its relationship to restriction, we write $M{\uparrow}^G$ for the induction of M to G. Again, some authors use $\mathrm{Ind}_H^G(M)$, and a few use M^G. This can be particularly confusing as M^G often means the set of G-invariant elements of M, and indeed will do when we talk about cohomology in Chap. 3. (Some people use $C_M(G)$ for the fixed-point set M^G.) Induction and restriction are both transitive; i.e., if $L \leq H \leq G$, M is a kL-module and N is a kG-module, then $(M{\uparrow}^H){\uparrow}^G = M{\uparrow}^G$ and $(N{\downarrow}_H){\downarrow}_L = N{\downarrow}_L$.

Induction and restriction are dual to one another by the *Nakayama relation*

$$\mathrm{Hom}_{kH}(N, M{\downarrow}_H) \cong \mathrm{Hom}_{kG}(N{\uparrow}^G, M),$$

where M is a kG-module and N is a kH-module. Since for group algebras $\mathrm{Hom}_{kG}(M, N) \cong \mathrm{Hom}_{kG}(N^*, M^*)$, and taking duals commutes with induction and restriction, we get the dual relation

$$\mathrm{Hom}_{kH}(M{\downarrow}_H, N) \cong \mathrm{Hom}_{kG}(M, N{\uparrow}^G).$$

Another construction to obtain one module from another is inflation. If G is a group and H is a normal subgroup of G, then the *inflation* of a $k(G/H)$-module M is the kG-module $\bar{M}$ whose underlying vector space is the same as M, and with group action $\cdot$ given by

$$m \cdot g = m(Hg)$$

for all $m \in M$ and $g \in G$, where $m(Hg)$ is the group action for M. It is easier to explain with representations: a representation $G/H \to \mathrm{GL}(M)$ yields a representation $G \to \mathrm{GL}(M)$ via composition: $G \to G/H \to GL(M)$, where the first map is the natural quotient map. Some authors,[2] notably [328], call this the 'lift' of M, but for us a lift is a different operation (see just after Example 3.2.7). A common notation for inflation is $\mathrm{Inf}_{G/H}^G(-)$. Often we conflate a module and its inflation.

[2] This included me until I wrote this book, because I learned representation theory from [328] and I never needed to discuss inflation and lifting in the same piece of work until now.

If G is a permutation group on a set X, and k is any field, we can construct the *permutation module*, the module kX, with basis as a vector space given by X, and given the obvious G-action, extended by linearity. In particular, G always possesses the regular permutation representation on G given by $x \mapsto xg$, and the permutation module in this case is called the *regular module*. This is simply the group algebra kG, thought of as a module over itself.

We now recount the standard representation theory over $\mathbb{C}$. If $\rho : G \to \mathrm{GL}_n(\mathbb{C})$ is a representation, then the *character* afforded by that representation is the function

$$\chi : G \to \mathbb{C}, \qquad \chi(x) = \mathrm{tr}(x\rho),$$

i.e., the trace of the matrix representing x. Since the traces of the matrices $A^{-1}BA$ and B are equal, the character of a representation does not depend on the basis chosen for the space. This also shows that characters are class functions, and indeed the irreducible complex characters, which are also called irreducible *ordinary* characters, form a basis for the vector space of class functions $G \to \mathbb{C}$. In particular, there are the same number of irreducible ordinary characters as conjugacy classes of G.

If χ and ϕ are two characters, then the inner product

$$\langle \chi, \phi \rangle = \frac{1}{|G|} \sum_{x \in G} \chi(x)\overline{\phi(x)} = \frac{1}{|G|} \sum_{x \in G} \chi(x)\phi(x^{-1})$$

(where $\bar{z}$ is the complex conjugate of the complex number z) turns the space of class functions into an inner product space, with the irreducible ordinary characters being an orthonormal basis for the space, i.e., the *row orthogonality relation*

$$\langle \chi, \phi \rangle = \frac{1}{|G|} \sum_{x \in G} \chi(x)\overline{\phi(x)} = \delta_{\chi,\phi}$$

for irreducible characters χ, ϕ holds. There is also the *column orthogonality relation*

$$\sum_{\chi \in \mathrm{Irr}(G)} \chi(x)\overline{\chi(y)} = \begin{cases} |C_G(x)| & x \text{ and } y \text{ are conjugate,} \\ 0 & \text{otherwise,} \end{cases}$$

where $\mathrm{Irr}(G)$ is the set of irreducible complex characters of G. Notice that $\langle \chi, \phi \rangle = \langle \phi, \chi \rangle$ for all complex characters χ and ϕ. The irreducible characters are often placed in a table, called the *character table*, with the irreducible characters of G labelling the rows, the conjugacy classes of G labelling the columns, and the value of the character on that conjugacy class as the entry in that row and column. Since the number of irreducible characters equals the number of conjugacy classes, this character table is square.

If the characteristic of the field k does not divide the order of the group, in particular if $\operatorname{char} k = 0$, then *Maschke's theorem* states that every kG-module is semisimple, i.e., if M is a kG-module then

$$M \cong M_1 \oplus M_2 \oplus \cdots \oplus M_r,$$

where the M_i are irreducible modules. If $k = \mathbb{C}$ therefore, there is a bijection between the isomorphism classes of kG-modules and the ordinary characters, given by taking the traces of the corresponding representation. If M is a kG-module with character χ, then write χ^* for the character of the dual module M^*.

When $k = \mathbb{C}$, the Nakayama relations above become *Frobenius reciprocity*: if χ is a character of G and ϕ is a character of H, then

$$\langle \chi{\downarrow_H}, \phi \rangle = \langle \chi, \phi{\uparrow^G} \rangle, \qquad \langle \phi{\uparrow^G}, \chi \rangle = \langle \phi, \chi{\downarrow_H} \rangle.$$

Applying Frobenius reciprocity to the regular character (the character of the regular module $\mathbb{C}G$), which is the trivial character for the trivial module induced to G, we see that, as a $\mathbb{C}G$-module,

$$\mathbb{C}G \cong \bigoplus_{V \in \operatorname{Irr}(G)} V^{\oplus \dim(V)},$$

where we have abused our notation slightly and allowed $\operatorname{Irr}(G)$ to denote both the irreducible ordinary characters and the irreducible $\mathbb{C}G$-modules. This decomposition of $\mathbb{C}G$ as a $\mathbb{C}G$-module is closely connected to the decomposition of $\mathbb{C}G$ as an algebra, namely

$$\mathbb{C}G \cong \bigoplus_{V \in \operatorname{Irr}(G)} M_{\dim(V)}(\mathbb{C}),$$

where $M_n(\mathbb{C})$ is the ring of $n \times n$ matrices over $\mathbb{C}$. This is called the *Artin–Wedderburn theorem*. The degree $\chi(1)$ for χ an ordinary character always divides $|G|$.

1.2 Group Theory

We should also talk a little bit about the groups that we will see in this book. Some of this is very standard undergraduate fare, some a bit more specialized.

If H is a subgroup of G, we will use $C_G(H)$ and $N_G(H)$ for the centralizer and normalizer of H in G respectively, $\operatorname{Aut}_G(H)$ for the set of *automorphisms* (isomorphisms from a group to itself) of H induced by elements of $N_G(H)$ (so $\operatorname{Aut}_G(H) \cong N_G(H)/C_G(H)$), and our conjugation will be the 'group theorists' convention' of left-to-right, so $x^g = g^{-1}xg$.[3] We write $Z(G)$ for the centre of G, and G' for the derived subgroup of G, generated by all elements $[g,h] = g^{-1}h^{-1}gh$ for $g, h \in G$. A group is *perfect* if $G = G'$. Two elements *commute* if $[g,h] = 1$, i.e., $gh = hg$.

If π is a set of primes, then a π*-group* is a group whose order is only divisible by primes in π, and if $\pi = \{p\}$ then we call a $\{p\}$-group simply a p*-group*. If π is a set of primes, write π' for all primes not in π. An element $g \in G$ is a π*-element* if the subgroup $\langle g \rangle$ of G is a π-group, and of course we again abbreviate and say p-element and p'-element.

The direct product $G \times H$ of two groups has both an external and an internal characterization: the external description is as pairs $(g,h) \in G \times H$ (here $G \times H$ is a set) with multiplication defined by $(g,h)(g',h') = (gg',hh')$ for $g, g' \in G$ and $h, h' \in H$. The internal description is that $E \cong G \times H$ if there are normal subgroups $G, H \trianglelefteq E$ such that E is generated by G and H and $G \cap H = 1$. A generalization of this is the *semidirect product* $G \rtimes H$. The internal characterization is easiest: here $E \cong G \rtimes H$ if there is a normal subgroup G and a subgroup H of E such that E is generated by G and H, and $G \cap H = 1$ (so we drop the normality condition on H). For the external characterization, we need a homomorphism $\phi : H \to \mathrm{Aut}(G)$, where $\mathrm{Aut}(G)$ denotes the set of all automorphisms of G. The semidirect product $G \rtimes_\phi H$ is the set of all pairs $(g,h) \in G \times H$ (here $G \times H$ is again a set) with multiplication defined this time by

$$(g,h)(g',h') = (g^{h'\phi}g', hh').$$

We assume that the reader is aware that every finite abelian group is the direct product of cyclic groups, and this gives us a large source of p-groups. In particular, it gives us *elementary abelian* p-groups, which are direct products of the cyclic group C_p of order p. The group $C_2 \times C_2$ is normally called the *Klein four group* and denoted V_4. If an elementary abelian group has order p^r, it has *rank* r. The p*-rank* of G is the largest rank of an elementary abelian subgroup of G, and the *sectional* p*-rank* of G is the maximum of the p-rank of all quotients H/K for $1 \leq K \trianglelefteq H \leq G$ (such groups H/K are called *sections* of G, hence the name).

We will need some non-abelian p-groups though, and so define three families of 2-groups that will appear in various sections of the book. The most basic are the *dihedral groups*

$$D_{2n} = \langle x, y \mid x^n = y^2 = 1,\ x^y = x^{-1} \rangle,$$

[3]Group theorists, and particularly finite group theorists, tend to compose group elements from left to right, so $(1,2)(1,3) = (1,2,3)$, but people who come from elsewhere in mathematics tend to do it the other way round, presumably guided by Euler's notation of $f(x)$ for a function. As with the previous footnote, those whose disagreement is profound can swap the '2' and '3' in '$(1,2,3)$'.

which have order $2n$, and are of course 2-groups if n is a power of 2. Another presentation of D_{2n} is

$$D_{2n} = \langle x, y \mid x^2 = y^2 = (xy)^n = 1 \rangle,$$

demonstrating that dihedral groups are the groups generated by two elements of order 2. Elements of order 2 in a group are called *involutions*.

The next easiest are *quaternion groups*,[4] which look like dihedral groups except that y has order 4, not 2. In order to make this work we have to make y^2 a power of x, so we get

$$Q_{2n} = \langle x, y \mid x^n = 1, y^2 = x^{n/2}, x^y = x^{-1} \rangle,$$

where we require that n is even, and again this is a 2-group if n is a power of 2. If instead of tweaking the order of y we tweak the action of y on x, we get the *semidihedral groups*,[5]

$$SD_{2^n} = \langle x, y \mid x^{2^{n-1}} = y^2 = 1, x^y = x^{2^{n-2}-1} \rangle.$$

A group G is nilpotent of class 1 if G is abelian, nilpotent of class c if $G/Z(G)$ is nilpotent of class $c-1$, and *nilpotent* if it is nilpotent of class c for some c. A finite group is nilpotent if and only if it is a direct product of p-groups for various primes p, so in particular p-groups are nilpotent. If G is of order p^n, then the nilpotency class of G lies between 1 and $n-1$, and it is said to be of *maximal class* if it has class $n-1$. The 2-groups of maximal class are the dihedral, semidihedral and quaternion 2-groups (and this is why they appear so often in statements of theorems). If G is a finite p-group and it does not have a subgroup $C_p \times C_p$, then either G is a cyclic group C_{p^n}, or $p = 2$ and G is a quaternion group Q_{2^n}.

We will use Sylow p-subgroups often, which are p-groups $P \leq G$ such that the index $|G : P|$ is not divisible by p, and write $\mathrm{Syl}_p(G)$ for the set of them. As we have seen, the set of all automorphisms of G is denoted by $\mathrm{Aut}(G)$; the normal subgroup of *inner automorphisms* $\mathrm{Inn}(G)$ of $\mathrm{Aut}(G)$ consists of conjugation maps $c_g : x \mapsto x^g$ for $g \in G$, and is isomorphic to $G/Z(G)$; $\mathrm{Out}(G)$, the *outer automorphism group*, is defined as the quotient group $\mathrm{Aut}(G)/\mathrm{Inn}(G)$. Of course, the outer automorphism group does not consist of automorphisms! It consists of cosets of automorphisms, but we often abuse our notation and call elements of both $\mathrm{Out}(G)$ and $\mathrm{Aut}(G) \setminus \mathrm{Inn}(G)$ outer automorphisms.

For π a set of primes again, write $O_\pi(G)$ for the largest normal π-subgroup of a finite group G. If $\pi = \{p\}$ we omit the braces and write $O_p(G)$ for $O_{\{p\}}(G)$, and $O_{p'}(G)$ for $O_{\{p\}'}(G)$. From the opposite direction, we write $O^\pi(G)$ for the smallest

[4] These are often called generalized quaternion groups, with ‘quaternion’ reserved for the group Q_8.

[5] These are also known, particularly in older papers, as ‘quasidihedral groups’.

normal subgroup whose quotient is a π-group, and write $O^p(G)$ and $O^{p'}(G)$ as with $O_p(G)$ and $O_{p'}(G)$.

We will also consider simple groups and their close relatives: a non-trivial group G is *simple* if it has no normal subgroups other than itself and 1; G is *quasisimple* if $G = G'$ and $G/Z(G)$ is simple; G is *almost simple* if there is a non-abelian simple group X such that $X \leq G \leq \mathrm{Aut}(X)$. (Since X is non-abelian and simple, $X/Z(X) = X$, so there is a natural isomorphism $X \cong \mathrm{Inn}(X)$. Thus we may embed X in $\mathrm{Aut}(X)$, and hence this definition makes sense.) If G is a finite, non-abelian, simple group then there is a unique finite quasisimple group $\hat{G}$, called the *universal central extension* of G, such that every quasisimple group X with $X/Z(X) \cong G$ is a quotient of $\hat{G}$ by a central subgroup. Such groups are often called *covers* of a simple group, with the order of $|Z(X)|$ being the multiple of the cover, (e.g., if $|Z(X)| = 3$, it is a triple cover).

There is a mutual extension of quasisimple and almost simple groups, which doesn't really have a name: one may think of a quasisimple group as a simple group with some centre placed underneath it, and an almost simple group as a simple group with some automorphisms on top of it. Often one needs to do both of these things, and have groups G such that there is a chain of subgroups

$$1 \leq Z \leq H \leq G$$

with $Z = Z(G)$, H quasisimple, and G/Z almost simple. These sorts of groups regularly appear when one tries to reduce questions about all finite groups to questions about a much smaller class of groups, which are 'close' to simple groups: the reduction can be to simple groups, to quasisimple groups, to almost simple groups, or to this less tightly defined class of groups. Sometimes we have to expand the collection of such groups still further, and include groups where H/Z is simple, but H need not be quasisimple, or where Z is central in H but not in G (so that the outer automorphisms of $H/Z(H)$ induced by elements of G can act non-trivially on $Z(H)$) or other subtle variations. It depends on the method of reduction and which theorems are used which type of groups are obtained at the end; such groups are often called *simple groups with decorations*, and sometimes *nearly simple groups*.[6]

As an example, take the group $\mathrm{PSL}_n(q)$, the projective special linear group of $n \times n$ matrices over the field $\mathbb{F}_q$. This is a simple group for $n \geq 2$ (except for $(n, q) = (2, 2), (2, 3)$); the group $\mathrm{SL}_n(q)$ is quasisimple and the group $\mathrm{PGL}_n(q) = \mathrm{GL}_n(q)/Z(\mathrm{GL}_n(q))$ is almost simple, with the same exceptions; the group $\mathrm{GL}_n(q)$ fits into the fourth category, sometimes without any modifications, but normally with the requirement that H is quasisimple dropped (and H is the subgroup $\mathrm{SL}_n(q) \cdot Z(\mathrm{GL}_n(q))$). With a slightly larger list of exceptions, $\mathrm{SL}_n(q)$ is the universal central extension of $\mathrm{PSL}_n(q)$.

[6]Both the case where $Z(G) = Z(H)$ and G acts non-trivially on $Z(H)$ have been called nearly simple in the literature.

A *composition series* for G is a chain of subgroups

$$1 = G_0 \trianglelefteq G_1 \trianglelefteq \cdots \trianglelefteq G_n = G$$

such that G_i/G_{i-1} is simple for all i, and the quotients G_i/G_{i-1} are called the *composition factors* of G. (If G is finite then it has a composition series, but if G is infinite, for example $\mathbb{Z}$, then it need not.) The *Jordan–Hölder theorem* states that the composition factors of G are well defined, in the sense that different composition series yield the same composition factors (with multiplicities), just as for modules.

A more specialized concept, that we will invoke only in Sect. 8.4, is that of isoclinism. Note that for a group G, if $x, y \in G$ and $z \in Z(G)$, then $[x, y] = [xz, y]$, so the commutator map can be seen as a map $G/Z(G) \times G/Z(G) \to G'$. Two groups G and H are *isoclinic* if there are isomorphisms $G/Z(G) \to H/Z(H)$ and $G' \to H'$ that commute with the commutator map above. For more information about isoclinic groups, see for example [122, Section 6.7].

A group G is *soluble* (often written 'solvable')[7] if it has a series

$$1 = G_0 \trianglelefteq G_1 \trianglelefteq \cdots \trianglelefteq G_n = G$$

such that G_i/G_{i-1} is abelian for all i. If G is finite, G is soluble if and only if all of the composition factors of G are cyclic of prime order. More generally, a finite group G is *p-soluble* if all composition factors are cyclic of prime order or a p'-group, i.e., a group of order not divisible by p. The simple alternating group A_5, of order 60, is therefore a 7-soluble group, as is the symmetric group S_5. It is easy to see that a finite group G is soluble if and only if it is p-soluble for all primes p. The famous *Feit–Thompson theorem* [219], also called the odd order theorem, states that any non-cyclic finite simple group has even order, or equivalently every finite 2-soluble group is soluble.

We should also mention the *classification of finite simple groups*, a vast project completed after decades of work. Broadly speaking, there are four classes of finite simple groups:

(i) cyclic groups C_p for p a prime;
(ii) alternating groups A_n for $n \geq 5$;
(iii) groups of Lie type such as $\mathrm{PSL}_n(q)$, classes of matrix groups over finite fields;
(iv) 26 sporadic groups, the largest of which is the Monster.

The first two classes are well known, but the third and fourth are less so. The third is really sixteen infinite series of groups: these split into classical and exceptional groups. The classical groups are the determinant 1 matrices that are stabilizers of certain types of forms on vector spaces: $\mathrm{SL}_n(q)$ is the stabilizer of the zero form, $\mathrm{SU}_n(q)$ is the stabilizer of a non-singular Hermitian form, $\mathrm{Sp}_{2n}(q)$ is the stabilizer of a non-singular alternating form, $\mathrm{SO}_{2n+1}(q)$ is the stabilizer of a non-singular

[7] Particularly by those of an American persuasion.

bilinear form for an odd-dimensional space, $\mathrm{SO}_{2n}^{\pm}(q)$ is the stabilizer of a +-type or −-type non-singular bilinear form, and there are extra considerations in characteristic 2 for the last case. These are not simple in general, but are quasisimple, except for the orthogonal groups, which have a quasisimple subgroup of index at most 2. Each of these is a 2-parameter family, requiring both the field $\mathbb{F}_q$ and the dimension of the vector space, n, $2n$ or $2n+1$.

The other groups are exceptional, and just need a field, as the vector space has fixed dimension. They are denoted by $G_2(q)$, $F_4(q)$, $E_6(q)$, $E_7(q)$, $E_8(q)$, ${}^2E_6(q)$, ${}^3D_4(q)$, ${}^2B_2(2^{2n+1})$, ${}^2G_2(2^{3n+1})$, and ${}^2F_4(2^{2n+1})$. The smallest is ${}^2B_2(2^{2n+1})$, called a Suzuki group, which acts on a 4-dimensional vector space over a field of order an odd power of 2, and the largest is $E_8(q)$, which acts on a 248-dimensional vector space over $\mathbb{F}_q$.

There is a general theory of groups of Lie type, enabling some aspects of the groups to be treated generally, but often for more difficult questions each family must be approached separately.

The final collection of finite simple groups are the sporadic simple groups. They range from the Mathieu group M_{11} of order 7920 to the Monster, of order

$$808017424794512875886459904961710757005754368000000000.$$

They include the Suzuki group, not to be confused with *a* Suzuki group, which we saw above. Apart from the Monster and the second largest sporadic group Baby Monster (which were conjectured to exist by Fischer, and constructed by Leon and Sims for the Baby Monster, and Griess for the Monster), they are all named after people, usually their discoverers, although a couple are more contentious.

The book [566] by Wilson has lots of information about the alternating, sporadic and Lie type groups, although he prefers a case-by-case construction of the Lie type groups. For more generic treatments, one normally must pass through the theory of algebraic groups, or 'groups with a BN-pair', which is another book in its own right. The recent book [417] by Malle and Testerman is a fairly short introduction to the general theory of these groups, suitable for graduate students.

Exercises

Exercise 1.1 Let G be a finite group and let ρ be the permutation representation of G on itself given by conjugation. Let ψ be the character of this representation. Show that

$$\psi = \sum_{\chi \in \mathrm{Irr}(G)} \chi \cdot \chi^*.$$

Exercise 1.2 Let G be a permutation group acting on a finite set X. Recall that G is 2-transitive on X if G is transitive on all ordered pairs (x, y) for $x, y \in X$ with $x \neq y$. Let χ denote the permutation character of the action of G on X. Show that G acts 2-transitively on X if and only if $\langle \chi, \chi \rangle = 2$.

Exercise 1.3 Let χ be the character of a kG-module M. Show that the characters of the kG-modules $S^2(M)$ and $\Lambda^2(M)$ are given by

$$\chi_{S^2(M)}(x) = \frac{\chi(x)^2 + \chi(x^2)}{2}$$

and

$$\chi_{\Lambda^2(M)}(x) = \frac{\chi(x)^2 - \chi(x^2)}{2}$$

respectively.

Exercise 1.4 We will use the previous exercise to prove facts about the Frobenius–Schur indicator. The *Frobenius–Schur indicator* of a character $\chi \in \operatorname{Irr}(G)$ is given by

$$\nu(\chi) = \frac{1}{|G|} \sum_{x \in G} \chi(x^2).$$

Show that $\nu(\chi) \in \{-1, 0, 1\}$, and that $\nu(\chi) = 0$ if and only if $\chi \neq \chi^*$.

(In fact, $\nu(\chi) = 1$ if χ is the character of a real representation, and $\nu(\chi) = -1$ if χ is real but cannot be afforded by a real representation.)

Exercise 1.5 Using the previous exercise, we now use the character table to count the number of elements of order 2, i.e., involutions, in a finite group G. Prove that the number of elements in G that square to the identity (including the identity!) is

$$\sum_{\chi \in \operatorname{Irr}(G)} \nu(\chi) \cdot \chi(1).$$

This is often called the *Frobenius–Schur count of involutions*.

Exercise 1.6 Assume that we have already proved that the irreducible characters form a basis for the space of class functions. We will show that the row and column orthogonality relations are equivalent. Let $x_1, \ldots, x_n$ denote conjugacy class representatives for a finite group G, and let $\chi_1, \ldots, \chi_n$ be the irreducible characters of G. Let X denote the character table of G, thought of as a matrix over $\mathbb{C}$. Let C denote the diagonal matrix whose ith entry is $|C_G(x_i)|$. Prove that the row orthogonality relation is equivalent to the statement $XC^{-1}\bar{X}^T = I$, and that the column orthogonality relation is equivalent to the statement $\bar{X}^T X = C$. Finally, show that these two statements about matrices are equivalent.

Exercise 1.7 Prove the following result of Burnside: if χ is an irreducible character of a finite group G, and $\chi(1) > 1$, then there exists $x \in G$ such that $\chi(x) = 0$.

Exercise 1.8 Let G be a finite group and let $\chi \in \mathrm{Irr}(G)$. Write

$$e_\chi = \frac{\chi(1)}{|G|} \sum_{g \in G} \chi(g^{-1})g.$$

Prove that

$$1 = \sum_{\chi \in \mathrm{Irr}(G)} e_\chi$$

and $e_\chi \cdot x = x \cdot e_\chi$ for all $x \in G$.

Prove that $e_\chi e_\psi = \delta_{\chi,\psi} e_\chi$ (where $\delta_{\chi,\psi}$ is the δ-function) if and only if the *generalized orthogonality relation* holds: for $\chi, \psi \in \mathrm{Irr}(G)$, we have

$$\frac{1}{|G|} \sum_{g \in G} \chi(g^{-1})\psi(gh) = \delta_{\chi,\psi} \frac{\chi(h)}{\chi(1)}.$$

The element e_χ is the *central idempotent* associated to χ.

Exercise 1.9 Use the regular representation of G, and the Artin–Wedderburn theorem, to show that if e_χ denotes the projection of 1 onto the submodule of $\mathbb{C}G$ corresponding to χ, then indeed

$$e_\chi = \frac{\chi(1)}{|G|} \sum_{g \in G} \chi(g^{-1})g,$$

as stated in the previous exercise.

Exercise 1.10 Prove the *Burnside–Brauer theorem*: let G be a finite group and let χ be a faithful character of G. Prove that, for any irreducible character $\psi \in \mathrm{Irr}(G)$, there exists some $n \geq 0$ such that ψ is a constituent of $\chi^{\otimes n}$. If χ takes m distinct values on G, i.e., the set $\{\chi(g) \mid g \in G\}$ has cardinality m, prove that the n above may be taken to be at most m.

Chapter 2
Blocks and Their Characters

Ordinary representation theory, that is, representation theory over the complex numbers, is a fairly good theory. Maschke's theorem tells us that every representation is a sum of irreducible representations, a given representation of a finite group G is determined by its character, and the group algebra $\mathbb{C}G$ is the direct sum of matrix algebras over $\mathbb{C}$, one for each irreducible representation, and of dimension equal to the degree of each representation.

All of this is wrong for fields k of characteristic $p > 0$, where p divides the order of the group. Not all representations are sums of irreducible representations, and hence a character does not necessarily determine a representation. Just as disconcertingly, the group algebra is not a sum of matrix algebras over k.

We can, however, write kG (for k an algebraically closed field of characteristic p) as a direct sum of *some* algebras. An indecomposable ideal summand of the group algebra kG will be called a block of kG, or a p-block of G. Blocks are the ideals from which our group algebra is built, so it is no surprise that they are fundamental in representation theory over fields of positive characteristic.

The analogue of a character of a representation in positive characteristic is a Brauer character: taking the trace of a matrix gives bad answers over fields of characteristic p, because for example $\chi(1)$ would no longer be the dimension of the matrix, but merely the dimension modulo p. To repair this, we note that the trace of a matrix is simply the sum of the eigenvalues of the matrix, the eigenvalues of a group element under a representation are roots of unity, and we simply take the corresponding roots of unity in the complex numbers. This will only work if the order of the group element is prime to the characteristic of the field; this is the right thing to do though, because the number of isomorphism classes of simple kG-modules is equal to the number of conjugacy classes of elements whose order is prime to p (called p-regular elements), so we again obtain a square character table, and the irreducible Brauer characters form a basis for such class functions, just as in the complex case.

D. A. Craven, *Representation Theory of Finite Groups: a Guidebook*, Universitext,
https://doi.org/10.1007/978-3-030-21792-1_2

Since we have Brauer characters and ordinary characters, both class functions on p-regular elements, and the Brauer characters form a basis for the class functions on p-regular elements, we can write every ordinary character as a linear combination of Brauer characters. It turns out that the coefficients in such a linear combination are always non-negative integers, called the decomposition numbers of a group, and we put them into a matrix. Given the ordinary character table, knowledge of the decomposition matrix of a group is equivalent to knowledge of the Brauer character table.

What goes wrong though, and why determining the Brauer characters is much harder than the ordinary characters, is that there is no natural bilinear form on the class functions via which the irreducible Brauer characters form an orthonormal basis. Of course, one may define such a form once one has the irreducible Brauer characters, but this requires you to know all irreducible Brauer characters first, and that's not the right way round for it to be of any use.

As not all indecomposable kG-modules are simple whenever $p \mid |G|$, knowledge of the Brauer character does not imply knowledge of the kG-module, but it is a good start. Much effort has been expended computing decomposition numbers, and we will talk more about these in later chapters, but suffice it to say it is among the most difficult problems in the representation theory of finite groups.

Each simple module, and indeed each ordinary character, belongs to a unique block. Grouping the Brauer and ordinary characters together by blocks makes the decomposition matrix into, as it happens, a block-diagonal matrix. Furthermore, the decomposition into blocks is the finest partition of the characters that makes the decomposition matrix block diagonal, yielding another possible definition of the blocks of a finite group.

If $|G| = p^a m$ for $p \nmid m$ (so that the Sylow p-subgroups of G have order p^a) then the defect of a block is the smallest non-negative integer d such that p^{a-d} divides $\chi(1)$ for all ordinary characters in that block. The defect group of a block is more difficult to define, but it is a p-subgroup of G of order p^d, well defined up to conjugacy. The defect group of a block has a profound, and still somewhat mysterious, impact on the structure of a block. For example, the number of ordinary characters in a block is bounded in terms of the defect; conjecturally there should be at most p^d ordinary characters in a block, but the best we can do in general is the Brauer–Feit theorem [59] (Theorem 4.7.1), that gives an upper bound of $p^{2d}/4+1$.

The chapter then begins to explore the connection between the block theory of a group G and the block theory of normalizers $N_G(P)$ of p-subgroups P of G, by stating the Brauer correspondence. This is a bijection between the blocks of kG with defect group D and the blocks of $kN_G(D)$ with defect group D. Many conjectures from representation theory aim to link the structures of corresponding blocks, and we will see a broad outline of these in Chap. 4.

This chapter will cover much of the basic theory of blocks that we will need throughout this book. We postpone until Chap. 4 a discussion about the various conjectures that aim to constrain the structure of a block in terms of its defect group (but we still tease the reader throughout this chapter with a taste of the delights to come), and we postpone until Chap. 3 work on kG-modules, as opposed to characters.

We will use the books of Navarro [447] (which has a character-theoretic leaning) and Linckelmann [394] (which has a module-theoretic leaning) as general references for this chapter. Feit's book [217] is now back in print, and can also be used as a source for much of this theory. Another popular choice is the textbook of Alperin [6], which has some results from block theory in it, but that book deals more with module theory.

2.1 Blocks and Block Idempotents

Let G be a finite group and let k be an algebraically closed field of characteristic p. Write

$$kG = B_1 \oplus B_2 \oplus \cdots \oplus B_r,$$

where each B_i is an indecomposable, two-sided ideal summand of kG. The B_i were introduced by Brauer and Nesbitt [60] in 1941, and are called the *blocks* of kG; a *p-block* of G is a block of kG for k an algebraically closed field of characteristic p. We will show that the B_i are unique up to ordering.

This is not the only definition of a block: there is also one involving elements. We will talk about that now, but we need a couple of definitions first.

Definition 2.1.1 Let A be a ring. An element $e \in A$ is an *idempotent* if $e^2 = e$. An element e is *central* if, for all $a \in A$, $ae = ea$. A non-zero idempotent e is *primitive* if, whenever $e = e_1 + e_2$ is an expression of e as a sum of idempotents such that $e_1e_2 = 0$, either $e_1 = 0$ or $e_2 = 0$.

A *primitive central idempotent* of A is a primitive idempotent of the centre $Z(A)$.

The block decomposition yields a decomposition of the identity element of kG into its projections onto the blocks, i.e.,

$$1 = e_1 + e_2 + \cdots + e_r,$$

where $e_i \in B_i$. Since $e_i \in B_i$, and B_i is an ideal, we have that $e_ie_j \in B_i \cap B_j = 0$ for $i \neq j$. Hence

$$1 = 1^2 = \left(\sum_{i=1}^{r} e_i\right)^2 = \sum_{i=1}^{r} e_i^2.$$

Thus $e_i^2 = e_i$ for all i, so the e_i are idempotents.

Furthermore, since $x \cdot 1 = x = 1 \cdot x$, we have

$$\sum_{i=1}^{r} xe_i = \sum_{i=1}^{r} e_i x.$$

We know that $e_i \in B_i$, and each B_i is an ideal of kG, so both $e_i x$ and xe_i lie in B_i. In particular, both of these expressions must give the projection of x onto B_i, and hence $xe_i = e_i x$. This means that the e_i are actually central idempotents of kG. We really see here that $B_i = kGe_i$: since $kG \cdot 1 = kG$,

$$kG = kG \cdot 1 = \sum_{i=1}^{r} kGe_i.$$

However, $kGe_i \leq B_i$, so the sum must be direct and $kGe_i = B_i$.

We finally show that the e_i are primitive. Suppose that $e_i = e + e'$ with e, e' central idempotents such that $ee' = 0$. We get that

$$B_i = kGe_i = kGe \oplus kGe',$$

and so one of kGe and kGe' is zero, as needed.

The element e_i is called the *block idempotent* of B_i. We have proved that all block idempotents are primitive central idempotents. Now we will prove the converse. Suppose that e is a primitive central idempotent. We have

$$e \cdot 1 = e \cdot \sum_{i=1}^{r} e_i = \sum_{i=1}^{r} ee_i;$$

since e and e_i are central idempotents so is ee_i, and $(ee_i)(ee_j) = e(e_i e_j) = 0$, so since e is primitive, all but one of the ee_i must be zero. Therefore $e = ee_i$ for some $1 \leq i \leq r$. On the other hand,

$$(e_i - e)^2 = e_i^2 + e^2 - 2e_i e = e_i - e.$$

We see that e and $e_i - e$ are both central idempotents, and $e(e_i - e) = 0$, so one of e and $e_i - e$ is zero as e is primitive. Since $e \neq 0$, we get that $e_i = e$, as needed.

Theorem 2.1.2 *An element of kG is a block idempotent if and only if it is a primitive central idempotent.*

Of course, we also get the following corollary.

Corollary 2.1.3 *The map $e \mapsto kGe$ establishes a one-to-one correspondence between the primitive central idempotents of kG and the blocks of kG.*

Since $B_i = kGe_i$, there is a split in the community over whether to call the ideal or the idempotent the block of kG. Broadly speaking, originally the ideal was always called the block, in early papers, but referring to the idempotent as a block is a more modern usage, reflecting the greater importance that idempotents have found in the theory as time has gone on. Nevertheless, many authors, including this one, still refer to the ideal as the block. Of course it doesn't really matter, since we can characterize the blocks either via the ideal route or via the primitive central idempotent route.

The simple, and in fact all indecomposable, kG-modules are distributed among the blocks of kG. We can see this most easily using the block idempotents: if M is an arbitrary (finite-dimensional) kG-module then $M \cdot 1 = M$, so

$$M = M \cdot 1 = \sum_{i=1}^{r} M \cdot e_i.$$

Since $e_i e_j = 0$ for $i \neq j$, this sum is in fact direct. Thus we can write

$$M = \bigoplus_{i=1}^{r} M \cdot e_i.$$

If M is indecomposable then there is some i such that $M \cdot e_i = M$ and $M \cdot e_j = 0$ for all $j \neq i$; we say that M *belongs to* the block B_i. Of course, if M belongs to a block so does every submodule and quotient of M, and hence all composition factors of M.

Writing kG as a module over itself, the regular module, we see that kG decomposes as a sum of the kGe_i, i.e., if B is a block of kG then B, viewed as a kG-module, belongs to B. In particular, we have proved the following.

Theorem 2.1.4 *The simple kG-modules are partitioned among the blocks of kG, and every block possesses at least one simple kG-module. Furthermore, if M is an indecomposable kG-module, then all composition factors of M belong to the same block.*

2.2 Brauer Characters

To go from a representation over $\mathbb{C}$ to a character one simply takes the trace of the matrices giving the representation. This isn't going to work over a field k of characteristic p, because if M is a module of dimension divisible by p then the character χ of M would satisfy $\chi(1) = 0$. We need a better way.

The trace of a matrix over $\mathbb{C}$ is the sum of the eigenvalues of that matrix. If k has characteristic p, and x in G has order n, then any representation of G yields a matrix for x whose eigenvalues are nth roots of unity. If n is prime to p then this matrix is conjugate to a diagonal matrix (i.e., the underlying vector space has a basis consisting of eigenvectors) with entries its eigenvalues, just as in the complex case. (Here we definitely require k to be algebraically closed.) Elements of G whose order is prime to p are called *p-regular* elements, and the set of them is denoted by G_{reg}. Complementing this definition, a *p-singular* element is one of order divisible by p.

However, rather than add these eigenvalues in k, we map them into $\mathbb{C}$. Formally, if $|G| = p^a m$ for some m with $p \nmid m$, then the eigenvalues of matrices representing any p-regular element are always mth roots of unity. Thus we choose a primitive mth root of unity $\zeta \in k$ and a primitive mth root of unity $\hat{\zeta} \in \mathbb{C}$, let $f : \langle \zeta \rangle \to \langle \hat{\zeta} \rangle$ be the obvious bijection, and take the sum

$$\sum_{\alpha \in E} f(\alpha),$$

where E is the collection (with multiplicities) of all eigenvalues of the matrix representing a p-regular element x. If $\rho : G \to \mathrm{GL}_n(k)$ is a representation then the *Brauer character* of ρ is the map $\psi : G_{\mathrm{reg}} \to \mathbb{C}$ that sends an element x to the sum above. (The Brauer character is often also called the *modular character*. The two terms are used interchangeably by authors, often even the same author.)

For example, if $p = 2$ and x has order 3, then x might have eigenvalues $1, 1, 1, \zeta, \zeta, \zeta^2$, where $\zeta^3 = 1$, in which case the Brauer character evaluates to $3 + 2\hat{\zeta} + \hat{\zeta}^2 = 2 + \hat{\zeta}$. A different choice of root of unity gives a different answer—in this case $2 + \hat{\zeta}^2$—but they are related (see Exercise 2.1).

Again, Brauer characters are class functions, but this time only defined on p-regular elements. The next theorem is the correct generalization to arbitrary algebraically closed fields of the statement that the ordinary characters of irreducible representations form a basis for all class functions.

Theorem 2.2.1 (Brauer [47]) *Let G be a finite group and let p be any prime.*

(i) *If M_1 and M_2 are two simple kG-modules with Brauer characters ψ_1 and ψ_2, then $\psi_1 = \psi_2$ if and only if $M_1 \cong M_2$.*
(ii) *The set of Brauer characters of the isomorphism classes of simple kG-modules forms a basis for the vector space of all class functions $G_{\mathrm{reg}} \to \mathbb{C}$. In particular, the number of isomorphism classes of simple kG-modules is equal to the number of conjugacy classes of p-regular elements.*

(See, for example, [447, Corollary 2.10] or [394, Theorem 5.13.13].) Just as for characteristic 0, for this theorem to work k really does need to be algebraically closed. Of course, if $p \nmid |G|$ then we get the statement from ordinary representation theory back, that the number of irreducible ordinary characters is equal to the number of conjugacy classes. As $k(G)$ denotes the number of irreducible ordinary characters of G, we write $l(G)$ for the number of irreducible Brauer characters of G.

At this point, one might think that although the number of irreducible characters changes depending on p, the character theories over all algebraically closed fields look similar. Given an arbitrary kG-module M, we let ϕ denote its Brauer character. Then ϕ may be written as a sum

$$\phi = \sum_i a_i \phi_i$$

where $a_i \in \mathbb{Z}_{\geq 0}$ and the ϕ_i are the irreducible Brauer characters. The integer a_i is the multiplicity of the simple module S_i as a composition factor of M, where S_i has Brauer character ϕ_i. This is true regardless of the characteristic of k.

There is one vital difference between k of characteristic 0 and of characteristic p: as we saw in Sect. 1.1, there is a natural inner product on the ordinary characters of a finite group, given by

$$\langle \chi, \phi \rangle = \frac{1}{|G|} \sum_{g \in G} \chi(g)\overline{\phi(g)},$$

which turns the irreducible characters into an orthonormal basis of the vector space of class functions over $\mathbb{C}$. There is no such natural inner product on the set of Brauer characters, so although, for example, one may take the tensor product of two Brauer characters just as one takes the tensor product of two ordinary characters, one has no idea a priori how to decompose this product into irreducibles, or even if it is actually irreducible.

This is a massive problem. The Monster simple group's character table (which the intrepid reader may find in [122]) was computed even before its existence was known, but as of the time of writing, the Brauer characters are known only for $p = 17, 19, 23, 31$, leaving $p = 2, 3, 5, 7, 11, 13, 29, 41, 47, 59, 71$ still to find.

As another example, the Murnaghan–Nakayama rule (Theorem 8.1.8) is an algorithm that can compute the ordinary character table of any symmetric group. However, even the degrees of the irreducible Brauer characters are not known for all primes (and no fast algorithm for computing them is known), never mind the full character table.

Example 2.2.2 If G is a finite p-group and k has characteristic p, then there is a single irreducible Brauer character, namely the trivial character.

Example 2.2.3 Let G be a 2-transitive permutation group of degree n (e.g., A_n for $n \geq 4$ and S_n for $n \geq 3$). Over the complex numbers, the permutation module is the direct sum of the trivial module and a simple module of dimension $n - 1$. In characteristic p, if $p \nmid n$ then again the module is the direct sum of a trivial module and another module, which is often simple. (It is simple for $G = A_n, S_n$, but if G is the sporadic simple Mathieu group M_{23} and $p = 2$ for example, it is the sum of two 11-dimensional modules. See [437] for more examples.)

If $p \mid n$ though then there is a trivial module and a trivial quotient, but they do not come from a trivial summand. For S_n and A_n (except for A_4) we have three composition factors, two of which are trivial. Moreover, the permutation module is indecomposable, completely different to the case where $p \nmid n$.

The sporadic simple Conway group Co_3, which is 2-transitive on a set of 276 points, has a more complicated permutation module structure for $p = 2, 3$: it has five composition factors, of dimensions 1, 1, 22, 22, 230 and 1, 1, 22, 126, 126 for $p = 2$ and $p = 3$ respectively. The first module is 'uniserial' (see Definition 3.1.2 below), so has a unique simple submodule, of dimension 1, but the second has simple submodules of dimensions 1 and 126.

Example 2.2.4 Let $G = A_5$, and let ζ denote a primitive 5th root of unity. The ordinary character table is as follows.

	1	(1, 2)(3, 4)	(1, 2, 3)	(1, 2, 3, 4, 5)	(1, 2, 3, 5, 4)
χ_1	1	1	1	1	1
χ_2	3	−1	0	$1+\zeta+\zeta^{-1}$	$1+\zeta^2+\zeta^{-2}$
χ_3	3	−1	0	$1+\zeta^2+\zeta^{-2}$	$1+\zeta+\zeta^{-1}$
χ_4	4	0	1	−1	−1
χ_5	5	1	−1	0	0

The 5-modular character table, on the other hand, is as follows.

	1	(1, 2)(3, 4)	(1, 2, 3)
ψ_1	1	1	1
ψ_2	3	−1	0
ψ_3	5	1	−1

In this last example we can see that the ordinary characters of G, when viewed simply as functions on G_{reg}, can be written as a linear combination of the Brauer characters. This is clear, as the Brauer characters form a basis for such class functions: but the coefficients in these linear combinations are non-negative integers. This is generally true.

Write $\mathrm{Irr}(G)$ for the irreducible ordinary characters of G, and if p is a given prime, write $\mathrm{IBr}(G)$ for the set of irreducible Brauer characters of G. If $\chi \in \mathrm{Irr}(G)$ is an ordinary character, then the restriction of χ to G_{reg} is a class function on G_{reg}, which is often denoted χ^0 (for example, in [447, p. 22]). We can write

$$\chi^0 = \sum_{\psi \in \mathrm{IBr}(G)} d_{\chi,\psi} \psi,$$

where $d_{\chi,\psi} \in \mathbb{C}$. The $d_{\chi,\psi}$ are called *decomposition numbers*. The $|\mathrm{Irr}(G)|$ by $|\mathrm{IBr}(G)|$ matrix whose entries are $d_{\chi,\psi}$ is called the *decomposition matrix* of G (at the prime p). The $d_{\chi,\psi}$ are not just elements of $\mathbb{C}$.

Theorem 2.2.5 *Decomposition numbers are non-negative integers.*

(See, for example, [394, Corollary 5.13.10] or [447, p. 23].) The columns of the decomposition matrix are labelled by irreducible Brauer characters, so order these by the p-blocks of the group. It turns out (see [217, p. 147] or [394, Proposition 4.16.9]) that if an irreducible ordinary character χ satisfies $d_{\chi,\psi} \neq 0$ for some $\psi \in \mathrm{IBr}(G)$, then the only $\phi \in \mathrm{IBr}(G)$ for which $d_{\chi,\phi} \neq 0$ are those in the same block as ψ. Just as we said that ψ belongs to a given block B, if $\chi \in \mathrm{Irr}(G)$ satisfies $d_{\chi,\psi} \neq 0$ for some ψ in B then we say that χ *belongs to* B as well. What we have just said is that ordinary characters belong to a unique block.

Write $\mathrm{Irr}(B)$ and $\mathrm{IBr}(B)$ for the irreducible ordinary and Brauer characters respectively that belong to the block B. We have partitioned both the ordinary and Brauer characters among the p-blocks of G. We get the following theorem.

Theorem 2.2.6 *If $d_{\chi,\psi} \neq 0$ then χ and ψ belong to the same p-block of G.*
Let

$$\mathrm{Irr}(G) \cup \mathrm{IBr}(G) = I_1 \cup I_2 \cup \cdots \cup I_r$$

be a partition of $\mathrm{Irr}(G) \cup \mathrm{IBr}(G)$ *into subsets such that whenever $d_{\chi,\psi} \neq 0$, χ and ψ lie in the same subset. Each I_i is the set of irreducible (ordinary and Brauer) characters belonging to a union of p-blocks of G. In particular, the decomposition of* $\mathrm{Irr}(G) \cup \mathrm{IBr}(G)$ *among the p-blocks of G is the finest such decomposition.*

In [447, Definition 3.1], this partition of $\mathrm{Irr}(G) \cup \mathrm{IBr}(G)$ is used as the definition of a p-block, and the equivalence with our definition (and therefore the proof of Theorem 2.2.6) is in [447, Theorem 3.3]. We can therefore determine the p-blocks of G by examining the decomposition matrix of G.

Example 2.2.7 The ordinary and modular character tables of A_5 for $p = 5$ were given in Example 2.2.4. From this it is easy to compute the decomposition matrix, and it is as follows.

	ψ_1	ψ_2	ψ_3
χ_1	1	.	.
χ_2	.	1	.
χ_3	.	1	.
χ_4	1	1	.
χ_5	.	.	1

(The entry 0 is often replaced by a . in order to detect patterns in the decomposition numbers. The location of zeros is very important, as will appear often in this book.)

We can see that there are two blocks—which is much easier with zeros replaced by full stops—one with χ_1 to χ_4, and ψ_1 and ψ_2 belonging to it, and one with χ_5 and ψ_3 belonging to it.

This is very nice, but we decided earlier that determining the ordinary characters was easy but determining the Brauer characters is hard. Since—once we know the ordinary character table—determining the Brauer characters and the decomposition numbers is the same, the obvious question is 'can we determine the blocks of a finite group without having to compute all of the Brauer characters?'

Fortunately, the answer is yes. If one wants to know how the ordinary characters are distributed into blocks, without knowing anything about the Brauer characters, then this can be done using central characters. A *central character* of G is formally a non-zero algebra homomorphism from $Z(kG)$ to k.

Let e be a primitive central idempotent of kG. If λ is a central character then

$$\lambda(e) = \lambda(e^2) = \lambda(e)^2,$$

so that $\lambda(e)$ is either 0 or 1. Furthermore, if e' is another primitive central idempotent with $e \neq e'$ then $ee' = 0$, so that $\lambda(e)\lambda(e') = 0$. Thus one can associate to every central character a unique block, namely the unique block whose idempotent does not lie in the kernel of λ.

Now let χ be an irreducible ordinary character of G. The centre of the group algebra $Z(kG)$ is spanned as a vector space by the class sums, i.e., $\widehat{C}$ for C a conjugacy class of G, where $\widehat{X}$ denotes the sum of the elements in X. We can construct a function

$$\lambda_\chi : Z(kG) \to \mathbb{C},$$

$$\widehat{x^G} \mapsto \frac{|x^G|\chi(x)}{\chi(1)},$$

where x is an element of G and, as usual, x^G denotes the conjugacy class containing x. To reduce notational overload, we write $\lambda_\chi(x)$ for $\lambda_\chi(\widehat{x^G})$. The image $\lambda_\chi(x)$ is an algebraic integer, rather than just an algebraic number.

We want to say that χ and ψ are in the same p-block if and only if λ_χ and λ_ψ are 'the same'; however, this clearly must depend on p, and so we interpret this as $\lambda_\chi \equiv \lambda_\psi$ modulo p. The correct way to view this is to view the character table over the p-adic rationals (or an extension of them), rather than the complex numbers: then λ_χ has image in the extension of the p-adic integers, and we can ask whether $\lambda_\chi - \lambda_\psi$ has image in the unique maximal ideal of (the extension of) the p-adic integers. As the quotient of (an extension of) the p-adic integers by its unique maximal ideal is a field of characteristic p, we get a map $\lambda_\chi : Z(kG) \to k$, as we desire. Brauer proved in [50] that two ordinary characters lie in the same block if and only if they have the same central character $Z(kG) \to k$.

It is possible, for example in [447] and [33, Section 1.9], to work entirely in the complex numbers. Here you choose a maximal ideal inside a ring of algebraic integers R (that contains all entries in the character table), containing pR, and check that the two central characters agree modulo this ideal. If the character values are actual integers, rather than algebraic integers, this simply means that the central characters agree modulo p.

Example 2.2.8 The character values of symmetric groups are all integers. To see this, note that every element $x \in S_n$ is rational, i.e., conjugate to every power x^i such that the orders of x and x^i are equal. In this case, as $\chi(x)$ is the sum of eigenvalues, $\chi(x^i)$ is the sum of those eigenvalues raised to the ith power, and $\chi(x) = \chi(x^i)$. From this we see that $\chi(x)$ is stable under any Galois automorphism of the field of values, and since character values are algebraic integers, they must be integers. The character table for S_5 is given below:

	1	(1, 2)	(1, 2)(3, 4)	(1, 2, 3)	(1, 2, 3, 4)	(1, 2, 3, 4, 5)	(1, 2, 3)(4, 5)
χ_1	1	1	1	1	1	1	1
χ_2	1	−1	1	1	−1	1	−1
χ_3	4	2	0	1	0	−1	−1
χ_4	4	−2	0	1	0	−1	1
χ_5	5	1	1	−1	−1	0	1
χ_6	5	−1	1	−1	1	0	−1
χ_7	6	0	−2	0	0	1	0

We can compute the central characters using the formula above. We write λ_i for the central character of χ_i, to avoid using double subscripts.

	1	(1, 2)	(1, 2)(3, 4)	(1, 2, 3)	(1, 2, 3, 4)	(1, 2, 3, 4, 5)	(1, 2, 3)(4, 5)
λ_1	1	10	15	20	30	24	20
λ_2	1	−10	15	20	−30	24	−20
λ_3	1	5	0	5	0	−6	−5
λ_4	1	−5	0	5	0	−6	5
λ_5	1	2	3	−4	−6	0	4
λ_6	1	−2	3	−4	6	0	−4
λ_7	1	0	−5	0	0	4	0

We read off the partitions of $\mathrm{Irr}(S_5)$ into p-blocks for $p = 2, 3, 5$ from this: the 2-blocks are $\{1, 2, 5, 6, 7\}$ and $\{3, 4\}$; the 3-blocks are $\{1, 4, 6\}$, $\{2, 3, 5\}$ and $\{7\}$; the 5-blocks are $\{1, 2, 3, 4, 7\}$, $\{5\}$ and $\{6\}$.

The Nakayama conjecture, Theorem 8.3.1, gives a purely combinatorial description of the distribution of ordinary characters into p-blocks for all symmetric groups.

This is all automated in computers, and so Magma and GAP[1] can easily compute the p-blocks of a finite group given its character table.

We can expand on this shift from the complex numbers to the p-adics, to produce a formal way to relate the ordinary and modular characters, rather than through the decomposition matrix, which also makes Theorem 2.2.5 a trivial consequence. The representation theories of a given finite group over $\mathbb{C}$ and any algebraically closed field of characteristic 0 are the same, so instead of $\mathbb{C}$ we take an extension of the p-adic rationals. For a given group G, we only need the presence of enough roots of unity (in fact $|G|$th roots of unity will do), not the whole algebraic closure, so we let K denote a finite extension of the p-adic rationals $\mathbb{Q}_p$ that contains enough roots of unity for all simple KG-modules to be absolutely simple, i.e., remain simple over the algebraic closure of K. Let $\mathcal{O}$ denote the ring of integers of K; then $\mathcal{O}$ possesses a unique maximal ideal $J(\mathcal{O})$ such that $k = \mathcal{O}/J(\mathcal{O})$ is a field of characteristic p, a finite extension of $\mathbb{F}_p$.

The idea is to conjugate representations of G over K until they become representations over $\mathcal{O}$, then reduce them modulo the maximal ideal $J(\mathcal{O})$ to form representations over k. This allows us to write every simple KG-module as a kG-module. The structure of the reduction does depend on the precise conjugate over $\mathcal{O}$ that we take, but the composition factors of the reduction do not. This gives the row of the decomposition matrix corresponding to that KG-module.

The triple $(K, \mathcal{O}, k)$ is called a *p-modular system*, and is the formal setup used in most papers on modular representation theory. If one is purely interested in the representation theory over k then one does not need p-modular systems, but particularly the categorical equivalences which we will meet in Chap. 3 use the ring $\mathcal{O}$.

Whether the triple $(K, \mathcal{O}, k)$ even exists is a good question. For our case above, when we deal with a single group, it is clear as we only need finite extensions of $\mathbb{Q}_p$ and $\mathbb{F}_p$. If we are dealing with infinitely many groups simultaneously, we might need to find a triple $(K, \mathcal{O}, k)$ such that K and k are algebraically closed. In other words, is there a ring $\mathcal{O}$ with a unique maximal ideal $J(\mathcal{O})$, such that both $\mathcal{O}/J(\mathcal{O})$ and the field of fractions of $\mathcal{O}$ are algebraically closed? The answer is 'yes': let K be any algebraically closed field of characteristic 0, and let $\mathcal{O}$ be a valuation ring inside K, i.e., a subring such that for all $\alpha \in K$, either $\alpha \in \mathcal{O}$ or $\alpha^{-1} \in \mathcal{O}$. Then K is clearly the field of fractions of $\mathcal{O}$, and $\mathcal{O}$ is a local ring with unique maximal ideal $J(\mathcal{O})$, and since K is algebraically closed it is an easy exercise to find a root of any non-constant polynomial in $k[X]$, where $k = \mathcal{O}/J(\mathcal{O})$.

[1] Magma and GAP are the two most widely used computer algebra systems at the moment. GAP is open source, and may be downloaded for free by anyone. Magma is commercial software, but they have an online calculator on their website that can be used for small computations.

However, although the ring $\mathcal{O}$ is a valuation ring, and is local, it is not a *discrete* valuation ring (see [394, Definition 4.2.1]), i.e., a principal ideal domain with a unique maximal ideal $J(\mathcal{O})$. In representation theory of finite groups, usually the local ring is taken to be a discrete valuation ring, rather than just a local ring. We will now show that we cannot choose a triple $(K, \mathcal{O}, k)$ where $\mathcal{O}$ is a discrete valuation ring and K is algebraically closed. Let α be the generator of the unique maximal ideal $J(\mathcal{O})$ of $\mathcal{O}$ (which exists as $\mathcal{O}$ is a principal ideal domain), and consider the equation $x^2 - \alpha$. As K is algebraically closed, this equation has a zero, say β, so that $\beta^2 = \alpha$. Since $\mathcal{O}$ is a discrete valuation ring, it is a valuation ring in K, so either β or β^{-1} lies in $\mathcal{O}$, say β. But α lies in $J(\mathcal{O})$, so in k we have that $\beta^2 = 0$; thus $\beta \in J(\mathcal{O})$. Hence $\beta = \alpha \cdot r$ for some $r \in \mathcal{O}$, and therefore $\alpha \cdot r = 1$. But generators of ideals cannot be units unless $J(\mathcal{O}) = \mathcal{O}$, which is a contradiction.

One hidden point here is whether the blocks of kG and $\mathcal{O}G$ are 'the same'. Of course, as $\mathcal{O}G$ is an $\mathcal{O}$-algebra, we can write $\mathcal{O}G$ as a sum of indecomposable, two-sided ideals. Again, we get a corresponding decomposition

$$1 = f_1 + f_2 + \cdots + f_s,$$

with each f_i belonging to a different block of $\mathcal{O}G$. (The f_i are necessarily central idempotents.) The surjective map $\mathcal{O} \to k$ yields a surjective map $\mathcal{O}G \to kG$, and the image of each f_i is a block idempotent of kG, so indeed everything works well. This idea is known as *lifting of idempotents*, and can be done modulo any nilpotent ideal (see, for example, [33, Theorem 1.7.3] for a particularly slick proof).

We are going to avoid using the ring $\mathcal{O}$ if at all possible, to keep things relatively simple. Unfortunately, sometimes we will have cause to mention it, but not for the rest of this chapter, at least.

There is, however, an easier way to determine the blocks if you know the character table of G. We produce a graph with vertices $\mathrm{Irr}(G)$, and connect two vertices χ and ψ if

$$\sum_{g \in G_{\mathrm{reg}}} \chi(g)\overline{\psi(g)} \neq 0,$$

where G_{reg} is the set of p-regular elements of G. The irreducible characters in a single p-block of G are exactly the characters in one connected component of this graph. This may be found in, for example, [447, Theorem 3.19]. Thus there is no need to take the maximal ideal at all, and one may simply read off the blocks from the character table without working with the ring $\mathcal{O}$ or any analogous ring.

This result first appeared in the work of Brauer and Feit [59], where they proved their upper bound on $k(B)$ (Theorem 4.7.1). In addition to this result, they proved that the diameter of this graph is always 2, i.e., there are always characters connected to all other characters in the block. These are the characters of 'height zero' in the block (see Definition 2.3.4 below). A proof can also be found in [447, Corollary 3.25].

The other important thing is that this shows that the distribution of ordinary characters into blocks does not depend on the choice of maximal ideal containing pR, which a priori it might in the definition above.

2.3 Defect

We have split the group algebra up into a direct sum of indecomposable, two-sided ideals, called blocks. All of the irreducible ordinary and Brauer characters have been partitioned among the blocks. But to get further in the theory we need to understand these blocks. We will start by attaching a non-negative integer d to each block, called its defect. This will enable us to move a bit further forward with our work, but to unleash the full force of local representation theory we will need to upgrade the defect to a defect group, a subgroup of G of order p^d, defined up to conjugacy. The defect group allows us to define the Brauer correspondence and give the main theorems, which show that at least some aspects of the representation theory of a finite group are locally controlled, that is, controlled by p-subgroups, their normalizers and centralizers.

We start with the defect. This has a simple definition.

Definition 2.3.1 Let G be a finite group. Let p^a denote the highest power of p dividing $|G|$. If B is a p-block of G, then the *defect* of B is the smallest non-negative integer d such that

$$p^{a-d} \mid \chi(1)$$

for all $\chi \in \mathrm{Irr}(B)$.

If B is the block to which the trivial character belongs (of course, from the definition of the decomposition matrix, it is easy to see that the trivial ordinary character and the trivial Brauer character lie in the same block) then B has defect a, where p^a is the order of a Sylow p-subgroup of G. This block—the unique block containing the trivial character—is called the *principal block*, and is in many respects the most important block in a finite group. The principal block has the largest possible defect, and is said to have *maximal defect*, or sometimes *full defect*.

On the other end of the scale are blocks of defect zero. These are completely understood, and complete the analogy with ordinary representation theory. If p does not divide the order of G, then every block must have defect zero, and since the ordinary and modular character tables are the same (see Exercise 2.6), the decomposition matrix is (up to ordering the rows) the identity matrix. Thus every block has a single ordinary character and a single modular character, and the decomposition matrix for that block is simply (1).

The next theorem gives us a complete understanding of blocks of defect zero.

Theorem 2.3.2 (Brauer [50]) *Let G be a finite group. Suppose that B is a p-block of G of defect zero. Then B has a single ordinary character, a single modular character, and the decomposition matrix of B is* (1). *Furthermore, B is isomorphic as an algebra to a matrix algebra over k of dimension $\chi(1)$, where χ is the unique (ordinary or modular, since they have the same degree) character belonging to B. Conversely, if χ is a character whose degree is divisible by the order of a Sylow p-subgroup of G, then χ belongs to a block of defect zero.*

Blocks of defect one are also close to being fully classified. There is a finite list of algebras, and every block of defect one is 'almost' isomorphic as an algebra to an algebra on that list. (In the theorem above, even for blocks of defect zero we see that there are infinitely many possible algebras, namely $M_n(k)$ for all n. The correct notion here is 'Morita equivalence' (see Sect. 3.3); all matrix algebras over k are Morita equivalent to one another, so up to Morita equivalence there is a single block of defect zero.) However it is still, for some primes, not completely clear exactly which algebras on this list can occur as blocks of a finite group and which cannot (see Sect. 5.3).

In this classification, which we discuss in Chap. 5, a bound on the number of ordinary characters in B can also be obtained: writing $k(B)$ for the number of irreducible ordinary characters that belong to B and $l(B)$ for the number of irreducible Brauer characters that belong to B, we have that $k(B) \leq p$ for blocks of defect one.

Notice that this means that $k(B) \leq p^d$ for blocks of defect d, for $d = 0, 1$. This isn't exactly a lot of evidence, but here is a conjecture, which I believe first appeared in [51].

Conjecture 2.3.3 (Brauer's $k(B)$-Conjecture) If B is a p-block of defect d in a finite group, then

$$k(B) \leq p^d.$$

We will give an account of the current progress on the $k(B)$-conjecture in Sect. 4.7.

What we also saw here was that characters of particular degree are guaranteed to be in blocks of defect zero. In particular, if B is a block of defect one, then every ordinary character in B has the same power of p dividing its degree. To see this, if $|G| = p^a m$ with $p \nmid m$, then by the definition of defect $p^{a-1} \mid \chi(1)$ for all $\chi \in \operatorname{Irr}(B)$, and if $p^a \mid \chi(1)$ then χ lies in a block of defect zero, hence not in B.

Definition 2.3.4 Let G be a finite group. Write $|G| = p^a m$ with $p \nmid m$. Let $\chi \in \operatorname{Irr}(G)$ lie in a p-block of defect d. The *height* of χ, denoted by $\operatorname{ht}(\chi)$, is defined to be the exact power of p dividing the integer

$$\frac{\chi(1)}{p^{a-d}}.$$

Since $p^{a-d} \mid \chi(1)$, the height is a non-negative integer. We have just seen that all characters in blocks of defect zero and one have height zero. The same is true for blocks of defect two, but not for blocks of defect three. It is not a coincidence that all p-groups of order p^2 are abelian, but not all p-groups of order p^3 are.

Example 2.3.5 Let G be a finite p-group. As we saw at the end of Sect. 1.1, the ordinary character degrees of G divide $|G|$, so are powers of p. Also, as there is a single irreducible modular character (we have seen this before in Example 2.2.2, and you are asked to prove it in Exercise 2.3), there can only be one p-block of G. Thus all ordinary characters belong to the principal block, and this has defect d, where $|G| = p^d$.

Notice that $k(B) = k(G) \leq |G|$, and all irreducible ordinary characters have height zero if and only if G is abelian.

Example 2.3.6 Let G be the sporadic simple Janko group J_1. This has an abelian Sylow 2-subgroup of order 8. The decomposition matrix of the principal 2-block was determined by Fong in [225], and is as follows.

	$\psi_1 = 1_1$	$\psi_2 = 20_1$	$\psi_3 = 56_1$	$\psi_4 = 56_2$	$\psi_5 = 76_1$
$\chi_1 = 1_1$	1	.	.	.	.
$\chi_2 = 77_1$	1	.	.	.	1
$\chi_3 = 77_2$	1	1	1	.	.
$\chi_4 = 77_3$	1	1	.	1	.
$\chi_5 = 133_1$	1	1	1	1	.
$\chi_6 = 133_2$	1	.	.	1	1
$\chi_7 = 133_3$	1	.	1	.	1
$\chi_8 = 209_1$	1	1	1	1	1

In this table, the numbers next to the characters are the degrees of the characters, so ψ_2 has degree 20, χ_2 has degree 77, and so on. The subscripts are so we can use the degrees as an alternative label for the characters, which is more informative than χ_i and ψ_i.

We see that all ordinary irreducible characters in the principal block (the degrees 1, 77, 133 and 209) have height zero. The irreducible Brauer characters of a finite group can have degree divisible by p, and need not even divide the order of the group. For example, $\mathrm{SL}_2(p)$ in characteristic p has an irreducible Brauer character of degree i for $1 \leq i \leq p$ (see Example 9.1.4).

Now let G be the group $\mathrm{PSL}_2(7)$, and again let k be a field of characteristic 2. This time the decomposition matrix of the principal 2-block is as follows.

	$\psi_1 = 1_1$	$\psi_2 = 3_1$	$\psi_3 = 3_2$
$\chi_1 = 1_1$	1	.	.
$\chi_2 = 3_1$	.	1	.
$\chi_3 = 3_2$	.	.	1
$\chi_4 = 6_1$	.	1	1
$\chi_5 = 7_1$	1	1	1

The Sylow 2-subgroups are again of order 8, but this time they are dihedral rather than abelian. This time we find an irreducible ordinary character of height one.

What these examples are telling us is that the defect of a block might not be a good enough indicator of character heights, and that we need to bring the structure of the Sylow 2-subgroup of the group into play. Thus to push further, we need to attach a group to every block, not merely an integer. The defect group of a block is a conjugacy class of p-subgroups of G of order p^d, where d is the defect of the block.

Conjecture 2.3.7 (Brauer's Height-Zero Conjecture) Let B be a block of a finite group. All irreducible ordinary characters in B have height zero if and only if the defect groups of B are abelian.

This conjecture dates back to 1955 [53]. The 'if' direction of this is now a theorem of Kessar and Malle [349], but the other direction remains open at this time. We will discuss the height-zero conjecture more in Sect. 4.1.

Thus the defect group is what we need to understand character heights better. Now we just have to define it. This isn't so easy, and we will have to use the block idempotents again.

Let Q be any p-subgroup of G. The *Brauer map* Br_Q is the map

$$\mathrm{Br}_Q : kG \to kC_G(Q), \qquad \mathrm{Br}_Q : \sum_{g\in G} a_g g \mapsto \sum_{g\in C_G(Q)} a_g g,$$

the projection onto $kC_G(Q)$.

When restricted to the Q-stable elements kG^Q of kG, i.e., those elements $\sigma \in kG$ such that $\sigma^x = \sigma$ for all $x \in Q$, this becomes an algebra homomorphism (Exercise 2.2). It is also surjective: note that kG^Q is spanned, as a vector space, by the sums of Q-orbits of elements of G, and of course if $x \in C_G(Q)$ then the Q-orbit of x is simply $\{x\}$. Hence $x \in kG^Q$ for all $x \in C_G(Q)$, and so Br_Q is surjective on restriction to kG^Q.

To show that it is an algebra homomorphism, we simply have to check that if $\sigma, \tau \in kG^Q$ then

$$\mathrm{Br}_Q(\sigma\tau) = \mathrm{Br}_Q(\sigma)\mathrm{Br}_Q(\tau),$$

and this is an easy exercise if one takes σ and τ to be the basis elements given above.

The definition of a defect group is now easy, but fairly opaque.

Definition 2.3.8 Let B be a p-block of G, with block idempotent e. A *defect group* of B is a p-subgroup D of G that is maximal under inclusion subject to the constraint that

$$\mathrm{Br}_D(e) \neq 0.$$

Since Br_1 is the identity map, defect groups exist for all blocks. Also, if D is a defect group and $g \in G$ then so is D^g. What is far from obvious is that the converse is true, that is, the defect groups of a block form a single conjugacy class of subgroups of G. They have order p^d, where d is the defect of the block. (See, for example, [394, Theorem 5.6.5 and Proposition 6.5.12] or [447, Theorems 4.3 and 4.6].)

Example 2.3.9 The principal block, the block containing the trivial module, has defect groups the Sylow p-subgroups of the group.

A block of defect zero has trivial defect group, and a block of defect one has defect group a cyclic group of order p. Blocks of defect two have abelian defect group.

Example 2.3.10 In the example of A_5 in characteristic 5, which we saw in Examples 2.2.4 and 2.2.7, there are two blocks: one, the principal block, has defect group C_5, and one has defect group 1.

Example 2.3.11 We return to J_1 from Example 2.3.6. The principal block has defect groups the Sylow 2-subgroups, which are isomorphic to $C_2 \times C_2 \times C_2$. There is also a block of defect one, so defect group C_2, with two ordinary characters of degree 76 and one modular character, also of degree 76. There are five blocks of defect zero, with character degrees 56, 56, 120, 120 and 120.

Before we move on, we can ask if there are any restrictions on which p-subgroups can be defect groups. We have seen that Sylow p-subgroups and the trivial subgroup can both be defect groups, and there are blocks with cyclic defect groups other than Sylow p-subgroups. It was proved by Green [264] (a short proof was given by Thompson [542]) that if D is a defect group of a p-block of a group G, then there exists $P \in \mathrm{Syl}_p(G)$ and $g \in G$ such that $D = P \cap P^g$; Green later improved this in [266] to $D = P \cap P^g$ for some $g \in C_G(D)$, and this reproves a result of Alperin [1] that defect groups are *tame intersections*, i.e., that $D = P \cap P^g$ such that $N_P(D)$ and $N_{P^g}(D)$ are Sylow p-subgroups of $N_G(D)$. Green actually proves in [266] that there is a p-regular element $z \in G$ such that $D \in \mathrm{Syl}_p(C_G(z))$, and $z = xy$ for two elements $x, y \in G$ such that $D = P \cap P^x = P \cap P^y$ for some $P \in \mathrm{Syl}_p(G)$.

2.4 Brauer Correspondence

The Brauer correspondence is the first step, both logically and historically, to our understanding of how the whole theory of local control of representation theory works. If D is a p-subgroup of G, the Brauer correspondence is a bijection between the p-blocks of G with defect group D and the p-blocks of $N_G(D)$ with defect group D.

Understanding the blocks of $N_G(D)$ is in some sense easier than understanding the blocks of G. At least, by induction it is. (Exercise 2.3 requires you to prove this theorem.)

Theorem 2.4.1 *If M is a simple kG-module then $O_p(G)$ lies in the kernel of M. Similarly, if B is a p-block of G and D is any defect group of B, then $O_p(G)$ lies in D.*

This seems like the right time to mention the companion to this. (Exercise 3.14 requires you to prove this theorem.)

Theorem 2.4.2 *If M is a simple kG-module in the principal block then $O_{p'}(G)$ lies in the kernel of M. In fact, the principal p-blocks of G and $G/O_{p'}(G)$ are isomorphic as k-algebras.*

We see in Exercise 3.8 that the simple modules, and the p-blocks, of G, are in a form of correspondence (not bijective in general on the level of blocks) with the simple modules and p-blocks of $G/O_p(G)$. Furthermore, the defect groups of corresponding blocks are D and a subgroup of $D/O_p(G)$. Thus in a general group, the p-blocks of $N_G(D)$ with defect group D possess simple modules that become blocks of defect zero in $N_G(D)/D$. But Theorem 2.3.2 establishes that we can determine blocks of defect zero easily by looking at the ordinary character degrees of $N_G(D)/D$. Thus this half of the Brauer correspondence should not be too hard to understand.

Let B be a block of kG with defect group D and block idempotent e. By definition of defect group, $\mathrm{Br}_D(e)$ is non-zero, and since e is central in G, it lies inside kG^D, so that

$$\mathrm{Br}_D(e^2) = \mathrm{Br}_D(e)^2;$$

hence $\mathrm{Br}_D(e)$ is a central idempotent of $kC_G(D)$. In fact, one may show that $\mathrm{Br}_D(e)$ is a central idempotent of $kN_G(D)$, and even more that it is primitive. Furthermore, every primitive central idempotent of $kN_G(D)$ with defect group D can be obtained as $\mathrm{Br}_D(e)$ for some block idempotent of G with defect group D.

This sets up the Brauer correspondence, also known as Brauer's first main theorem [50, 54].

Theorem 2.4.3 (Brauer's First Main Theorem) *The map $e \mapsto \mathrm{Br}_D(e)$ induces a bijection between primitive central idempotents of kG with defect group D and primitive central idempotents of $kN_G(D)$ with defect group D.*

(See, for example, [447, Chapter 4] or [394, Section 6.7].) The name of this theorem suggests that there is at least one more main theorem of Brauer. In fact, there are two more, but the second main theorem requires us to understand generalized decomposition numbers, which we will not do in this book. The third main theorem can be glibly stated as the fact that the Brauer correspondent of the principal block of kG is always the principal block of $kN_G(P)$ for P a Sylow p-subgroup of G.

Theorem 2.4.4 (Brauer's Third Main Theorem, Weak Version) *If G is a finite group and P is a Sylow p-subgroup of G, then the principal p-block of G and the principal p-block of $N_G(P)$ are Brauer correspondents.*

This was proved in [56, Theorem 3]. It is stronger than this, but to state its full strength we would need to generalize the Brauer correspondence. Rather than do this and achieve the full force of the third main theorem, we prefer to move on to other things and gain an insight into the module structure in the next chapter.

Exercises

Exercise 2.1 Prove that choosing a different primitive mth root of unity $\zeta \in k$ in Sect. 2.2 yields the same Brauer character table after reordering columns.

Note that it is definitely not true that one merely has to permute the rows, i.e., replacing ζ by a different root of unity sends Brauer characters to Brauer characters. The easiest example to see this is $G \cong \mathrm{SL}_2(32)$ for $p = 2$. Here there are five 2-dimensional modules up to isomorphism, and there is an element x in G of order 31 with eigenvalues on each of these modules θ, θ^{-1}, for various primitive 31st roots of unity θ. However, this means that only ten of the 31 primitive roots appear as eigenvalues of x on simple kG-modules of dimension 2. The others appear as eigenvalues of *powers* of x.

Exercise 2.2 Prove that the Brauer map Br_Q from Sect. 2.3 is an algebra homomorphism when restricted to Q-stable elements kG^Q.

Exercise 2.3 Prove directly, i.e., without appealing to Theorem 2.2.1, that if G is a finite p-group then G possesses a unique irreducible Brauer character over a field of characteristic p.

Deduce, using the weak version of Clifford's theorem, Theorem 7.1.1, that for any finite group G, $O_p(G)$ lies in the kernel of every Brauer character.

Exercise 2.4 Prove the following result of Brauer:

$$k(G) = \sum_x l(C_G(x)),$$

where the sum runs over all conjugacy classes of p-elements x. (Hint: every element $g \in G$ may be written uniquely as $g = g_1 g_2$, where g_1 is a p-element, g_2 is a p'-element, and g_1 and g_2 commute.)

Exercise 2.5 Let b be a block idempotent, and write

$$b = \sum_{g \in G} \alpha_g g.$$

Write $\alpha = \sum_{g \in G} \alpha_g$. Show that $\alpha = 0$ if b is non-principal, and that $\alpha = 1$ if α is principal.

Exercise 2.6 Let G be a finite group and let p be a prime not dividing $|G|$. Show that the ordinary and Brauer character tables are the same.

Exercise 2.7 This exercise will use an alternative definition of the defect group. If x is an element of G, then a defect group of x is a Sylow p-subgroup of $C_G(x)$. Thus to every conjugacy class we may associate a conjugacy class of defect groups. Let B be a block of kG, with block idempotent b and central character λ. An alternative definition of a defect group is a p-subgroup of smallest order amongst the defect groups of conjugacy classes whose sums are not in the kernel of λ.

(i) Show that with this alternative definition, defect groups exist.
(ii) Prove that the principal block has Sylow p-subgroups as defect groups, using this alternative definition.

Exercise 2.8 Recall that if χ is an ordinary character then the dual, denoted χ^* or $\bar{\chi}$, has character values the complex conjugate of χ.

Prove that the same holds true for the dual of a Brauer character. If $\chi \in \mathrm{Irr}(G)$ and $\psi \in \mathrm{IBr}(G)$, prove that $d_{\chi,\psi} = d_{\chi^*,\psi^*}$. Deduce that, for any $\chi, \psi \in \mathrm{Irr}(G) \cup \mathrm{IBr}(G)$, χ and ψ lie in the same block if and only if χ^* and ψ^* do.

Thus duality induces a bijection on characters that respects the block decomposition. A *real block* is one that is sent to itself under the duality operation. Note that if B contains a real character then B is real; if $p = 2$ then the converse holds, and real blocks have a real irreducible character, and also an irreducible Brauer character. This is a result of Gow and Willems [259] (see also [447, Theorem 3.33]). For odd primes, the presence of a real Brauer character in a block implies the existence of a real ordinary character in it [447, Theorem 3.35], but the existence of a real ordinary character in a block does not imply the existence of a real Brauer character ($\mathrm{GL}_2(3)$ at $p = 3$ is a counterexample). Navarro provided me with a counterexample to the statement that a real block need have a real irreducible character, ordinary or

Brauer: the smallest such group for $p = 3$ is of order 144, and is of the form $((C_3 \times C_3) \rtimes C_8) \rtimes C_2$.

Exercise 2.9 Let χ be an irreducible character of G. Prove that

$$|\chi(x)|^2 = \frac{\chi(1)}{|G|} \sum_{g \in G} \chi([x, g]),$$

for all $x \in G$. (Hint: use central characters and Schur's lemma.)

Chapter 3
Modules

The last chapter focused on blocks and characters. We occasionally mentioned modules, for instance when we showed that each indecomposable module belongs to a single block. This chapter delves into the module theory much more.

We start with projectives. The short definition of a projective indecomposable module is an indecomposable summand of the regular module, and every projective module is a sum of projective indecomposable modules. The projective indecomposable modules, or PIMs, are in one-to-one correspondence with the simple modules, and their composition factors are given by the Cartan matrix, which is simply the product of the decomposition matrix with its transpose.

Projectives are important because every module is a quotient of a projective module. The smallest projective module for which a given module M is a quotient is called the projective cover of M. Projective covers turn out to be of great importance when we introduce the stable module category, which broadly speaking is the module category with maps being taken modulo projectives.

Because group algebras are special among all algebras (they are 'symmetric', see [394, Section 2.11]), projective modules are also 'injective'. This means that not only is every module a quotient of a projective, it is also a submodule of a projective. In particular, a projective indecomposable module has a unique simple submodule and a unique simple quotient, and they are isomorphic. This yields the one-to-one correspondence with the simple modules.

We then develop the Green correspondence: this looks a bit like the Brauer correspondence, but for indecomposable modules rather than for blocks. To every indecomposable module we attach a conjugacy class of p-subgroups, called vertices, and set up a bijection between indecomposable kG-modules with vertex Q and indecomposable $kN_G(Q)$-modules with vertex Q.

We lastly look at categories associated with modules: the module category, stable category, and finally the derived category. Along the way we meet endotrivial

D. A. Craven, *Representation Theory of Finite Groups: a Guidebook*, Universitext,
https://doi.org/10.1007/978-3-030-21792-1_3

modules, which are modules M such that $M \otimes M^*$ is the sum of the trivial module and a projective module. The indecomposable endotrivial modules form an abelian group under tensor product, with inverse the dual, and the structure of this group is starting to become clear thanks to recent research.

We also consider extensions of modules, i.e., given modules A and B, modules E such that there is a surjective map $E \to A$ with kernel B. With this knowledge, we can introduce the stable and derived categories, and give some of their basic properties.

The ideas in this chapter mirror those in the previous one: Chap. 2 develops the numerical, character-theoretic information about finite group representation theory, and this chapter develops the structural, module-theoretic information.

We continue to use [394], and will also use [6] for this chapter, as [447] deals less with these concepts.

3.1 Projectives

Let G be a finite group and let k be a field of characteristic a prime p. The *module category* of kG, written mod-kG, is the category whose objects are all finite-dimensional (right) kG-modules, and whose morphisms are all kG-module homomorphisms. (The category of left kG-modules is usually denoted by kG-mod.) We will assume that k is algebraically closed, but this is often not necessary.

Just as free groups are the free objects in the category of groups, free modules are the free objects in the category of modules. The concept of a free module is given by a universal property, but there is a much simpler description of a free module. For R a ring, a *free R-module* is any module with an R-basis. Thus the free objects in mod-kG, i.e., the finite-dimensional free modules, are simply the modules $kG^{\oplus n}$ for all positive integers n.

A *projective module* is any summand of a free module, so free modules are projective, but there are projective modules that are not free unless G is a p-group. The reason for this is that the projective indecomposable kG-modules are, up to isomorphism, in one-to-one correspondence with the simple kG-modules. As G has a single simple module if and only if it is a p-group by Theorem 2.2.1, the module kG—which must be a sum of indecomposable projective modules, and every such one must occur inside it—is indecomposable if and only if G is a p-group.

Example 3.1.1 Let G be an elementary abelian p-group of order p^n, generated by $x_1, \dots, x_n$. Notice that kG can be identified with

$$k[X_1, \dots, X_n]/(X_1^p, X_2^p, \dots, X_n^p),$$

where $X_i = x_i - 1$. This is of course the projective indecomposable module for G. In general, we can identify the group algebra of a direct product of cyclic p-groups of orders $r_1, \ldots, r_n$ with the quotient

$$k[X_1, \ldots, X_n]/(X_1^{r_1}, \ldots, X_n^{r_n})$$

of the polynomial ring in n variables.

Before we can say more about projective modules, we need to talk a bit more about arbitrary modules and their structure; we need some language to discuss their submodule structure, for example. If a module is semisimple, that is a sum of simple modules, then it is easy to understand this structure, but in general most modules are not semisimple, so we need some notation for these cases.

The *socle* $\operatorname{soc}(M)$ of a module M is the sum of all simple submodules of M. It is a semisimple submodule of M, and it is the largest semisimple submodule of M in the sense that if N is a semisimple submodule of M then $N \leq \operatorname{soc}(M)$.

We can iterate this procedure, and define $\operatorname{soc}^i(M)$ to be the preimage in M of the socle of $M/\operatorname{soc}^{i-1}(M)$ (and $\operatorname{soc}^0(M) = 0$). This should be reminiscent of the upper central series from nilpotent groups, for example, defining the ith centre $Z^i(G)$ to be the preimage in G of $Z(G/Z^{i-1}(G))$.

Dual to this process is the *Jacobson radical* or simply the *radical*, $\operatorname{rad}(M)$, of a module M, which is the intersection of all kernels of maps $M \to N$ where N is a simple module. This time, it is $M/\operatorname{rad}(M)$ that is semisimple, and if M/N is semisimple then $\operatorname{rad}(M) \leq N$. We can iterate taking radicals like we did for taking socles, defining $\operatorname{rad}^i(M)$ to be $\operatorname{rad}(\operatorname{rad}^{i-1}(M))$ (and $\operatorname{rad}^0(M) = M$).

More generally, if A is an algebra, we define the Jacobson radical $J(A)$ of A to be the intersection of the annihilators of all simple right A-modules.

The socle series is the series

$$0 \leq \operatorname{soc}(M) \leq \operatorname{soc}^2(M) \leq \cdots \leq \operatorname{soc}^r(M) = M,$$

and similarly for the radical series. The *socle length* of M is the smallest integer r such that $\operatorname{soc}^r(M) = M$, and similarly for the *radical length*. The radical and socle lengths coincide (Exercise 3.10). The ith *radical layer* of a module M is the quotient $\operatorname{rad}^{i-1}(M)/\operatorname{rad}^i(M)$, and similarly the ith *socle layer* is the quotient $\operatorname{soc}^i(M)/\operatorname{soc}^{i-1}(M)$. The quotient $M/\operatorname{rad}(M)$ is often called the *top* of M (and by some authors the *head* of M).

We occasionally want to describe the socle or radical layers in the form of a picture, and we write

$$\begin{matrix} A \\ B \\ C \end{matrix}$$

to mean that a module has socle C, second socle layer B and third socle layer A.

Definition 3.1.2 A kG-module M is *uniserial* if any of the following equivalent conditions hold:

(i) there is a unique composition series for M;
(ii) $\mathrm{soc}^i(M)/\mathrm{soc}^{i-1}(M)$ is simple (or zero) for all $i \geq 1$;
(iii) $\mathrm{rad}^{i-1}(M)/\mathrm{rad}^i(M)$ is simple (or zero) for all $i \geq 1$;
(iv) every semisimple subquotient of M is simple, i.e., there is no subquotient $M_1 \oplus M_2$ for two non-zero modules M_i.

Of course, a semisimple module is uniserial if and only if it is simple.

Example 3.1.3 Let G be a cyclic p-group, of order p^n. As we saw in Example 3.1.1, the group algebra kG is identified with the polynomial ring $k[X]/(X^{p^n})$. This is a uniserial module, with ith radical layer being generated by X^{i-1}. We will show later, in Example 3.1.8, that the same holds for all indecomposable kG-modules.

The next result generalizes the characteristic 0 Artin–Wedderburn theorem, which states that $\mathbb{C}G$ is the direct sum of matrix algebras

$$\mathbb{C}G \cong \bigoplus_{\chi \in \mathrm{Irr}(G)} M_{\chi(1)}(\mathbb{C})$$

of degrees $\chi(1)$ for χ an irreducible complex character of G. Hence, as a module,

$$\mathbb{C}G \cong \bigoplus_{\chi \in \mathrm{Irr}(G)} M_\chi^{\oplus \chi(1)},$$

where M_χ is a module with character χ. We write kG as a sum of some indecomposable kG-modules, which are no longer necessarily simple, but still occur with multiplicity equal to the irreducible character degrees; this time the Brauer character degrees, rather than the complex character degrees.

Theorem 3.1.4 *Let G be a finite group. If $\psi \in \mathrm{IBr}(G)$, write M_ψ for a simple kG-module with Brauer character ψ. There exist indecomposable modules P_ψ for each $\psi \in \mathrm{IBr}(G)$ such that $P_\psi/\mathrm{rad}(P_\psi) \cong \mathrm{soc}(P_\psi) \cong M_\psi$, and*

$$kG \cong \bigoplus_{\psi \in \mathrm{IBr}(G)} P_\psi^{\oplus \psi(1)}.$$

(See, for example, [6, Theorem 5.3, Corollary 5.4 and Theorem 6.6].) This result needs k to be algebraically closed, as if k is not algebraically closed then ψ need not be an irreducible Brauer character. The indecomposable module P_ψ still exists, but the decomposition of kG has to be modified: now the sum is over all simple kG-modules, which will not in general be simple over the algebraic closure, and the number of times P_ψ appears is not $\psi(1)$, but $\phi(1)$, where ϕ is any irreducible Brauer character that is a constituent of ψ (they all have the same degree).

The module P_ψ in this theorem is a *projective indecomposable module*—sometimes *principal indecomposable module*—or PIM (which rather helpfully abbreviates both names). Comparing this with the block decomposition, it is clear that, as a kG-module, a block B is the sum of the projectives P_ψ for ψ belonging to B, each appearing with multiplicity $\psi(1)$. Thus if B is a block of defect zero with unique simple module M, then M is projective and B is simply $M^{\oplus \dim(M)}$ as a kG-module, since blocks of defect zero are simply matrix algebras (see Theorem 2.3.2).

Projectives can be quite mysterious objects the first time you meet them. I think part of the issue is that representation theory neophytes, and indeed I, prefer submodules to quotients. Choose a semisimple module V: we build larger and larger modules M with V as the socle of M. Of course, we might not be able to build any module other than V, in which case V itself is projective. (This is the case if $p \nmid |G|$.) If we can build a larger module, then we keep going until we can no longer build a larger module.

Without some theory it isn't obvious that we would ever reach that point. Why can modules not be continually piled on top of V, making ever larger modules with the same socle? It turns out that there is a unique projective module X with socle V, and *every* module M with socle V lies inside X.[1] This tells you two things: the first is that one cannot simply build ever larger modules with the same socle. The second is that, if I can build two modules M_1 and M_2 with socle V, then there is some larger module $M_{1,2}$, with socle V again, that contains (isomorphic copies of) both M_1 and M_2 and is the (not direct) sum of them. In other words, adding the modules in M_1 on top of V does not stop us adding the modules in M_2 on top of V, which again is not obvious without any theory.

In particular, if M is a module properly containing a projective submodule X, then $\mathrm{soc}(M)$ must be strictly larger than $\mathrm{soc}(X)$. However, more is true. To give the best version of this next result needs the definition of Ext^i, which is Proposition 3.4.1. The version written here is often enough for most purposes.

Proposition 3.1.5 *Let G be a finite group and let M be a kG-module. If X is a submodule or quotient of M and X is projective, then X is a summand of M.*

The composition factors of the projective indecomposable modules are of tremendous importance in representation theory. The multiplicities of a given simple module in a given indecomposable module are called the *Cartan numbers*, and they are arranged into a matrix, naturally called the *Cartan matrix*. Given an ordering on the set of isomorphism classes of simple modules, $S_1, \dots, S_n$, the Cartan number $c_{i,j}$ is the multiplicity of the simple module S_i in the projective indecomposable module with socle S_j. If one orders the S_i block by block, the Cartan matrix becomes block diagonal.

[1]One consequence of trying to produce consistent notation is that you run into difficulties. I'm trying to use P for a Sylow p-subgroup of G, but of course P is a good choice for a projective module. I could use S for a Sylow p-subgroup instead, but then I want to use S for a source of a module (see the next section). I don't want to use M or V because these are general modules. Thus I end up using X for a projective module.

Theorem 3.1.6 *Let $S_1, \ldots, S_n$ be the simple kG-modules up to isomorphism.*

(i) *If S_i and S_j lie in different blocks then $c_{i,j} = 0$.*
(ii) *$c_{i,j} = c_{j,i}$, so that the Cartan matrix is symmetric.*
(iii) *If the columns of the decomposition matrix D of kG are ordered $S_1, \ldots, S_n$, then $D^T D = C$, where C is the Cartan matrix of kG.*

(See, for example, [394, Proposition 4.10.4 and Theorem 4.12.3] or [447, p. 25 and p. 50].)

By this theorem, we can talk about the Cartan matrix of a block of a finite group, and therefore we can ask how the Cartan matrix of a block might depend on the defect group of the block. We have already mentioned the Brauer–Feit theorem (Theorem 4.7.1 below), which states that we can bound the number of ordinary characters that belong to a block, and hence the number of simple modules, in terms of the defect group of the block. Hence if B is a block with defect group D, then the Cartan matrix has a bounded number of entries.

Conjecture 3.1.7 (Weak Donovan Conjecture) If B is a p-block with defect group D, then the entries of the decomposition matrix, or equivalently the Cartan matrix, of B are bounded in terms of $|D|$.

Another way to say this is that there are only finitely many possible decomposition matrices or Cartan matrices for blocks with a fixed defect group. Brauer originally proposed [55, Problem 22] that the entries of the Cartan matrix should be bounded by $|D|$, but this has been shown to be false in general by Landrock [381]. The much weaker version here can be found, for example, in [4, Conjecture K]. We will say more about this conjecture in Sect. 4.5, and in particular for which groups it is known.

If M is a simple module and X is the projective indecomposable module with socle M, then since $X/\text{rad}(X) \cong M$, there is an obvious surjective homomorphism from X to M. In fact, since every module is a quotient of a free module, if M is an arbitrary finite-dimensional module then there is a unique (up to isomorphism) projective module X such that X surjects onto M, and if X' is any other projective module that surjects onto M, then X' has X as a summand. In other words, X can be chosen so that the kernel of the map $X \to M$ has no projective summands. We call X the *projective cover* of M. There is no standard notation for the projective cover of a module, and we use $\mathcal{P}(M)$ to denote this module. It is not that difficult to check that if M is a module and $M/\text{rad}(M) = M_1 \oplus \cdots \oplus M_r$ is a decomposition into simple summands, then the projective cover $\mathcal{P}(M)$ is simply the sum of the $\mathcal{P}(M_i)$.

Example 3.1.8 Let us return to Example 3.1.3, so let G be a cyclic group of order p^n. Let M be a kG-module of dimension d. This is equivalent to specifying a single matrix $A \in \text{GL}_d(k)$ such that A^{p^n} is the identity matrix.

As we know, A is conjugate in $\mathrm{GL}_d(k)$ to its Jordan normal form, which is a direct sum of block matrices of the form

$$\begin{pmatrix} 1 & 1 & 0 & \cdots & 0 \\ 0 & 1 & 1 & \ddots & \vdots \\ \vdots & \ddots & \ddots & \ddots & 0 \\ \vdots & & \ddots & \ddots & 1 \\ 0 & \cdots & \cdots & 0 & 1 \end{pmatrix}.$$

Of course, if M is indecomposable then the Jordan normal form of A must have a single block, and $A^{p^n} = 1$ if and only if $d \leq p^n$.

It is easy to see that the quotient $k[X]/(X^d)$ is isomorphic to M if M is indecomposable and of dimension d. Thus $k[X]/(X^{p^n}) \cong kG$ is the projective cover of M.

Not only is the projective cover as defined above unique up to isomorphism (some authors allow the addition of arbitrary projectives to it, which can be technically useful in some situations), the map $\mathcal{P}(M) \to M$ is also unique, up to applying an automorphism of $\mathcal{P}(M)$. Thus it makes sense to talk of the kernel of this map, which is denoted by $\Omega(M)$, and called the *Heller translate* or *syzygy* of M (introduced by Heller in [277]). The map $M \mapsto \Omega(M)$ induces a bijection on the set of modules with no projective summand, if one defines $\Omega(0) = 0$. We use the notation $\Omega^0(M)$ to denote the largest summand of M that does not contain any projective summands, which exists and is unique up to isomorphism (but not unique as a submodule in general) by the Krull–Schmidt theorem. As $M \mapsto \Omega(M)$ is a bijection, we can define $\Omega^i(M)$ for any integer i, not just positive integers. Another construction of $\Omega^{-1}(M)$ is to take the injective hull of M, rather than a projective cover. The *injective hull* of M is the smallest projective module $\mathcal{I}(M)$ such that M is a submodule of $\mathcal{I}(M)$: we then define $\Omega^{-1}(M)$ to be the cokernel (codomain modulo image) of the map $M \to \mathcal{I}(M)$. The notion of a projective cover is dual to the notion of an injective hull in the category-theoretic sense, in that they can be defined by diagrams, and reversing the arrows goes from one to the other. They are also dual in the sense that if M is a kG-module then $\mathcal{P}(M)^* = \mathcal{I}(M^*)$.

(The reason it is called an injective hull, is that technically $\mathcal{I}(M)$ is the smallest *injective* module containing M as a submodule. In representations of finite groups, injectives are the same as projectives, but in other categories this need not be the case, and one may not be able to take projective covers, or take injective hulls, or even take either of them if you are particularly unlucky. For example, the category of finite abelian groups does not have either projective covers or injective hulls.)

The map $\Omega(-)$ satisfies a number of very useful properties, such as $\Omega(M \oplus N) \cong \Omega(M) \oplus \Omega(N)$, and

$$\Omega^0(\Omega(M) \otimes N) \cong \Omega(M \otimes N).$$

(The second of these is Exercise 3.6.) In the next section we relate the indecomposable modules for a group to those of *p-local* subgroups, i.e., normalizers and centralizers of p-subgroups of a finite group (sometimes this also includes p-subgroups themselves), so the fact that

$$\Omega^0(\Omega(M)\downarrow_H) \cong \Omega(M\downarrow_H)$$

will clearly play a role. (The Ω^0 is needed because $\Omega(M)\downarrow_H$ will in general have projective summands, while $\Omega(M\downarrow_H)$ does not.)

Also of interest in later parts of this book is the notion of a periodic module. A module is *periodic* if there exists a positive integer i such that $M \cong \Omega^i(M)$, with the smallest such i being the module's *period*.

The possible periods for a periodic module depend, naturally given the statement about restriction above, on the defect group of the block to which the module belongs. Thus we often focus on periodicity for modules for p-groups, and then extend this to all groups using the methods from the next section. So for now we focus on p-groups.

Notice that, from the interaction of the tensor product and the Heller operator above, every indecomposable module is periodic if and only if the trivial module is. It turns out by work of Carlson [94, Corollary 5.6] that if M is a periodic module then $\dim(M)$ is divisible by p^{r-1} (but not if and only if), where r is the p-rank of the p-group G, i.e., the rank of a maximal elementary abelian subgroup. The p-groups of p-rank 1 are cyclic and quaternion, and $\Omega^2(k) = k$ for G cyclic and $\Omega^4(k) = k$ for G quaternion.

A free module for G can be thought of as a periodic module of period 0, so one might believe that there is a relationship between free and periodic modules, and indeed there is: they fit into a general schema of 'complexity', which measures the growth of a projective resolution for a given module. Free modules have complexity 0 and periodic modules have complexity 1. A *projective resolution* for a module M is a (generally infinite) sequence

$$\cdots \to P_n \to P_{n-1} \to \cdots \to P_1 \to P_0 \to M \to 0,$$

where the image of each map is the kernel of the next map in the sequence (so in particular the map $P_0 \to M$ is surjective) and the P_i are all projective. If the P_i are all of smallest dimension out of all projective resolutions, then this is a *minimal resolution*. (It is intuitively clear that a minimal resolution is intimately connected to the Heller operator, and for example the kernel of the map $P_0 \to M$ is $\Omega(M)$.) We will work more with projective resolutions in Sect. 3.4.

If there exist non-negative integers a and c such that

$$\dim(P_n) \leq an^{c-1}$$

for all $n \geq 1$ in a projective resolution for M, and c is minimal with this property, then M is said to have *complexity* c. It is far from clear that the complexity is finite, but it is. The complexity of the trivial module is equal to the p-rank of G, and this is an upper bound for the complexity of any module. Projective modules have complexity 0, and periodic modules have complexity 1.

A theorem of Chouinard [112] states that a module is free if and only if its restriction to every elementary abelian subgroup is free: this was also proved by Carlson [95] for periodic modules, and then generalized by Alperin and Evens [10] at around the same time to all complexities. Carlson used this result, together with the fact that if G is abelian then all periodic modules are of period 1 or 2, to show that, if G is a non-abelian p-group of order p^n, and every maximal abelian subgroup of G has order greater than p^m, then the period of every periodic module must divide $2p^{n-m}$. A few years earlier, Carlson proved in [94] that the dimension of a periodic module is always divisible by p^{r-1}, where the p-group has p-rank r.

This is just the start of homological algebra, and we won't have time to talk much about it in this book, or the deep role that the elementary abelian subgroups play in it, which is touched on in these results above. However, we will have to develop some more homological algebra to be able to really begin to understand what a module category looks like.

3.2 Vertices, Sources and the Green Correspondence

Let G be a finite group, and let p be a prime, with k an algebraically closed field of characteristic p. For this section we might as well assume that p divides $|G|$, because otherwise the content of this section gets really rather boring.

Having defined projective modules, we now define relative projective modules.

Definition 3.2.1 A kG-module M is *relatively H-projective* for some subgroup $H \leq G$ if there exists a kH-module N such that M is a summand of $N{\uparrow}^G$.

Since the permutation module of G on the cosets of H is simply $k_H{\uparrow}^G$ (where k_H is the trivial module for H), the regular module is just $k_1{\uparrow}^G$, where k_1 is the trivial module for the trivial subgroup of G. As projective modules are summands of sums of the regular module, this shows that projective modules are the same as relatively 1-projective modules.

The proof of Maschke's theorem, slightly but carefully altered, shows that if P is a Sylow p-subgroup of G then every module M is projective relative to P (Exercise 3.2). In fact, this is the 'correct' generalization of Maschke's theorem.

Theorem 3.2.2 ((General) Maschke's Theorem) *Every kG-module is projective relative to a Sylow p-subgroup of G.*

If the Sylow p-subgroup of G is trivial then every module is projective, so in particular every module must be semisimple, since projective submodules must always be summands (see Proposition 3.1.5 above).

Suppose that M is relatively H-projective for a subgroup H of G, and that Q is a Sylow p-subgroup of H. There exists a kH-module N such that M is a summand of $N{\uparrow}^G$. However, N itself must be relatively Q-projective by Theorem 3.2.2, so N itself is a summand of $N_1{\uparrow}^H$ for some kQ-module N_1. Thus M is a summand of $N_1{\uparrow}^H{\uparrow}^G = N_1{\uparrow}^G$, and M is relatively Q-projective. Of course, if M is relatively H-projective and $H \leq K \leq G$ then M is also relatively K-projective.

Thus being relatively projective is really a property of p-subgroups. Just as with defect groups, we are going to pick out one of these. A *vertex* of an indecomposable module M is a p-subgroup Q such that M is relatively Q-projective but not relatively R-projective for any $R < Q$.

Since every module is projective relative to a Sylow p-subgroup, vertices always exist. The main result about vertices is that they are always conjugate in G, just like defect groups.

Thus let M be an indecomposable module for G, and let Q be its vertex. This means that there exists some indecomposable kQ-module S such that $S{\uparrow}^G$ has M as a summand. A *source* of M is any indecomposable kQ-module S such that $M \mid S{\uparrow}^G$. The next question is, are sources determined up to isomorphism?

Example 3.2.3 Let $G = A_4$, and let $p = 2$. Let P denote the Sylow 2-subgroup of G, generated by x and y of order 2. Let N be the permutation module of P on the cosets of $\langle y\rangle$, and let $M = N{\uparrow}^G$. This 6-dimensional module is indecomposable, and as a kP-module it is the sum of three permutation modules, on the cosets of $\langle x\rangle$, $\langle y\rangle$ and $\langle xy\rangle$ (all conjugate). Thus we see that the vertices of M are $\langle x\rangle$, $\langle y\rangle$ and $\langle xy\rangle$, and the source is the trivial module.

Now let N' be a 4-dimensional indecomposable module that is not invariant under the outer automorphism group of P. (See Sect. 6.1 for a classification of the indecomposable modules for Klein four groups.) If M' denotes the induction of N' to G, then the action of P on M' is the sum of the three non-isomorphic modules, each of the same dimension as N'. Each of these is a source of M'.

So the answer to our question above is clearly 'no', but all the sources involved here were 'conjugated' by elements of $N_G(Q)$. If S is a kQ-module and $g \in N_G(Q)$, then define the *conjugate module* S^g to have the same elements as S, but action given by

$$s \cdot x = s(x^g)$$

for $s \in S$ and $x \in Q$, where the right-hand side is the original action of Q on S. (This is often denoted $S \otimes g$ in the literature, for example in [33].) This yields a very similar-looking module, and the main result on sources is that these are determined up to $N_G(Q)$-conjugacy, just not isomorphism. We collect this into a theorem.

Theorem 3.2.4 (Green [263]) *Let M be an indecomposable kG-module.*

(i) *If Q and $\bar{Q}$ are vertices of M, then Q and $\bar{Q}$ are conjugate in G.*
(ii) *If S and $\bar{S}$ are both sources of M with the same vertex Q, then $\bar{S} \cong S^g$ for some $g \in N_G(Q)$.*

As well as introducing the concepts of vertex and source, and proving this result, in [263] Green also introduced the now-standard representation-theoretic definition of relative projectivity (although it was foreshadowed by work of Donald Higman in [281], which introduced Higman's criterion (Exercise 3.1), and formally defined in more generality by Higman [282] and Hochschild [297]), proved Green's indecomposability criterion, Theorem 3.3.1, and proved three more titbits about vertices, which are interesting enough to be given their own numbers.

Proposition 3.2.5 (Green [263]) *Let G be a finite group and let B be a p-block of G, with defect group D.*

(i) *If an indecomposable kG-module M lies in B, then M is relatively D-projective.*
(ii) *There is a simple B-module with vertex D.*
(iii) *If a kG-module M has vertex Q, contained in a Sylow p-subgroup P of G, then $\dim(M)$ is divisible by $|P : Q|$.*

If Q is a non-cyclic p-group then there are infinitely many non-isomorphic indecomposable kQ-modules, so by inducing them to a finite group G, we see that any finite group contains infinitely many indecomposable modules with a given (non-cyclic) vertex. If we restrict to the class of simple modules, rather than indecomposable modules, then things look very different.

Conjecture 3.2.6 (Feit's Conjecture [216]) Let Q be a finite p-group. There exists a finite set $\mathcal{M}$ of indecomposable kQ-modules such that, if G is a finite group and M is a simple kG-module with vertex Q, then a source of M is isomorphic to a member of $\mathcal{M}$.

This conjecture didn't quite appear in [216, 2.8], but a weaker version just involving blocks of maximal defect. This version is what people tend to call Feit's conjecture now. For a given finite group G and p-subgroup Q, an indecomposable kQ-module S is the source of only finitely many indecomposable kG-modules (each such module is a summand of $S{\uparrow}^G$, which is finite-dimensional), so this is quite a restrictive statement. We will discuss this more in Sect. 4.6.

Example 3.2.7 Suppose that M is a kG-module with vertex Q and *trivial source*, i.e., the source is the trivial module. This means that M is a summand of the module $(k_Q){\uparrow}^G$, where k_Q is the trivial module on Q; but this is just the permutation module on the cosets of Q, so a trivial-source module is a summand of the permutation module on the cosets of its vertex.

Conversely, if M is a summand of the permutation module on the cosets of H, then M is a summand of $(k_H){\uparrow}^G$, so M is relatively H-projective. If Q denotes a Sylow p-subgroup of H, then k_H is a summand of $(k_Q){\uparrow}^H$, and therefore M is a summand of $(k_Q){\uparrow}^G$, i.e., it is not only relatively Q-projective, but also with trivial 'source'. We place source in inverted commas because we have not shown that Q is a vertex of M, but even though the vertex could be smaller than Q, the source will still be trivial. (See Exercise 3.4.)

What is far less easy to see is that if ϕ is the Brauer character of M, which has trivial source, then ϕ is the reduction modulo p of an ordinary character: such a character is called *liftable*. In fact, more is true.

Proposition 3.2.8 *If M is a trivial-source kG-module, then there is a trivial-source $\mathcal{O}G$-module, unique up to isomorphism, whose reduction modulo p is M.*

(The ring $\mathcal{O}$ was introduced in Sect. 2.2 as the bridge between k and a field of characteristic 0.) That trivial-source kG-modules have a unique lift to trivial-source $\mathcal{O}G$-modules is a result of Scott [518], (for a quick proof, see [33, Section 3.11]) and we will use it in Sect. 8.2.

The main result on vertices and sources is the Green correspondence. This establishes a bijection between indecomposable kG-modules with vertex a p-subgroup Q of G and indecomposable $kN_G(Q)$-modules with vertex Q. This bijection is given by induction and restriction; we induce a module, and look for a particular summand of the resulting module, and similarly, we restrict a module and take a particular summand.

The way we will find this summand is to eliminate all the other summands. If Q is a p-subgroup of G and $N_G(Q) \leq H \leq G$, then define two sets of subgroups of G:

$$\mathfrak{X} = \{Q \cap Q^g \mid g \in G \setminus H\}, \qquad \mathfrak{Y} = \{H \cap Q^g \mid g \in G \setminus H\}.$$

Notice that $\mathfrak{X}$ consists of proper subgroups of Q, whereas $\mathfrak{Y}$ consists of conjugates of proper subgroups of Q, and perhaps conjugates of Q itself, but not Q. For example, if G is a group where all involutions are conjugate and $|Q| = 2$, then $N_G(Q)$ contains a Sylow 2-subgroup of G, hence clearly there are conjugates of Q inside $\mathfrak{Y}$ whenever the Sylow 2-subgroup of G is not cyclic or quaternion. (In these cases a Sylow 2-subgroup possesses a single involution, as mentioned in Chap. 1.)

The Green correspondence is as follows.

Theorem 3.2.9 (Green Correspondence [265]) *Let G be a finite group and let Q be a p-subgroup of G, with $N_G(Q) \leq H \leq G$.*

If M is an indecomposable kG-module with vertex Q, then

$$M{\downarrow_H} \cong f(M) \oplus Y,$$

where $f(M)$ is indecomposable and with vertex Q, and Y is a sum of indecomposable modules, each of which is projective relative to a subgroup in $\mathfrak{Y}$. Conversely, if N is an indecomposable kH-module with vertex Q, then

$$N{\uparrow^G} \cong g(N) \oplus X,$$

where $g(N)$ is indecomposable and with vertex Q, and X is a sum of indecomposable modules, each of which is projective relative to a subgroup in $\mathfrak{X}$.

We have $g(f(M)) \cong M$ and $f(g(N)) \cong N$ for all indecomposable kG-modules M and indecomposable kH-modules N.[2]

Green's theorem is a powerful generalization of a result of Conlon [120], which proved that if N is an indecomposable module in a block B' of $N_G(D)$ with defect group D, and B is its Brauer correspondent in G, then there is a summand M of $N{\uparrow}^G$ in B with N a summand of $M{\downarrow}_H$. If N has vertex D as well, then the Green correspondence states that M must have vertex D, and we have pinned down the block to which the Green correspondent belongs.

In other words, we have the Green correspondence between modules of G and modules of $N_G(Q)$ for various Q, and a Brauer correspondence between blocks of G and blocks of $N_G(Q)$ for various Q. One should expect that they are compatible, and indeed they are.

Proposition 3.2.10 (Conlon–Green [120, 265]) *Let G be a finite group, and let B be a p-block of G with defect group D. Writing B' for its Brauer correspondent in $N_G(D)$, if M is an indecomposable B-module with vertex D, then the Green correspondent of M belongs to B'.*

The Heller translate $\Omega(-)$ we introduced in the last section interacts well with sources and Green correspondents, because of the nice interplay with the restriction map. If S is a source of an indecomposable module M, then $\Omega(M)$ has the same vertex as M, and $\Omega(S)$ is a source of $\Omega(M)$. Similarly, if M has Green correspondent $f(M)$, then $f(\Omega(M)) \cong \Omega(f(M))$. These facts come in handy when trying to understand blocks with cyclic defect group, where there are only finitely many indecomposable modules. We will talk a lot more about blocks with cyclic defect group in Chap. 5.

The Green correspondence allows us to relate a module for G with a module for $N_G(Q)$ for some Q, but how do we detect the vertex Q of the module? There are a couple of situations where some information can be obtained. First, as we mentioned before, Green proved that if B is a block of kG with defect group D, then every module in B is relatively D-projective. In certain circumstances one can do better.

Theorem 3.2.11 (Knörr [365]) *Let G be a finite group and let M be a simple kG-module belonging to a block B with defect group D. There is a vertex Q of M such that Q is a subgroup of D containing $C_D(Q)$. In particular, if D is abelian then $Q = D$.*

This shows that blocks with abelian defect groups are easy, at least for simple modules. If the vertex is cyclic, then we get even more.

[2] We want to say that f and g are bijections and $f = g^{-1}$, but we defined f and g on *all* indecomposable kG- and kH-modules, not on isomorphism classes, so all we know is that $f = g^{-1}$ *up to isomorphism*. We see that [6, Theorem 11.1] and [265], for example, avoided this trap, but [33, Theorem 3.12.2], and an earlier version of this book, fell into it.

Theorem 3.2.12 (Erdmann [190]) *Let G be a finite group and let M be a simple kG-module, belonging to a block B with defect group D. If a vertex of M is cyclic, then D is a vertex of M (so in particular D is cyclic).*

This cannot be extended further, as if $G = \mathrm{PSL}_2(q)$ for $q \equiv 1 \bmod 4$ then from [191] we see that G has two non-trivial simple modules in the principal 2-block with Klein four vertex. Since the defect group is dihedral of order 2^n for arbitrarily large n, we see that the previous theorem cannot be extended in any way to Klein four vertex. However, if p is odd then this appears not to be the case, and we should be able to bound the order of a defect group D of a block B in terms of the order of a vertex Q of a simple B-module, i.e., there is some global function $f : \mathbb{N} \to \mathbb{N}$ such that for odd primes p we have $|D| \leq f(|Q|)$. We will talk about this in Sect. 4.6.

These are general theorems that give you some information about vertices of simple modules. In practice, one can restrict a given module and look at its summands, using the Burry–Carlson–Puig theorem.

Theorem 3.2.13 (Burry–Carlson–Puig [82, 467]) *Let G be a finite group with a subgroup H, let M be an indecomposable kG-module, and let Q be a p-subgroup of G, with $N_G(Q) \leq H \leq G$. If the restriction $M{\downarrow_H}$ contains a summand N with vertex Q then M has vertex Q, and M and N are Green correspondents.*

So if you already have a guess for the vertex of a module, this theorem can really help you out.

This seems like an appropriate place to collect some useful results about induction and restriction. The most basic of these is the Nakayama relation from Chap. 1, which we remind the reader states that

$$\mathrm{Hom}_{kG}(M, N{\uparrow^G}) \cong \mathrm{Hom}_{kH}(M{\downarrow_H}, N),$$

where $H \leq G$, M is a kG-module and N is a kH-module. (There is a similar relation with the terms swapped.) The most useful, and the basis of proving much of the Green correspondence, is the Mackey formula, which tells you how to induce and then restrict.

Theorem 3.2.14 (Mackey Formula [410]) *Let G be a finite group and let H and L be subgroups of G. Let M be a kH-module. We have that*

$$M{\uparrow^G}{\downarrow_L} \cong \bigoplus_{t \in T} M^t{\downarrow_{H^t \cap L}}{\uparrow^L},$$

where T is a set of (H, L)-double coset representatives in G.

We can also relate tensor products and induction, using the Nakayama relation, to get for example (Exercise 3.5) that if M is a kG-module and N is a kH-module, then

$$(M{\downarrow_H} \otimes N){\uparrow^G} \cong M \otimes (N{\uparrow^G}),$$

and using the Mackey formula we get that if M is a kH-module and N is a kL-module then

$$(M\uparrow^G) \otimes (N\uparrow^G) \cong \bigoplus_{t\in T} (M^t\downarrow_{H^t\cap L} \otimes N\downarrow_{H^t\cap L})\uparrow^G.$$

(See, for example, [33, Corollary 3.3.5].)

The equation $(M\downarrow_H \otimes N)\uparrow^G \cong M \otimes (N\uparrow^G)$ implies that the tensor product of a projective module and any other module is always projective. To see this, note that it first follows from the statement that the tensor product of a free module and any other module is always free. Then we have that $kG \cong k_1\uparrow^G$, where k_1 is the trivial module for the trivial group. Letting $H = 1$ and $N = k_1$, the equation yields that $M\otimes k_1\uparrow^G$ is isomorphic to some module for the trivial group induced to G. Since the only modules for the trivial group are trivial, we see that $M \otimes (k_1\uparrow^G) \cong M \otimes kG$ is always free. Hence if M is any kG-module and N is a projective kG-module, $M \otimes N$ is always projective.

In fact, more is true.

Proposition 3.2.15 *Let G be a finite group and let H be a subgroup of G. If M and N are kG-modules with M relatively H-projective, then $M \otimes N$ is relatively H-projective.*

It is a simple consequence of our previous fact about tensor products and induction: if there is some kH-module S such that $M \mid S\uparrow^G$, then

$$(M \otimes N) \mid \left((S\uparrow^G) \otimes N\right) \cong (S \otimes N\downarrow_H)\uparrow^G.$$

Trying to piece all this information together about vertices, tensor products, complexity and so forth, we get the idea of a support variety $V_G(M)$, where G is an arbitrary finite group and M is a kG-module. In order to define this, and at least be able to talk sensibly about it, we would need a deeper understanding about group cohomology than is provided by this book, but with what we have so far we can define $V_G(M)$ for G an elementary abelian p-group. This description, called the rank variety (for elementary abelian p-groups, the support variety is isomorphic to the rank variety), was developed by Carlson [96]. (For a description of the general case, see for example [34, Section 5.7].)

Let $G = \langle x_1, \ldots, x_n\rangle$ be an elementary abelian p-group of rank n, and let $X_i = x_i - 1$ be elements of kG, in fact of the Jacobson radical $J(kG)$ of kG. Write V_G for the subspace of kG spanned by the X_i, and notice that, if $v \in V_G$ then $v^p = 0$, so that $1 + v$ has order p whenever $v \neq 0$.

This means that $\langle 1 + v\rangle$ is a subgroup of kG of order p, and of course the group algebra $k(\langle 1 + v\rangle)$ is a subalgebra of kG, so we may restrict kG-modules to get $k(\langle 1 + v\rangle)$-modules. Such a subgroup $\langle 1 + v\rangle$ is called a *cyclic shifted subgroup*.

Cyclic shifted subgroups were introduced by Dade [145], and feature in the statement of Dade's lemma, which allows us to detect freedom of a module on elements. (Since we are dealing with p-groups, there is only one indecomposable projective, and so projective and free are the same thing.)

Theorem 3.2.16 (Dade's Lemma [145]) *If P is an elementary abelian p-group, and M is a finite-dimensional kP-module, then M is free if and only if the restriction of M to every cyclic shifted subgroup is free.*

If M is a kP-module for an elementary abelian p-group P, and M is not free, then there are some cyclic shifted subgroups for which the restriction fails to be free. Carlson asked what the collection of those shifted subgroups looks like, and found that it had the structure of a variety.

Definition 3.2.17 Let G be an elementary abelian p-group and let M be a kG-module. The *rank variety* $V_G(M)$ consists of 0 and all points v in V_G such that $M\downarrow_{\langle 1+v\rangle}$ is not free.

The rank variety satisfies some very nice properties, which we will give in one big proposition.

Proposition 3.2.18 *Let M, M_1 and M_2 be kG-modules for an elementary abelian p-group G.*

(i) $V_G(M) = \{0\}$ *if and only if M is projective.*
(ii) *More generally,* $\dim(V_G(M))$ *is the complexity of M.*
(iii) *If $p \nmid \dim(M)$ then* $V_G(M) = V_G$.
(iv) $V_G(M_1 \oplus M_2) = V_G(M_1) \cup V_G(M_2)$.
(v) $V_G(M_1 \otimes M_2) = V_G(M_1) \cap V_G(M_2)$.
(vi) $V_G(M) = V_G(M^*) = V_G(\Omega(M))$.

Some of these are easy to see: if M is projective, i.e., a free module, then the restriction of M to $\langle 1+v\rangle$ is free; if the restriction of M_1 to $\langle 1+v\rangle$ is not free, then the same holds for $M_1 \oplus M_2$, and so on. This proposition is also true for the support variety $V_G(M)$, when G is not elementary abelian, which can be thought of as piecing together $V_G(E)$ for E running over the conjugacy classes of (non-trivial) elementary abelian subgroups of G. For the variety V_G, i.e., $V_G(k)$, this decomposition result is *Quillen stratification* [472, 473], and it was generalized to $V_G(M)$ by Avrunin and Scott [24].

The third part of this proposition tells us that rank varieties are only really of interest for modules of dimension a multiple of p. If $G = C_p \times C_p$, there should be a characterization of those indecomposable modules M whose rank variety is 1-dimensional, i.e., such that M is periodic. This was given in [126, Conjecture E], and is my personal favourite out of the conjectures I have made. Call a module *algebraic* if it satisfies a polynomial, under direct sum and tensor product, or equivalently, only finitely many isomorphism classes of summands appear in all successive tensor powers M, $M^{\otimes 2}$, $M^{\otimes 3}$, and so on.

Conjecture 3.2.19 Let k be an algebraically closed field of characteristic p, let $G = C_p \times C_p$, and let M be an indecomposable kG-module. If $\dim(M)$ is divisible by p, then M is algebraic if and only if it is periodic.

The reason why I think this is intriguing is that neither direction of this conjecture seems clear, and it relates the tensor product and cohomological structures of $C_p \times C_p$ in a way not seen before. There are partial versions of this conjecture for all p-groups, but it is for $C_p \times C_p$ where it takes its most attractive form. (For $p = 2$ this result holds, as we can see from the description of the indecomposable modules for V_4, their tensor products and their Heller translates, in Sect. 6.1.)

We have given a few ideas concerning how the modules of a finite group relate to the modules of various subgroups of a group. In order to progress further we need a better language to talk about the collection of all modules belonging to a finite-dimensional algebra, concepts like the module categories, equivalences of categories, and so on. The next section develops this language so we can talk about when two blocks 'look the same'.

3.3 The Module Category

Category theory is the language used to talk about much of modern mathematics, and representation theory is no exception. The *module category*, denoted mod-kG, has as objects all finite-dimensional kG-modules and as morphisms all kG-module homomorphisms.

We can define concepts such as being free and projective in terms of universal properties in the standard category-theoretic sense. A module is projective if, given modules M and N, a surjective homomorphism $\phi : M \to N$ and any homomorphism $\psi : X \to N$, there exists a map $\theta : X \to M$ such that the diagram

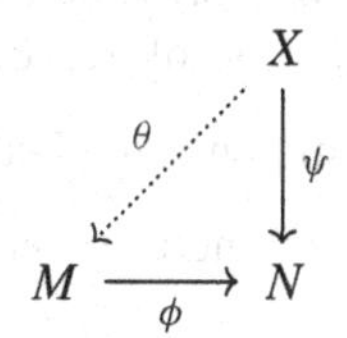

commutes, i.e., $\theta\phi = \psi$. This is equivalent to our definition of it being a summand of a free module (Exercise 3.13).

Let B be a p-block of a finite group G, with defect group D. As we have said before, for a given kD-module S, there are only finitely many summands of $S{\uparrow}^G$, so there are only finitely many indecomposable modules with a given source. Thus it makes sense to examine modules for a given p-group, just as we did when discussing periodicity. The first thing to do is to understand modules for a p-group P with vertex smaller than P.

Theorem 3.3.1 (Green's Indecomposability Criterion [263]) *Let G be a finite group, let H be a subnormal subgroup of G of p-power index, and let M be an indecomposable kH-module. The induction $M{\uparrow}^G$ is indecomposable.*

(This theorem definitely requires k to be algebraically closed.) If P is a cyclic p-group of order p^n, then Example 3.1.8 shows that every indecomposable module is uniserial, and therefore there are only finitely many indecomposable modules for cyclic p-groups.

If P is non-cyclic then there is a quotient of P that has an elementary abelian subgroup $C_p \times C_p$. There are infinitely many indecomposable modules for $C_p \times C_p$: an easy way to see this is to note that the trivial module is non-periodic and so $\Omega^i(k)$ for $i \in \mathbb{Z}$ form infinitely many indecomposable modules. It is possible to construct infinitely many indecomposable modules directly whenever k is infinite, by sending x and y, generators of $C_p \times C_p$, to

$$\begin{pmatrix} 1 & 1 \\ 0 & 1 \end{pmatrix} \quad \text{and} \quad \begin{pmatrix} 1 & \lambda \\ 0 & 1 \end{pmatrix}$$

for $\lambda \in k$. (These are all non-isomorphic: in order for two of these to be isomorphic, one must find a matrix in $\mathrm{GL}_2(k)$ that conjugates the first matrix to itself, i.e., centralizes it. But any matrix that centralizes the first matrix must centralize the second.) However, all we have constructed are indecomposable modules parametrized by k. We can, and will in Sect. 6.1, give a construction of all indecomposable modules for the Klein four group, and they fit into a set of 0- and 1-parameter families, indexed by elements of k.

On the other hand, the modules for $C_p \times C_p$ over a field of odd characteristic p are given by all pairs of commuting matrices of order p. This is a complete mess, and we cannot hope to classify these up to isomorphism.

Thus there are (at least) three types of behaviour: finite; infinite, but reasonable; a complete mess. These three types of representation type are called finite, tame and wild, and we formally define the concepts of 'reasonable' and 'complete mess' now.

Definition 3.3.2 Let A be a finite-dimensional k-algebra.

(i) A has *finite representation type* if there are only finitely many indecomposable A-modules.
(ii) A has *tame representation type* if there are infinitely many indecomposable A-modules and, for each $n \in \mathbb{N}$, the indecomposable A-modules of dimension n fall into finitely many 1-parameter families indexed by elements of k, together with finitely many other modules.
(iii) A has *wild representation type* if, for every finite-dimensional k-algebra A', there is an embedding of the module category of A' into the module category of A preserving isomorphism and indecomposability of A'-modules, specifically a faithful, exact functor from mod-A' to mod-A that respects the action of k on homomorphisms between A'-modules, and preserving isomorphisms and indecomposability.

We will see this 'k-linearity', respecting the action of k on homomorphisms, again soon in this section, so we do not go into detail yet. You can still get a good idea of how bad wild is from this definition without it, but k-linearity is important for us later (see in particular Definition 4.5.1).

Of course, if A does not have finite representation type then it has infinite representation type, but just because the indecomposable modules for A cannot be nicely parametrized does not mean that we can embed every other representation theory into it. However, the trichotomy theorem of Drozd [171] (see also the work of Crawley-Boevey [135]) proves that any finite-dimensional algebra A has one of the three types above.

Related to this is the first Brauer–Thrall conjecture, which states that, for a given finite-dimensional (in fact, artinian will do) algebra A, if there is some integer n such that every indecomposable A-module has at most n composition factors, then A has finite representation type. This was proved by Roiter [494] in 1968. (The second Brauer–Thrall conjecture states that if A is of infinite type and infinite then there are infinitely many integers n such that there are infinitely many indecomposable modules with n composition factors; over an algebraically closed field this was proved by Bautista [27] in 1985, but the general case remains open.)

We have seen above that if A is a block of a finite group then A has finite representation type if and only if A has cyclic defect group, but what about tame and wild types?

Theorem 3.3.3 *A p-block of a finite group has tame representation type if and only if the defect group is one of dihedral, semidihedral or quaternion, i.e., $p = 2$ and the defect group is either Klein four or is non-abelian and has maximal class.*

It was proved by Kruglyak [375] and Brenner [63] that the p-group P has wild type if $|P : P'| > 4$ and P is non-cyclic; Bondarenko and Drozd [41] finished it off, proving that if $|P : P'| \leq 4$ and P is non-cyclic (these are precisely the dihedral, semidihedral and quaternion 2-groups) then P has tame representation type.

There is a nice parametrization of the indecomposable modules for dihedral 2-groups (proved independently by Bondarenko [40] and Ringel [484]) and given in Sect. 6.2, and a less nice parametrization of the indecomposable modules for semidihedral 2-groups [41] (this is how Bondarenko and Drozd completed the proof of Theorem 3.3.3), with another description by Crawley-Boevey [136], but for quaternion 2-groups the only current method to obtain indecomposable modules is to restrict those from the semidihedral 2-groups down to the subgroup of index 2. This shows that there are only tamely many, but does not provide a construction. This deficiency in the theory was remarked upon in [33, Section 4.11], and in the couple of decades since its publication, it still hasn't been solved.

So if the block does not have finite or tame representation type then we don't have any hope of completely understanding the full module category of the block. This is because to understand the modules for one wild k-algebra means understanding the modules for every other k-algebra. Thus the entire module category of every finite group in characteristic 3 is embedded in the module category of $k(C_3 \times C_3)$. It also contains a famous problem in linear algebra that is considered 'impossible', namely classifying all ordered n-tuples of $m \times m$ matrices up to simultaneous conjugation, i.e., whether there is a single $m \times m$ matrix that simultaneously conjugates a given ordered collection of $m \times m$ matrices into another given ordered collection.

Given this hopeless task, we can either consider only some of the modules belonging to the block, or we can test if two blocks have equivalent module categories without having a full description of them. For the latter, we have the concept of a Morita equivalence.

Definition 3.3.4 Two rings B and B' are *Morita equivalent* if the module categories of B and B' are equivalent.

(In many textbooks you will find that two rings are Morita equivalent if the module categories are equivalent, meaning *all* modules, not just finitely generated ones. However, as part of Morita's theorem (Theorem 3.3.8) below it turns out that this distinction is unnecessary.)

Although this is the correct definition for arbitrary rings, we need to be slightly more careful for group algebras. Group algebras, and in general all k-algebras, are not just rings but carry a structure given by multiplication by elements of k. Thus, for X, Y and Z in the module category of a k-algebra, $\mathrm{Hom}(X, Y)$ carries the structure of a k-module and the composition $\mathrm{Hom}(X, Y) \times \mathrm{Hom}(Y, Z) \to \mathrm{Hom}(X, Z)$ is k-bilinear.

A priori, a Morita equivalence need not respect this extra 'k-linear' structure, so if B and B' are k-algebras then our definition is slightly different.

Definition 3.3.5 Two k-algebras B and B' are *Morita equivalent* if the module categories of B and B' are equivalent as k-linear categories.

Thus the equivalence of categories should respect multiplication by elements of k on $\mathrm{Hom}(X, Y)$. This turns out to not just be semantics, as there are Morita equivalences between blocks of finite groups algebras *as rings* that are not Morita equivalences *as k-algebras*, and in fact in some cases the two blocks are not Morita equivalent as k-algebras (see Sect. 4.5, particularly Definition 4.5.1). It is a fundamental part of Donovan's conjecture, which we state below.

We give a quick example, which will be important in the next chapter.

Example 3.3.6 If R is a ring, then R and the matrix ring $M_n(R)$ are Morita equivalent. (See, for example [394, Corollary 2.8.8].) All p-blocks of defect zero are Morita equivalent, both as rings and as k-algebras.

In order to state some of Morita's theorem, we need the idea of a progenerator.

Definition 3.3.7 Let A be a k-algebra. A (right) A-module M is called a (right) *progenerator* of A if

(i) M is finitely generated,
(ii) M is projective, and
(iii) as a right A-module, A is a summand of $M^{\oplus n}$ for some $n > 0$.

Left progenerators are defined similarly.

For blocks of finite groups, progenerators are simply finite sums of projective indecomposable modules that include at least one copy of each projective in the block. The *opposite ring* of a ring R, denoted R^{opp}, is the ring with the same underlying set, the same addition, but multiplication is reversed, i.e., the product of x and y in R^{opp} is yx, not xy.

Theorem 3.3.8 (Morita's Theorem [432]) *Let A and B be finite-dimensional k-algebras. The following are equivalent:*

(i) *A and B are Morita equivalent;*
(ii) *the categories of all (including non-finitely generated) A-modules and B-modules are equivalent;*
(iii) *M is a progenerator for A, with B and $\mathrm{End}_A(M)^{\mathrm{opp}}$ isomorphic as k-algebras;*
(iv) *there is an (A, B)-bimodule M and a (B, A)-bimodule N (in fact, N is isomorphic to $\mathrm{Hom}_A(M, A)$) such that $M \otimes_B N \cong A$ and $N \otimes_A M \cong B$, as (A, A)- and (B, B)-bimodules respectively.*

(See, for example, [394, Theorem 2.8.2].)

This whole lot is a bit of a mouthful, but the third condition states that we should be able to test Morita equivalence in a fairly straightforward way, and gives us a recipe to construct Morita equivalences in practice, as well as in theory.

Now that we have Morita equivalences, we can strengthen the weak Donovan conjecture into the full-strength Donovan conjecture.

Conjecture 3.3.9 (Donovan Conjecture) Let D be a finite p-group. There exists a finite list $\mathcal{A}$ of k-algebras such that, if B is a p-block of a finite group, and B has defect group D, then B is Morita equivalent to one of the algebras in $\mathcal{A}$.

In other words, up to Morita equivalence, there are only finitely many blocks of finite groups with defect group D.

This appears, for example, in [4, Conjecture M]. We will talk about Donovan's conjecture in Sect. 4.5.

A few paragraphs ago, we said that because the module category is in general too massive to understand, at least whenever p divides $|G|$, we can either consider Morita equivalences or consider only some of the indecomposable modules. In the other direction, that of finding small collections of 'interesting' modules, one collection stands out as being useful in a variety of contexts.

If M is a kG-module, then M is *endopermutation* if $M \otimes M^* \cong \mathrm{End}_k(M)$ is isomorphic to a permutation module. Similarly, M is an *endotrivial* module if $M \otimes M^*$ is the sum of a projective module and a single trivial module. Certainly trivial modules are permutation modules, and the free module is also a permutation module, so for p-groups endotrivial modules are endopermutation. However, in general, projective modules are not necessarily permutation modules.

If M and N are endopermutation modules for a finite p-group, then $M \oplus N$ is usually not an endopermutation module, because

$$(M \oplus N) \otimes (M \oplus N)^* \cong (M \otimes M^*) \oplus (N \otimes N^*) \oplus (M \otimes N^*) \oplus (M^* \otimes N),$$

and so it is an endopermutation module if and only if $M \otimes N^* = \mathrm{Hom}(N, M)$ is a permutation module. If M and N are indecomposable and have the same vertex, then $\mathrm{Hom}(N, M)$ is a permutation module if and only if $M \cong N$ [144, Theorem 3.8].

It is easier to work with endotrivial modules, and they still show us most of the ideas. If M and N are endotrivial modules then $M \otimes N$ is also endotrivial, because

$$(M \otimes N) \otimes (M \otimes N)^* \cong (M \otimes M^*) \otimes (N \otimes N^*) \cong (X_1 \oplus k) \otimes (X_2 \oplus k) \cong X \oplus k,$$

where X_1, X_2 and X are all projective modules. If M is endotrivial and $M = M_1 \oplus M_2$, then either $M_1 \otimes M_1^*$ or $M_2 \otimes M_2^*$ is projective, and this implies that either M_1 or M_2 is projective. (It cannot be that both are projective, as then we cannot have a trivial summand.) Thus any endotrivial module is the sum of a projective and an indecomposable endotrivial module, and this allows us to place a group structure on the class of indecomposable endotrivial modules, called the *Dade group*. The trivial module is the identity and the dual is the inverse. We can also use our nice formula combining the tensor product and Heller translate to see that

$$\Omega(M) \otimes \Omega(M)^* \cong \Omega(M) \otimes \Omega^{-1}(M^*) \cong M \otimes M^* \cong k$$

if one is working modulo projectives. (This previews the idea of the stable category—where we always work modulo projectives—which will be formally introduced in Sect. 3.5.)

Of course, at some point we should note that the trivial module is a summand of $M \otimes M^*$. This is true if and only if $p \nmid \dim(M)$. This forms part of a very useful lemma of Benson and Carlson.

Lemma 3.3.10 (Benson–Carlson [35]) *Let M and N be indecomposable kG-modules for a finite group G.*

(i) *$M \otimes N$ has a trivial summand if and only if $p \nmid \dim(M)$ and $N \cong M^*$.*
(ii) *If $p \mid \dim(M)$ then p divides the dimension of every summand of $M \otimes N$.*

(This definitely requires k to be algebraically closed.) The structure of the Dade group $T(P)$ of endotrivial modules for a p-group P has been completely determined. In the original paper [144] where endopermutation and endotrivial modules were defined, Dade computed $T(P)$ for P an abelian p-group: $T(P)$ is cyclic, generated by $\Omega(k)$. (He also computed the whole group of endopermutation modules in that paper.) The Dade group is a finitely generated abelian group by a result of Puig [470], and as such is the sum of a torsion part and a torsion-free part. The rank of the torsion-free part was computed by Alperin [8], and depends on the p-rank of the p-group: if it has p-rank 1 then $|T(P)|$ is finite; if it has p-rank 2 then the torsion-free rank of $T(P)$ is the number of conjugacy classes of maximal subgroups $C_p \times C_p$; if the p-rank of P is 3 or more then it is the number of conjugacy classes of maximal subgroups $C_p \times C_p$ plus 1.

Hence we need to understand the torsion part of $T(P)$. Carlson and Thévenaz, in a series of papers [104–106], completed the classification of torsion endotrivial modules, and hence of all endotrivial modules. If p is odd, then $T(P)$ is actually a torsion-free group if P is non-cyclic, and if P is cyclic then $|T(P)|$ has order 2, consisting of k and $\Omega(k)$. If $p = 2$ then again $T(P)$ is torsion-free if P is not cyclic, quaternion or semidihedral; if P is cyclic of order 2 then $T(P) = 1$, if it cyclic of order at least 4 then again $|T(P)| = 2$, and if P is semidihedral then $T(P) \cong \mathbb{Z} \oplus C_2$. Finally, if P is quaternion then $T(P) \cong C_4 \oplus C_2$.

The Dade group $D(P)$ of all endopermutation modules for a finite p-group is slightly more complicated to define. First, we consider only *capped* endopermutation modules, i.e., endopermutation modules with an indecomposable summand of vertex P (necessarily unique up to isomorphism), which is called the *cap* of the module. The tensor product of two caps is also a capped endopermutation module, and by taking its cap, we obtain a group structure on the set of caps. All endotrivial modules are capped endopermutation modules, so we have an embedding of $T(P)$ into $D(P)$. Using the classification of $T(P)$ Bouc [45] determined the complete structure of $D(P)$, using the computation of the torsion-free part by Bouc and Thévenaz in [46].

More recently, work has shifted to understanding the endotrivial modules for general groups G, not just p-groups. Let P be a Sylow p-subgroup of G: by taking restrictions of modules from G to P, we see that there is a homomorphism $\phi : T(G) \to T(P)$. We want to understand the kernel $T(G, P)$ of this map, i.e., all trivial-source, endotrivial modules for G, and also the image of ϕ in $T(P)$.

In 2006, Carlson, Mazza and Nakano [98] generalized Alperin's result on the torsion-free rank of $T(G)$ from p-groups to all finite groups G, and started the process of analysing $T(G)$ for G a simple group, determining $T(G)$ for G a group of Lie type in characteristic p. One of the key obstructions to understanding $T(G)$ rather than $T(P)$ is the following: a kG-module is endotrivial if and only if its restriction to all elementary abelian p-subgroups is endotrivial, and if a kG-module is endotrivial then its Green correspondent in $N_G(P)$ is (endotrivial modules always have Sylow p-subgroups as vertices), but if M is an endotrivial module for $N_G(P)$ then its Green correspondent in G need not be. From [103], we at least know that $T(N_G(P))$ is understood completely.

Suppose first that a Sylow p-subgroup of G is cyclic, let Q be a subgroup of order p in G, and let $H = N_G(Q)$. Mazza and Thévenaz proved [425] that $T(G)$ is isomorphic to $T(H)$ via induction and restriction, and furthermore, $T(H)$ is generated by 1-dimensional kH-modules and $\Omega(k)$. This completely determines $T(G)$ in these cases, leaving only groups with non-cyclic Sylow p-subgroups.

If $p = 2$ and the Sylow 2-subgroups P of G are quaternion or semidihedral, then there are torsion elements in $T(P)$, and in [102] Carlson, Mazza and Thévenaz prove that the map $\phi : T(G) \to T(P)$ is surjective in these cases. Again, they give a complete description of $T(G)$ for these groups. As cyclic, semidihedral and quaternion groups are the only ones to have torsion in $T(P)$, this means that, as an abstract abelian group, $\mathrm{im}(\phi)$ is understood; actual generators in general are not, however, known. It was conjectured that $\mathrm{im}(\phi)$ is simply the G-stable elements of $T(P)$, i.e., those that are stable under conjugation by elements of G. Unfortunately, Barthel, Grodal and Hunt have recently found a counterexample for $G = \mathrm{PSL}_3(7)$.

By a case-by-case analysis, if P is non-cyclic then $T(G) = T(G, P) \oplus \mathrm{im}(\phi)$: this was checked for the quaternion and semidihedral cases directly in [102], and follows if $T(P)$ is torsion-free by general facts about abelian groups. Thus we focus our attention on $T(G, P)$.

A number of papers have attacked specific simple groups, for example groups of Lie type in characteristic p in [98], $\mathrm{GL}_n(q)$ and $\mathrm{SL}_n(q)$ where $p \nmid q$ in [100, 101], and symmetric and alternating groups in [97, 99].

In a different direction, Balmer proved in [25] that $T(G, P)$ is isomorphic to the group of weak homomorphisms: a map $\chi : G \to k^\times$ is a *weak homomorphism* if $\chi(g) = 1$ whenever $g \in P$ or $P \cap P^g = 1$, and if $g, h \in G$ with $P \cap P^h \cap P^{gh} \neq 1$, then $\chi(gh) = \chi(g)\chi(h)$. If χ and ψ are weak homomorphisms then so is $\chi\psi$, where $(\chi\psi)(g) = \chi(g)\psi(g)$, and this turns the set of weak homomorphisms into an abelian group, isomorphic to $T(G, P)$. In particular, this shows that if G has trivial intersection Sylow p-subgroups then $T(G, P)$ is isomorphic to the group of 1-dimensional representations (over k) of $N_G(P)$.

Recently, Grodal has given in [268] a homotopical description of $T(G, P)$ as the 1-cohomology group of a certain category, called the *orbit category*: the category whose objects are the sets G/Q for Q a non-trivial p-subgroup of G, and whose morphisms are G-homomorphisms.

In many cases, there is an isomorphism between $T(G, P)$ and $\mathrm{Hom}_{kG}(G_0, k^\times)$, where G_0 is the group generated by the subgroups $N_G(Q)$ for $1 < Q \leq P$. This gives a topological method for constructing $T(G, P)$ which is very powerful in general. For example, in [268] Grodal shows that if G is the Monster then $T(G, P)$ is trivial if $p = 2, 3, 5, 7, 11, 13$ (i.e., the Sylow p-subgroup is non-cyclic), and $T(G, P) = \mathrm{Hom}(N_G(P)/P, k^\times)$ if $p > 13$. His general methods also reprove many of the results from the literature mentioned above, which fundamentally rely on a certain subgroup complex being simply connected. For example, if $G = \mathrm{GL}_n(q)$ for $p \nmid q$ and the p-rank of G is least 3, then $T(G, P) = \mathrm{Hom}(G, k^\times)$, which was known by [100], but if $G = \mathrm{Sp}_{2n}(q)$ and the order of p modulo q is odd, and G has p-rank at least 3, then again $T(G, P) = 1$, and this result is new.

Translated into finite group theory, the criterion is fairly easy to understand: let K_G be the subgroup of $N_G(P)$ generated by all elements $g \in N_G(P)$ that may be written as $g = x_1 x_2 \dots x_n$, where for each $1 \leq i \leq n$ we have $x_i \in O^{p'}(N_G(Q_i))$ for some $1 < Q_i < P$, and there exists some $1 \neq z \in P$ such that $z^{x_1 \dots x_i}$ lies inside Q_i for all $1 \leq i \leq n$. Then $T(G, P)$ is the abelianization (i.e., group modulo its derived subgroup) of $N_G(P)/K_G$. Using this criterion, the author was able to swiftly deduce the structure of $T(G, P)$ for all sporadic simple groups [130], and vastly reduce the length of the proof for symmetric and alternating groups, which took up two papers originally [97, 99]. (There is also a short proof in [268] for the Monster simple group using the original homotopy-theoretic methods.) This new method of determining $T(G, P)$ has been used by Carlson, Grodal, Mazza and Nakano to determine $T(G, P)$ for all groups of Lie type; the results have been announced but not yet published.

These new ideas have already improved our understanding of $T(G)$, and there is hope that a more complete description of it will result from the progress that is now underway.

Endopermutation modules appear in constructing equivalences of categories. For example, later in this chapter we will see stable equivalences of Morita type, and if tensoring by a module induces a stable equivalence of Morita type from a module category to itself, then that module is in fact endotrivial (see, for example, [394, Proposition 7.5.12]). Also, the sources of simple modules in nilpotent blocks, which we will see in Sect. 6.3, are endopermutation modules.

3.4 Extensions

If you have done any extension theory of groups, some of this is going to look very familiar, and it might also make results like the Schur–Zassenhaus theorem make more sense, as it puts cohomology in its 'correct' framework, as the special case of $\mathrm{Ext}^1(M, N)$ for modules M and N.

If M and N are two kG-modules, then an *extension* of N by M is a module E such that E possesses a submodule M with $E/M \cong N$,[3] or in other words, there is a 'short exact sequence'

$$0 \to M \to E \to N \to 0.$$

This is not the first time we have met exact sequences: projection resolutions were exact, if not short. A sequence of modules with homomorphisms between them is exact at a particular module M if the image of the map entering M is equal to the kernel of the map leaving M. We are going to need this later when looking at the derived category, so let's introduce a few terms right now. A *sequence* or *complex* is a chain of modules and homomorphisms

$$\cdots \to M_{i-1} \to M_i \to M_{i+1} \to \cdots$$

where the M_i are modules for all $i \in \mathbb{Z}$, and there are maps $\phi_i : M_i \to M_{i+1}$ such that $\mathrm{im}(\phi_{i-1}) \leq \ker(\phi_i)$ for all $i \in \mathbb{Z}$. If all but finitely many of the M_i are the zero module then the sequence is *bounded*. The sequence is *exact* at M_i if $\mathrm{im}(\phi_{i-1}) = \ker(\phi_i)$, and exact if it is exact at M_i for all $i \in \mathbb{Z}$. It is a *short exact sequence* if there exists some $i \in \mathbb{Z}$ such that $M_j = 0$ for $j < i-1$ and $j > i+1$.

[3]Warning! The statement 'E is an extension of A by B' means different things to different people. If A and B are modules, then A is the quotient, and if A and B are groups then B is the quotient.

We often ignore the zero terms in the sequence, or include only one zero either side of the non-zero terms. Notice that a short exact sequence is the shortest that an exact sequence can be without being $0 \to M \to M \to 0$.

The quotient module $\ker(\phi_i)/\mathrm{im}(\phi_{i-1})$ is called the *cohomology* at M_i. We refer to M_i as the term at *degree* i. Of course, the degree of a term is a relative concept, since everything may be shifted, but the degree is important when you have more than one complex.

An extension is another word for short exact sequence. An extension

$$0 \to M \to E \xrightarrow{\phi} N \to 0$$

splits if there exists a module homomorphism $\psi : N \to E$ such that $\psi\phi = \mathrm{id}_N$, or in other words there is a complement to the module M in E. In the category of groups, that just means that the group is a semidirect product, as not all subgroups are normal subgroups, but for modules, like abelian groups (which are just $\mathbb{Z}$-modules) a split extension is simply a direct sum.

The extensions of N by M are given by a vector space, which is labelled $\mathrm{Ext}^1_{kG}(N, M)$, although people sometimes drop the subscript kG if it is obvious which group is being considered. Formally, this is the derived functor of $\mathrm{Hom}_{kG}(-, M)$ applied to N, but that probably doesn't give you any information about it. We take a projective resolution of N:

$$\cdots \to P_n \to \cdots \to P_1 \to P_0 \to N \to 0,$$

and then apply the functor above to it, cutting off the N, to get

$$0 \to \mathrm{Hom}_{kG}(P_0, M) \to \mathrm{Hom}_{kG}(P_1, M) \to \cdots \to \mathrm{Hom}_{kG}(P_n, M) \to \cdots :$$

$\mathrm{Ext}^n_{kG}(N, M)$ is the cohomology of this complex at degree n (where P_n is the term in degree $-n$ in the original complex), so at $\mathrm{Hom}_{kG}(P_n, M)$.

(In [33, Proposition 2.5.5], it is shown that we can either take a derived functor of $\mathrm{Hom}_{kG}(N, -)$ and apply it to M, or take a derived functor of $\mathrm{Hom}_{kG}(-, M)$ and apply it to N, and we obtain the same answer. The only difference is one is a left derived functor and one is a right derived functor. This means we need an *injective* resolution rather than a projective resolution if we want to do it the other way around.)

The reader that's really on the ball might guess that this means that there is a relationship between Ext^1, Hom and the Heller translate $\Omega(-)$, and yes, there is: a surjective map

$$\mathrm{Hom}_{kG}(\Omega(N), M) \to \mathrm{Ext}^1_{kG}(N, M).$$

There is a kernel to this map, called $\mathrm{PHom}_{kG}(N, M)$, which we will see in the next section. This gives you another way to work out at least the dimension of Ext^1, if you know the projectives. Another thing it shows is that, if M and N are simple modules, the dimension of $\mathrm{Ext}^1_{kG}(N, M)$ is the multiplicity of M as a composition factor of $\mathrm{rad}(\mathcal{P}(N))/\mathrm{rad}^2(\mathcal{P}(N))$.

Proposition 3.4.1 *If X is a projective kG-module and M is any kG-module, then $\mathrm{Ext}^n_{kG}(X, M) = 0$ for all $n \geq 0$. In particular, any projective submodule or quotient is a summand.*

The projective modules are far too big to actually compute, so in practice people don't do that. There are computer algorithms[4] that can produce $\mathrm{Ext}^1_{kG}(N, M)$ for any two modules, if the groups and modules involved aren't too big, and there are theoretical tools that can help.

But how does $\mathrm{Ext}^1_{kG}(N, M)$ actually translate into extensions? We have a short exact sequence

$$0 \to M \to E \to N \to 0,$$

and we fit a projective resolution of N above it, then force maps to the sequence from the resolution. (These maps do exist because of the universal property of projective modules, which we saw in the previous section. For a construction, see for example [33, Theorem 2.4.2].) This gives us some commutative diagram

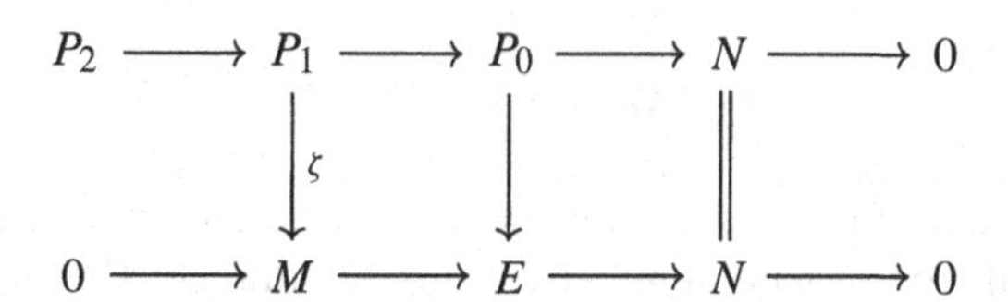

and the element of $\mathrm{Ext}^1_{kG}(M, N)$ that this corresponds to is the class of ζ (as Ext^1 is a quotient space).

If we have two extensions of N by M, say E and E', then they are *equivalent* if there is a commutative diagram

$$\begin{array}{ccccccccc} 0 & \longrightarrow & M & \longrightarrow & E' & \longrightarrow & N & \longrightarrow & 0 \\ & & \| & & \downarrow & & \| & & \\ 0 & \longrightarrow & M & \longrightarrow & E & \longrightarrow & N & \longrightarrow & 0. \end{array}$$

If there is such a diagram, then a result known as the Five Lemma from homological algebra (see, for example, [409, Lemma VIII.4.1]) implies that the map $E' \to E$ is an isomorphism. By chaining together equivalences we can see that being equivalent

[4]For example the command `Ext(M,N)` in Magma computes $\mathrm{Ext}^1_{kG}(M, N)$ as a k-vector space, and given an element of that space, the program can compute the corresponding extension.

is helpfully an equivalence relation on the collection of all short exact sequences. The equivalence classes of this relation form a set, which we may also denote by $\mathrm{Ext}^1_{kG}(N, M)$. In order to turn this into a vector space we need to know how to add two extensions, and it is not at all obvious that we can in any sense add two extensions. If we map everything back into the projective resolution then we can do it, but what if we start off with two short exact sequences?

Given the two extensions E and E' above, we first form the direct sum $E \oplus E'$: this has a submodule $M \oplus M$ with quotient $N \oplus N$, and we need a module with a submodule (isomorphic to) M and a quotient (isomorphic to) N. Thus we need to take a submodule and a quotient, peeling off one copy of M from the bottom and one copy of N from the top of the module. To remove a copy of N, take the submodule Y of $E \oplus E'$ consisting of all pairs (e, e') whose image in N is the same under the maps $\phi : E \to N$ and $\phi' : E' \to N$, i.e., $e\phi = e'\phi'$. This removes one copy of N from the top, and now we need to identify a copy of M as a submodule to quotient out by. If $\psi : M \to E$ and $\psi' : M \to E'$ are our two maps from the short exact sequence, then the elements $(m\psi, 0) - (0, m\psi')$ for $m \in M$ form a submodule $\bar{M}$ of $E \oplus E'$, and indeed $\bar{M}$ is isomorphic to M. We now take the quotient $\bar{E} = Y/\bar{M}$, which is an extension of N by M again. We need to furnish the reader with maps from M to $\bar{E}$ and from $\bar{E}$ to N: the map $m \mapsto (m\psi, 0) \sim (0, m\psi')$ is injective, and the map $(e, e') \to e\phi = e'\phi'$ gives us a surjective map, hence a short exact sequence

$$0 \to M \to \bar{E} \to N \to 0,$$

called the *Baer sum* of the two short exact sequences. The enthusiastic reader can now check that there is a natural map from this definition of Ext^1 to the definition involving projective resolutions, and the Baer sum maps to the standard sum of homomorphisms.

It is also fairly easy to see with this construction that this is commutative, associative, and that the direct sum forms the identity element.

As well as being able to add extensions, we want to be able to compose them as well. We cannot do this with any two extensions, just as we cannot multiply any two matrices: the 'inner terms' must agree. In this case, this means that the image of one must be the kernel of the other. If A, B and C are modules, and we have short exact sequences

$$0 \to A \to E_1 \to B \to 0, \qquad 0 \to B \to E_2 \to C \to 0,$$

then we can chain these together, to get an exact sequence with four non-zero terms, rather than just three:

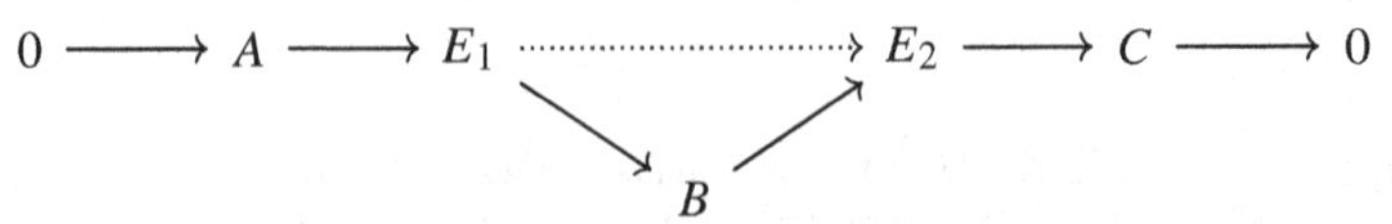

$$\begin{array}{ccccccccccc} 0 & \longrightarrow & A & \longrightarrow & E_1 & \cdots\cdots\cdots\rightarrow & & E_2 & \longrightarrow & C & \longrightarrow 0 \\ & & & & & \searrow & & \nearrow & & & \\ & & & & & & B & & & & \end{array}$$

This is an element of $\mathrm{Ext}^2_{kG}(C, A)$, as it happens. In general, $\mathrm{Ext}^n_{kG}(C, A)$ can be thought of as exact sequences with $n + 2$ non-zero terms, starting in A and ending in C, although with a somewhat non-trivial equivalence relation. So as long as we set $A = C$, we can place a product structure on the *Ext algebra*

$$\mathrm{Ext}_{kG}(A, A) = \bigoplus_{i \geq 0} \mathrm{Ext}^i_{kG}(A, A)$$

(where $\mathrm{Ext}^0_{kG}(A, A) = \mathrm{Hom}_{kG}(A, A)$), that goes from $\mathrm{Ext}^i_{kG}(A, A) \otimes \mathrm{Ext}^j_{kG}(A, A)$ to $\mathrm{Ext}^{i+j}_{kG}(A, A)$, doing as we have done above. This is called the *Yoneda product*, and together with a generalization of the Baer sum to $\mathrm{Ext}^n_{kG}(A, A)$ (given by the category-theoretic notions of pullbacks and pushouts, which are essentially what we did above) we turn $\mathrm{Ext}_{kG}(A, A)$ into a ring. This ring is graded (see Sect. 7.2), and for $A = k$ it is in fact *graded commutative*, i.e., $xy = (-1)^{|x|\,|y|}(yx)$, where x and y are homogenous elements (see Sect. 7.2) and $|x|$ and $|y|$ are the degrees of x and y.

A lot of energy has been expended by the community to compute Ext^1 between simple modules for various classes of groups. It suffices to go block by block, since we know from Sect. 2.1 that if M and N lie in different blocks then the only extension between them is the direct sum, so $\mathrm{Ext}^1_{kG}(N, M) = 0$ (in fact $\mathrm{Ext}^i_{kG}(N, M) = 0$, as we can see by writing $\mathrm{Ext}^i_{kG}(N, M)$ as a quotient of $\mathrm{Hom}_{kG}(\Omega^i(N), M)$, which is 0 as projectives only contain composition factors from their block). If a block has cyclic defect groups then Ext^1 between simple modules has dimension at most 1, as we will see in Chap. 5. For small-rank groups of Lie type there is a pretty decent theory. One can find a fairly authoritative survey of the current state of things in [307, Chapter 12].

The other main source of results in the literature is in group cohomology. Group cohomology is a special case of extension theory, because

$$H^n(G, M) = \mathrm{Ext}^n_{kG}(k, M).$$

Thus questions about group cohomology can be rephrased in terms of questions about extensions between simple modules. The most stringent conjecture is *Guralnick's conjecture* [270, Conjecture 2], which states that $\dim(H^1(G, M))$ is universally bounded by some constant n, for G an arbitrary finite group and M an arbitrary faithful, simple kG-module (this needs k to be algebraically closed, or M to be absolutely simple). For p-soluble groups there are no faithful simple modules with non-zero cohomology by Theorems 2.4.1 and 2.4.2, and [22, Theorem 3] shows that $H^1(G, M)$ for a general finite group G is bounded above by $H^1(H, N)$ for some simple subgroup H of G and simple submodule N of $M{\downarrow_H}$. Thus we may assume that G is a simple group. Originally, the only known examples of $H^1(G, M)$ for simple modules M for simple groups G had dimensions 0, 1 and 2, but examples with dimension 3 were found by Scott [519] and Bray–Wilson [62]. Scott and

Sprowl, in unpublished work in 2012, found examples in $\mathrm{SL}_7(q)$ with $H^1(G, M)$ having dimension 5, and Guralnick's conjecture started to look shaky.

On the other hand, back in 1986, Guralnick proved [270] that the dimension of $H^1(G, M)$ is at most $2\dim(M)/3$, for all finite groups G and all faithful simple modules M. He also conjectured in that paper that $\dim(H^1(G, M))$ should be bounded by $\dim(M)/2$, which was later proved by Guralnick and Hoffman in [271]. Even better, in [272], Guralnick and Tiep proved that for a Lie type group G of rank d and Weyl group W, in characteristic $r \neq p$, the dimension of $H^1(G, M)$ for any absolutely simple kG-module M is at most $|W| + d$, and later improved the bound to $|W|^{1/2} + d - 1$ in [273]. This bound is growing, of course, but it is decoupled from the dimension of M.

In [273], Guralnick and Tiep also prove that the dimension of $H^1(G, M)$ (for M an absolutely irreducible kG-module) is bounded by a constant depending only on p and the sectional p-rank of G. This was motivated by a consequence of Donovan's conjecture (Conjecture 3.3.9) mentioned by Kessar: as there are conjecturally only finitely many p-blocks with a given defect group D up to Morita equivalence, if M and N are simple kG-modules then $\mathrm{Ext}^i_{kG}(M, N)$ should be universally bounded in terms of D and i.

However, Guralnick's conjecture, while still technically open, was blown out of the water by results of Lübeck in [399]. Announced at the same conference as the Scott–Sprowl results above, he found some examples of modules with much larger 1-cohomology groups. In the later paper, he produced an example of a simple module for a simple group $E_6(q)$ with q a power of p, with 1-cohomology of dimension 3537142. Now the consensus is that it is definitely false, but the method that constructs these large dimensions of cohomology, which is to compute so-called Kazhdan–Lusztig polynomials, appears difficult to do theoretically for infinitely many ranks, and of course computational results can only produce finitely many polynomials.

3.5 The Stable and Derived Categories

When we considered endotrivial modules, we noted that things look much nicer if we 'modded out' by projectives. Also, the formula for Heller translates and tensor products, modding out by projectives, is simply

$$\Omega^n(M) \otimes \Omega^m(N) \cong \Omega^{n+m}(M \otimes N),$$

which is a nice and succinct way of writing it. We also saw that, modulo projectives, $\mathrm{Ext}^1_{kG}(N, M)$ and $\mathrm{Hom}_{kG}(\Omega(N), M)$ are isomorphic, but now we need to mod out by maps that factor through a projective, rather than just ignoring projective summands of modules. The idea of modding out by projectives is not just to make formulae look nice, but plays a central role in modern representation theory.

The *stable category*, denoted $\underline{\text{mod}}\text{-}kG$, is the category whose objects are all finite-dimensional kG-modules, and whose morphisms are given by the factor sets $\text{Hom}_{kG}(M, N)/\text{PHom}_{kG}(M, N)$, where $\text{PHom}_{kG}(M, N)$ consists of all maps $\phi : M \to N$ that can be factorized as $\phi = \alpha\beta$, where $\alpha \in \text{Hom}_{kG}(M, X)$ and $\beta \in \text{Hom}_{kG}(X, N)$ with X projective, i.e., all maps that factor through a projective. Thus any projective module is isomorphic to 0 in the stable category.

We saw above that the stable category makes formulae for the Heller translate nicer, but it does a lot more than that, it makes it into a functor. First, *Schanuel's lemma* (see, for example, [33, Lemma 1.5.3]), which was glossed over during the definition of $\Omega(-)$, states that if

$$0 \to M \to X \to N \to 0 \quad \text{and} \quad 0 \to M' \to X' \to N \to 0$$

are short exact sequences with X, X' projective, then $M \oplus X' \cong M' \oplus X$. (We couldn't have used this language then because we didn't know about short exact sequences at the time.) This shows that if $\Omega(M)$ were defined as the kernel of *any* surjective map from a projective module to M, then $\Omega(M)$ is well defined up to isomorphism in the stable category. Moreover, given the diagram

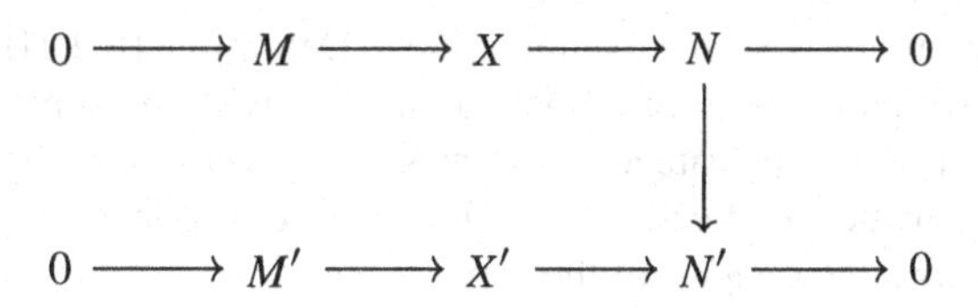

one gets a composition $X \to N'$, and hence a map $X \to X'$ from the universal property of projective modules (see the start of Sect. 3.3), which we can restrict to obtain a map $M \to M'$ that is unique up to maps that factor through a projective module, i.e., unique up to isomorphism in the stable category. Thus $\Omega(-)$ becomes a functor on the stable category $\underline{\text{mod}}\text{-}kG$.

Stable categories are nice, but they don't seem to necessarily encode all that much information about the module category. Even the one thing you might think they do preserve—the number of simple modules—is only a conjecture.

Conjecture 3.5.1 (Alperin–Auslander Conjecture[5]*)* If B and B' are stably equivalent p-blocks, then $l(B) = l(B')$.

Whereas we saw from Morita's theorem that Morita equivalences are induced by bimodules, this is not necessarily true for stable equivalences, but *nearly* true. We start with a definition, trying to add something from Morita equivalences to stable equivalences.

---

[5] I believe that this first appears in [23] for general finite-dimensional k-algebras, and Alperin's name is possibly attached because of his publicizing it. By [479] and [496] it was known in group representation theory circles as the Alperin–Auslander conjecture. Thanks to Jeremy Rickard for this historical information.

Definition 3.5.2 (Broué [69]) Let A and B be k-algebras, and let M and N be (A, B)- and (B, A)-bimodules respectively that are both finitely generated projective modules as both left and right modules. The modules M and N induce a *stable equivalence of Morita type* between A and B if $M \otimes_B N \cong A$ and $N \otimes_A M \cong B$ in the stable categories $\underline{\text{mod}}\text{-}(A \otimes_k A^{\text{opp}})$ and $\underline{\text{mod}}\text{-}(B \otimes_k B^{\text{opp}})$ respectively.

In fact, [479, Theorem 3.2] states that all stable equivalences induced by exact functors between blocks are isomorphic (as functors) to stable equivalences of Morita type, and this is usually the case in examples, so this isn't a big restriction. However, in certain applications, particularly to Broué's conjecture, replacing a stable equivalence of Morita type by one induced by a complex of bimodules is the 'correct' thing to do from a theoretical point of view.

I haven't been entirely honest with you so far. The stable category has the structure of a *triangulated category*, and all of the statements about stable equivalences are as triangulated categories, not just plain categorical equivalences. This matters, just as it matters that Morita equivalences are as k-linear categories, rather than just categories. Whereas the concept of a k-linear category was easy to define, the concept of a triangulated category is not.

The actual definition (see, for example, [550, Definition 10.2.1]) takes about a page to write down and will be of no help at all, so I will give a brief overview of what it means. A triangulated category $\mathcal{A}$ comes with a *translation functor* $T : \mathcal{A} \to \mathcal{A}$, which is an equivalence of categories. There is also a collection of sequences in $\mathcal{A}$, called *distinguished triangles*, of the form

$$A \to B \to C \to A[1],$$

where $A[1] = T(A)$. The category is triangulated if the set of distinguished triangles satisfies a collection of axioms: one states that any map $A \to B$ can be completed to a (distinguished) triangle (the term C in such a triangle is called a *mapping cone* of the map); another states that one may 'rotate' triangles, so that

$$A \xrightarrow{u} B \xrightarrow{v} C \xrightarrow{w} A[1]$$

is distinguished if and only if

$$B \xrightarrow{v} C \xrightarrow{w} A[1] \xrightarrow{-T(u)} B[1]$$

is distinguished. (The mapping cone of the identity map is 0, and the triangle $A \to A \to 0 \to A[1]$ is always distinguished.) The most complicated axiom is the 'octahedral axiom', which is so-called because the traditional commutative diagram associated to it is of an octahedron.

The stable category is an example of a triangulated category, with the distinguished triangles coming from all short exact sequences, and translation functor Ω^{-1}. A short exact sequence

$$0 \to A \to B \to C \to 0$$

yields an element of $\mathrm{Ext}^1_{kG}(C, A)$, and as we saw in Sect. 3.4, $\mathrm{Ext}^1_{kG}(C, A)$ is isomorphic to $\mathrm{Hom}_{kG}(\Omega(C), A)$ *in the stable category*, i.e., $\underline{\mathrm{Hom}}_{kG}(\Omega(C), A)$: we therefore obtain a sequence

$$0 \to A \to B \to C \to \Omega^{-1}(A).$$

These are the distinguished triangles, one for every short exact sequence.

It is not just the stable category that is triangulated, so is the derived category. The derived category is a bit tricky to define. The most categorically formal method invokes model categories, which we will not. We outline two approaches here. There are three categories, each with the same objects, but with increasingly complicated morphisms: the category of chain complexes, the homotopy category of chain complexes, and finally the derived category.

Let A be some finite-dimensional k-algebra, for example a group algebra or a block. The objects in these categories are all complexes with objects in mod-A. (Recall from the previous section that a complex is a chain of modules and homomorphisms such that the composition of any two homomorphisms in the chain is zero.) We write $Y = (Y_i)$ for a chain complex (i.e., the terms in the complex are Y_i for $i \in \mathbb{Z}$) and the maps between them are often referred to as a *differential*, and simply written ∂ or d. The condition on the differential can be stated as $\partial^2 = 0$.

The *category of chain complexes* is easiest. The morphisms are simply *chain maps*, maps

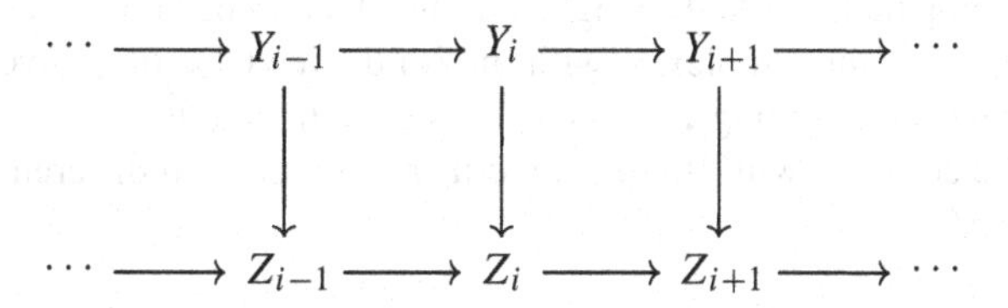

such that all squares commute. This is a perfectly fine category, but is too rigid for obtaining equivalences. The next stage is to define a homotopy between two chain maps. This is analogous to homotopies in algebraic topology: let $Y = (Y_i)$ and $Z = (Z_i)$ be two complexes, with differentials ∂_Y and ∂_Z respectively. Let (f_i) and (g_i) be chain maps from (Y_i) to (Z_i). A *chain homotopy* from (f_i) to (g_i) is a collection of homomorphisms $h_i : Y_i \to Z_{i-1}$ such that

$$f_i - g_i = h_i \circ \partial_Z + \partial_Y \circ h_{i+1}.$$

The property of being chain homotopic is an equivalence relation (Exercise 3.15). The *homotopy category of chain complexes* $K(\text{mod-}A)$ has morphism sets $\text{Hom}(Y, Z)/\sim$, where $\text{Hom}(Y, Z)$ is the set of chain maps from the complex Y to the complex Z, and $\sim$ is the homotopy equivalence relation.

The obvious question is 'Why this construction?' As we saw in Sect. 3.4, complexes have cohomology, and it turns out that a chain map induces a map on cohomology of those complexes (Exercise 3.16). Homotopic chain maps induce the same map on cohomology. In the category of chain complexes, an isomorphism between complexes Y and Z means that they are isomorphic complexes, i.e., there is an inverse map such that the composition is the identity. In the homotopy category, Y and Z are isomorphic if there are chain maps from Y and Z and back again whose composition is homotopic to the identity. In particular, an isomorphism $Y \to Z$ in this category induces an isomorphism on cohomology.

The *derived category* $D(\text{mod-}A)$ pushes this to its conclusion, forcing any map that induces an isomorphism on cohomology, called a *quasi-isomorphism*, to be an isomorphism. As we said, the most rigorous way to construct the derived category requires the notion of a model category, but that pushes us too far off-course. Informally, one thinks of the derived category as having objects all complexes, and morphisms chain maps subject to the requirement that quasi-isomorphisms are isomorphisms in this category. A more formal way to describe the morphisms from Y to Z in $D(\text{mod-}A)$ is as so-called 'roofs'

$$Y \leftarrow Y' \rightarrow Z$$

in $K(\text{mod-}A)$, where the first map is a quasi-isomorphism and the second is any map in $K(\text{mod-}A)$. Composing roofs is non-trivial: it involves finding a third roof that allows you to chain the two roofs together.

The derived category is triangulated because the homotopy category is triangulated, with translation functor given by shifting the degree of the complex by 1. In order to define the distinguished triangles, we need mapping cones. Let $f : Y \to Z$ be a morphism of chain complexes, with the ith degree term of Y given by Y_i and the ith degree term of Z being Z_i, and differentials ∂_Y and ∂_Z. The *mapping cone*, $\text{Cone}(f)$, is the complex with ith degree term $Y_{i+1} \oplus Z_i$, and differential given by the 2×2 matrix

$$\begin{pmatrix} -\partial_Y & f \\ 0 & \partial_Z \end{pmatrix}.$$

There are natural maps $Z \to \text{Cone}(f)$ and $\text{Cone}(f) \to Y[1]$ (where $Y[1]$ is the complex Y but shifted by one degree, so that the ith term of $Y[1]$ is Y_{i+1}), so we obtain a chain of maps

$$Y \xrightarrow{f} Z \to \text{Cone}(f) \to Y[1] :$$

these are the distinguished triangles in $K(\text{mod-}A)$, and in $D(\text{mod-}A)$. They exist in the category of chain complexes as well, but the identity map $Y \to Y$ does not extend to a triangle $Y \to Y \to 0 \to Y[1]$, but merely to something homotopic to it (Exercise 3.17), hence distinguished in $K(\text{mod-}A)$.

The *bounded derived category*, denoted $D^b(\text{mod-}A)$, is simply the subcategory of $D(\text{mod-}A)$ consisting of all complexes that are bounded, so that they are 0 for all but a finite number of terms. Two algebras are *derived equivalent* if their (bounded) derived categories are equivalent as triangulated categories.

For each bounded complex in $D(\text{mod-}A)$, we can select an isomorphic complex whose terms are all projectives. We do this in the case where the sequence is one term, concentrated in degree zero. Suppose that $0 \to M \to 0$ is the sequence, and write

$$\cdots \to P_n \to P_{n-1} \to \cdots \to P_1 \to P_0 \to M$$

for a projective resolution of M. We chop off the term M, just as we did in Sect. 3.4, and compare it to our sequence $0 \to M \to 0$. This yields the diagram

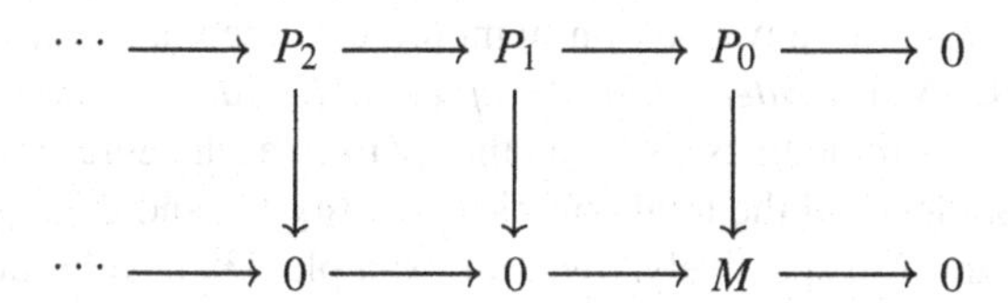

where all maps downwards are zero except for the last one, which is our original map $P_0 \to M$. This is a quasi-isomorphism because the cohomologies of the two sequences are 0 away from degree zero, and exactly M at degree zero, and furthermore the map induces an isomorphism of the two cohomology groups. The only problem is that the projective resolution lies in $D(\text{mod-}A)$, not $D^b(\text{mod-}A)$.

Another way to describe $D^b(\text{mod-}A)$ is therefore the subcategory of $K(\text{mod-}A)$ whose objects are all complexes of projectives (often denoted $K(\text{proj-}A)$), that are zero for sufficiently large degrees, and have no cohomology for all sufficiently small degrees. (This is the approach taken in, for example, [394].)

The theory of derived categories is far greater and more detailed than we can hope to cover in this book, so we give just some of the most important facts about derived categories, and particularly bounded derived categories in the rest of the section.

Morita equivalences, as stated in Theorem 3.3.8, are given by bimodules. Stable equivalences (of Morita type) are also given by bimodules. In general, derived equivalences are given by *complexes* of bimodules, rather than a single bimodule.

The complexes of bimodules we need are tilting complexes: a *tilting complex* of A-modules is a bounded complex C of projective A-modules such that there are no maps in $K^b(\text{mod-}A)$ between C and $C[n]$ for all $n \neq 0$. To avoid any trivial tilting complexes, we require that the summands of C as a complex generate the homotopy category $K^b(\text{proj-}A)$ of projective A-modules.

Note the similarity between this and progenerators from Morita's theorem: the projective module has been replaced by a complex of projective modules, and the statement that B is a summand of $M^{\oplus n}$ for some n is replaced by a generation statement on the homotopy category.

Rickard proved the analogue of Morita's theorem for derived equivalences in [477].

Theorem 3.5.3 *Let A and B be p-blocks of finite groups. The following are equivalent:*

(i) *A and B are derived equivalent;*
(ii) *$K^b(\text{mod-}A)$ and $K^b(\text{mod-}B)$ are equivalent (as triangulated categories);*
(iii) *B is isomorphic to* $\mathrm{End}(X)$*, where X is a tilting complex of A-modules.*

Nowadays people talk of 'two-sided' tilting complexes, in analogy with the fourth equivalent condition in Morita's theorem (Theorem 3.3.8). Part of this definition is the requirement to take the tensor product of two complexes, which is the *total complex*. For reasons of space we do not give the definition here, and instead refer the reader to any book on homological algebra, or for example [394, Definition 1.17.9]. A *two-sided tilting complex* or *Rickard complex* is a bounded complex C of (A, B)-bimodules such that the terms in each degree are projective as left and right modules, and the total complexes $C \otimes_B C^*$ and $C^* \otimes_A C$ are quasi-isomorphic to A and B respectively. (See, for example, [394, Definition 2.21.4].)

The stable category is a quotient of the derived category. For this reason, a derived equivalence (as triangulated categories) between two blocks implies a stable equivalence between two blocks, so if two stably equivalent blocks are in fact derived equivalent, then the Alperin–Auslander conjecture is true for that pair of blocks.

Proposition 3.5.4 *If A and B are derived equivalent p-blocks then $l(A) = l(B)$.*

Derived equivalences also preserve the quantity $k(B)$, which is the dimension of the centre of B. However, working just over the field k, some character-theoretic information about B cannot be easily seen, and to obtain this one needs to move to the ring $\mathcal{O}$ from Sect. 2.2. The definitions and results above all translate through to $\mathcal{O}$, and the character-theoretic shadow of a derived equivalence is a 'perfect isometry'. (See Sect. 4.4 for more details.)

For this reason we introduce splendid derived equivalences. The word 'splendid' is actually an abbreviation, for '**spl**it-**end**omorphism two-sided tilting complex of summands of permutation modules **i**nduced from **d**iagonal subgroups'. The responsibility for this is Rickard's [478].

Definition 3.5.5 Let G_1 and G_2 be finite groups, with common Sylow p-subgroup P. Let ΔP denote the diagonal subgroup $\{(x, x) \mid x \in P\}$ of $G_1 \times G_2$. Let B_i be the principal block of kG_i. A two-sided tilting complex X of (B_1, B_2)-bimodules is *splendid* if all terms of X are (B_1, B_2)-bimodules with vertex ΔP and trivial source.

Because trivial-source modules are liftable to $\mathcal{O}$ (Proposition 3.2.8) splendid derived equivalences are liftable to $\mathcal{O}$, so it suffices to work over k in this definition. Of course, if the block is not principal, we need to be a bit more careful. In [478], Rickard stuck to principal blocks, but afterwards Linckelmann wrote down a general definition [392, 393].

When looking for derived equivalences, one should in general be trying to find splendid (derived) equivalences. Not only does it make lifting to $\mathcal{O}$ easier, they are the 'correct' type of derived equivalence in the realm of finite group representation theory. Since [478], any equivalence induced by trivial-source modules is now called 'splendid', so we will see splendid Morita equivalences in Sect. 6.3. The splendour of the Morita equivalence repairs many of the issues with Morita equivalences that we identify in Sect. 4.5.

Rickard in [480] produced a theorem that allows you to prove that two blocks are derived equivalent if you already know that they are stably equivalent. In essence, you 'lift' the stable equivalence to an equivalence of derived categories.

Theorem 3.5.6 *Let A and B be p-blocks of finite groups, and write $S_1, \dots, S_n$ for the simple A-modules up to isomorphism. Let F : mod-$A \to$ mod-B be an exact functor, and suppose that F induces a stable equivalence. Let $X_1, \dots, X_n$ be objects of D^b(mod-B) that satisfy the following properties:*

(i) *for each i, X_i is stably equivalent to $F(S_i)$;*
(ii) *for each i, j, $\mathrm{Hom}(X_i, X_j[l]) = 0$ whenever $l < 0$;*
(iii) *for each i, j, $\dim(\mathrm{Hom}(X_i, X_j)) = \delta_{i,j}$;*
(iv) *the X_i generate D^b(mod-B) as a triangulated category.*

Then A and B are derived equivalent.

This is useful for constructing derived equivalences, particularly for sporadic groups, and also for checking that your guess for the images of simple modules under a putative derived equivalence is correct.

Having stated that splendid equivalences should be what we want, we then only say that the two derived categories are equivalent. However, if we start off with a 'splendid' stable equivalence of Morita type then we obtain a splendid equivalence [299, Theorem 4.4.2].

One special type of derived equivalence is a perverse equivalence. Defined by Chuang and Rouquier as part of their proof [115] of Broué's conjecture for symmetric groups (see Sects. 4.4 and 8.3), perverse equivalences are a special type of derived equivalence, defined purely combinatorially. Informally, one builds up the derived equivalence one simple module at a time.

Suppose that A and B are two derived equivalent (symmetric) algebras, for example two blocks of finite groups. The numbers of isomorphism classes of simple A- and B-modules are the same: we fix a bijection f from the isomorphism classes of simple A-modules to the classes of simple B-modules. The simple A-modules can be placed in D^b(mod-A) via $0 \to M \to 0$, so we can consider the images of the simple A-modules in D^b(mod-B). These sequences have lengths and cohomology.

The derived equivalence F is *perverse* if there is a function π on the simple modules, that takes values in the natural numbers, such that, for any simple module M, the cohomology of the complex $F(M)$ satisfies the following properties:

(i) it is zero outside of the degrees $-\pi(M), \dots, -1$;
(ii) $f(M)$ appears exactly once, in degree $-\pi(M)$;
(iii) for some other simple A-module N, $f(N)$ appears in the cohomology only in degrees between $-\pi(M)$ and $-\pi(N) - 1$ inclusive.

This definition is a bit dry, so here's an informal explanation of what a perverse equivalence is meant to do. If our two algebras were Morita equivalent, we could send each simple A-module M to a complex $0 \to N \to 0$ of B-modules. This isn't possible if the two algebras aren't Morita equivalent, so instead we will set up Morita equivalences between 'subquotients' of the derived categories. Ordering the simple modules $M_1, \dots, M_n$ by increasing π-function, the module M_1 gets sent to a complex $0 \to N_1 \to 0$ for some simple B-module N_1. The next module, M_2, gets sent to $0 \to N_2 \to 0$ in the 'quotient' by the subspace generated by the image of M_1. By the subspace, we mean the set of all complexes whose cohomology contains only those composition factors N_1, and even then only in degrees between $-\pi(M_1)$ and -1. In general, M_i gets sent to a complex $0 \to N_i \to 0$ modulo cohomology coming from all N_j for $j < i$.

This is a rough explanation, and is not technically correct, for example the sequence $0 \to N_i \to 0$ must be shifted by $-\pi(M_i)$. An analogy would be that a Morita equivalence is the identity matrix, and a perverse equivalence is a unitriangular matrix. Composing equivalences looks like multiplying matrices: multiplying two identity matrices yields an identity matrix, and multiplying two unitriangular matrices yields a unitriangular matrix. However, if we compose two perverse equivalences with different orderings $M_1, \dots, M_n$, then we multiply two matrices that are unitriangular with respect to different bases, and then we can get an arbitrary matrix (or arbitrary derived equivalence). However, it is not known if, for any two derived equivalent algebras, there is a chain of perverse equivalences that demonstrate that they are derived equivalent.

This is not the most general definition of perverse equivalence, as seen for example in [116]. However, it is the version that works best for finite groups, as certain derived equivalences for group algebras are known to be perverse (for example, between blocks of symmetric groups [115] and certain classical groups [179], and also blocks with cyclic defect group [116]), and certain important conjectural derived equivalences should be perverse. The cohomology of the images $F(M)$ of the simple A-modules allows us to compute the Cartan matrix of B in terms of the Cartan matrix of A. If B and A are blocks, we can also extract the decomposition matrix for B from the equivalence and knowledge of the decomposition matrix of A.

We won't say all that much about perverse equivalences here, except to make a few remarks. First, given a finite-dimensional algebra A, and a perversity function π on the simple A-modules, there is always some algebra B and a perverse equivalence from A to B with perversity function π. Furthermore, B is unique up to Morita equivalence. This means, for example, that in a finite group situation, if two blocks are Morita equivalent, and there are two perverse equivalences to other blocks with the same perversity functions, then the two target blocks are also Morita equivalent. This is one way in which the theory of perverse equivalences impinges on finite group representation theory: one may use it to prove a Morita equivalence between two blocks of finite groups indirectly, without ever having to compare them.

It also allows you some control on the decomposition numbers of a block. This area is still inchoate, but results of Rouquier and myself show that one may lift perverse equivalences to certain characteristic 0 objects. This enables us to compute the perverse equivalence in characteristic 0, and then apply it to infinitely many different prime characteristics simultaneously. This method allows us to compare the decomposition matrices, not only of different finite groups for the same prime, but different finite groups at different primes. It offers a glimpse of a Morita theory of blocks that allows us to view the characteristic of the field as a parameter rather than a constant, and potentially heralds a revolution in the representation theory of finite groups.

Exercises

Exercise 3.1 In [281], Higman introduced what is now known as *Higman's criterion*. Let G be a finite group, H be a subgroup of G, and M be an indecomposable kG-module. Let T be a right transversal to H in G. Prove that the following are equivalent:

(i) M is relatively H-projective;
(ii) M is a summand of $M{\downarrow_H}{\uparrow^G}$;
(iii) if N is any kG-module and $\phi : N \to M$ is a surjective homomorphism such that ϕ splits on restriction to H, then ϕ splits;
(iv) given any two kG-modules X and Y, a surjective homomorphism $\phi : X \to Y$, and a homomorphism $\alpha : M \to Y$, if there exists a kH-module homomorphism $\beta : M \to X$ such that $\beta\phi = \alpha$, then there exists a kG-module homomorphism with this property;
(v) there exists a kH-module endomorphism f such that

$$\sum_{g \in T} f^g = \mathrm{id}_M.$$

Higman's criterion is the equivalence of the first and last of these statements.

Exercise 3.2 Modify the proof of Maschke's theorem (see for example [328, Chapter 8]) to prove that every module is projective relative to a Sylow p-subgroup.

Hint: use an equivalent condition from Exercise 3.1.

Exercise 3.3 A module is *cyclic* if it is generated by one element. Show directly, i.e., not as a consequence of the freedom of kG, that any cyclic module is a quotient of kG. Prove that if G is finite, then finite-dimensional and finitely generated modules are the same.

Exercise 3.4 Let G be a finite group, let H be a subgroup of G, and let Q be a Sylow p-subgroup of H.

(i) Show that k_H is a summand of $k_Q{\uparrow}^H$.
(ii) Suppose that M is a summand of the permutation module $k_H{\uparrow}^G$. Show that M has trivial source, has vertex contained in Q, and show that M need not have vertex exactly Q. (Hint: the Mackey formula.)

Exercise 3.5 If H is a subgroup of G, M is a kG-module and N is a kH-module, prove that

$$(M{\downarrow}_H \otimes N){\uparrow}^G \cong M \otimes (N{\uparrow}^G).$$

Exercise 3.6 Let M and N be kG-modules. Prove that

$$\Omega^0(\Omega(M) \otimes N) \cong \Omega(M \otimes N).$$

Exercise 3.7 Let G be an elementary abelian p-group of order p^n, and let M be the module $kG/\mathrm{rad}^2(kG)$, of dimension $n+1$. Show that for $i < p$, the module $S^i(M)$ can be naturally identified with $kG/\mathrm{rad}^{i+1}(kG)$. Deduce that, if $n > 1$, then M is non-algebraic.

Exercise 3.8 Let G be a finite group with a normal p-subgroup N. Theorem 2.4.1 states that N lies in the kernel of every simple kG-module, so there is a canonical bijection between the simple kG-modules and the simple $k(G/N)$-modules.

Show that if N is central then this canonical bijection preserves the partition of simple modules among the p-blocks, and hence induces a bijection between the p-blocks of G and G/N. If N is not central, show that the canonical bijection sends a p-block of G to a union of p-blocks of G/N.

Let B be a p-block of G, and let $\bar{B}_1, \dots, \bar{B}_r$ denote the p-blocks of G/N in correspondence with B. Show that, if D denotes a defect group of B, then there is a defect group of $\bar{B}_i$ contained in D/N, and D/N is a defect group for at least one of the B_i.

Exercise 3.9 This more difficult exercise will guide you through the proof of the Green correspondence, Theorem 3.2.9. Let G be a finite group, Q a p-subgroup of G, and let H be a subgroup of G containing $N_G(Q)$.

(i) Let M be a kG-module with vertex Q. Prove that there exist kH-modules V and V', both with vertex Q, such that $M \mid V{\uparrow}^G$ and $V' \mid M{\downarrow}_H$.
(ii) If M is a relatively Q-projective kH-module, show that

$$M{\uparrow}^G{\downarrow}_H \cong M \oplus W,$$

where W is a direct sum of a collection of modules each of which is projective relative to a member of $\mathfrak{Y}$.

(iii) If M is an indecomposable kG-module with vertex Q, show that $M{\downarrow_H}$ is the direct sum of an indecomposable module V and a collection of modules, each of which is projective relative to a member of $\mathfrak{Y}$. The module V has vertex Q and $M \mid V{\uparrow^G}$.

(iv) If N is an indecomposable kH-module with vertex Q, then $N{\uparrow^G}$ is the direct sum of an indecomposable module U and a collection of modules, each of which is projective relative to a member of $\mathfrak{X}$. The module U has vertex Q and $N \mid U{\downarrow_H}$.

(v) Deduce the Green correspondence from the last two statements.

Exercise 3.10 Let M be a finite-dimensional module for a finite group G. Suppose that M has socle length s and radical length r. Prove that

$$\mathrm{soc}^i(M) \geq \mathrm{rad}^{r-i}(M)$$

for all i, and in particular that $r = s$.

Exercise 3.11 Prove *Fitting's lemma*: let M be an indecomposable kG-module, and let f be an endomorphism of M. Prove that f is either an automorphism or $f^n = 0$ for some $n > 0$. (Hint: iterate f.)

Exercise 3.12 Prove that the blocks of kG are the indecomposable summands of kG as a (kG, kG)-bimodule, i.e., a $k(G \times G)$-module. Prove that a defect group of a block of kG is a vertex of it viewed as a $k(G \times G)$-module.

Exercise 3.13 Prove the equivalence of the definition of a projective module as a summand of a free module and the universal property given at the start of Sect. 3.3.

Exercise 3.14 Let G be a finite group and let H be a normal p'-subgroup of G. Prove that the inflation to G of a projective $k(G/H)$-module is projective, and therefore there is a Morita equivalence between any block of $k(G/H)$ and their inflations to G. Deduce Theorem 2.4.2.

Exercise 3.15 Prove that being chain homotopic is an equivalence relation.

Exercise 3.16 Prove that a chain map induces a map on cohomology of the complexes.

Exercise 3.17 Show that the mapping cone of the identity map is homotopic to the zero complex.

Exercise 3.18 If $M_1 \to M_2 \to M_3$ is a short exact sequence of A-modules, show that there is a distinguished triangle $M_1 \to M_2 \to M_3 \to M_1[1]$ in $D(A)$, where M_i is embedded in $D(A)$ via $0 \to M_i \to 0$.

Chapter 4
The Local-Global Principle

We have already seen in Chap. 2 that the p-subgroups of a group have a deep influence on the modular representation theory in characteristic p, and there were hints that this influence bleeds through into the characteristic 0 case with Brauer's height-zero conjecture. In 1971 this local control of global invariants gained a new facet, with the McKay conjecture, and that kick-started a new direction for the subject.

Broadly speaking, there are two types of local-global conjecture: bounding results and constructive results. Bounding results state that certain global invariants, like $k(G)$ (or $k(B)$) and $l(G)$ (or $l(B)$) are bounded in terms of the defect group or the vertex, for example. Examples of these are Donovan's conjecture and Feit's conjecture, and in a different sense Brauer's $k(B)$-conjecture. Constructive results aim to give precisely an invariant of the whole group or whole block in terms of information associated to normalizers of p-subgroups of the group, and their blocks. Examples of these are Brauer's height-zero conjecture, Broué's conjecture, and the McKay conjecture.

These can be further subdivided into numerical conjectures and structural conjectures. For example, the McKay conjecture aims to count the number of irreducible characters of p'-degree in G, whereas Donovan's conjecture claims that there are only finitely many blocks with a given defect group up to Morita equivalence, which is a statement about the module category rather than a number.

In this chapter we will explore some of the deep and interconnected web of conjectures and results that comprise the local-global principle. At the time of writing, the true picture remains obscure, our understanding being akin to seeing a sprinkling of mountain tops, with the full range hidden by clouds.

In time the clouds will drift away and we will see everything, but for now we can do little more than name the mountain tops; this chapter will serve as an atlas for the world we have so far explored.

Brauer's height-zero conjecture has already been introduced in Sect. 2.3, and states that character heights depend on whether the defect group is abelian or not.

D. A. Craven, *Representation Theory of Finite Groups: a Guidebook*, Universitext,
https://doi.org/10.1007/978-3-030-21792-1_4

The (Alperin–)McKay conjecture gives a general description of characters of height zero in a block, and Alperin's weight conjecture states that one can determine the number $l(B)$ of simple modules in terms of local information about the block B. Broué's conjecture posits derived equivalences, but only when the defect group is abelian. Donovan's and Feit's conjecture aim to constrain the possible Morita classes of blocks, and sources of simple modules, respectively. Finally, Brauer's $k(B)$-conjecture bounds the number of ordinary characters lying in a block B in terms of the defect group of B.

Each of these states that some property of the representation theory of a group G can be described in terms of, or at least restrained by, the local structure of G. Pieced together, they paint a picture of a rigid relationship, but one that is still out of focus.

4.1 Brauer's Height-Zero Conjecture

We first met the height-zero conjecture back in Sect. 2.3.

Conjecture 4.1.1 (Brauer's Height-Zero Conjecture) Let B be a block of a finite group G, with defect group D. All ordinary irreducible characters of B have height zero if and only if D is abelian.

It was first stated in 1955 [53], and a series of papers in the last few years means we have a good understanding of what needs to be done to prove it. Since an irreducible ordinary character has degree divisible by the order of a Sylow p-subgroup if and only if it lies in a block of defect zero, blocks of defect one certainly have all characters of height zero. For general blocks with cyclic defect group, the height-zero conjecture was proved by Rothschild in [495], using the Brauer tree (see Chap. 5).

Reynolds [474] proved the result when the defect group D is normal in G in the late 1950s but the result did not come out until 1963. For the 'if' part, i.e., that if the defect group is abelian then all characters have height zero, Fong [224] (trailed in [223]) proved this for p-soluble groups. (In this paper he also proved the other half of the height-zero conjecture for the principal block of a p-soluble group.) For a full reduction of the 'if' direction to quasisimple groups we had to wait until 1988, when Berger and Knörr [37] gave the reduction.

Since the simple groups are classified, general statements about the parametrization of blocks of quasisimple groups, and the irreducible characters that belong to them, should prove the result. Indeed, for some classes of simple groups—as an obvious example the sporadic groups—it is very easy to prove the result. For symmetric groups, their degrees and associated blocks are given by the hook formula and Nakayama conjecture (Theorems 8.1.3 and 8.3.1), so it becomes a question of combinatorics. For the groups of Lie type in characteristic p the result is easy because the defect group is either a Sylow p-subgroup (and then the group is $\mathrm{SL}_2(q)$, as all other groups have non-abelian Sylow p-subgroups) or the trivial group (see Theorem 9.1.9), and the result is clearly true for blocks of defect zero.

As with many of the local-global conjectures, the hard case is the groups of Lie type in characteristic different from p, which we will talk about in Chap. 9. When p is small, i.e., 2, 3 and 5, the representation theory of groups of Lie type behaves very differently to the theory at larger primes. Many of these notions will be defined and described in more detail in Chap. 9, so we won't say much here: for most blocks of groups of Lie type, the work of Cabanes and Enguehard [86, 87] proves the result, leaving a few, very difficult cases. The last piece in the puzzle was the classification of quasi-isolated blocks of exceptional groups of Lie type in small primes. This was accomplished by Kessar and Malle [349].

Thus we have a proper, complete theorem in this section.

Theorem 4.1.2 (Kessar–Malle) *If B is a block of a finite group with abelian defect group, then all irreducible ordinary characters in B have height zero.*

(Although we have attributed this to Kessar and Malle, there are of course a large number of contributors to it in the preceding papers.)

We now should talk about the 'only if' direction. If we were naïvely approaching this, we would say that this looks easier than the 'if' direction: after all, for 'only if', we merely need to find one character of positive height, whereas for the 'if' direction we need to prove something about all characters, and if we are given a single group then this is true. This makes, for example, a proof for sporadic simple groups easy, but in the end no easier than for the 'if' direction.

In reality, it is significantly harder to make progress in this direction. For soluble groups and $p \geq 5$, the result was proved by Wolf in [569]. A few years later, Gluck joined Wolf and proved the result for all soluble groups [255], and then later pushed it to p-soluble groups [254]. Unlike Fong's result for the 'if' direction, the 'only if' proof requires the classification of finite simple groups.

The main tool in this is the Gluck–Wolf theorem, which gives a criterion for the Sylow p-subgroups of a finite group to be abelian.

Theorem 4.1.3 (Gluck–Wolf Theorem [254]) *Let G be a finite group, let N be a normal subgroup such that G/N is p-soluble, and let χ be an irreducible ordinary character of N. If $p \nmid (\phi(1)/\chi(1))$ for all irreducible characters ϕ of G such that $\langle \phi\downarrow_N, \chi\rangle \neq 0$, then the Sylow p-subgroups of G/N are abelian.*

Like many of the other local-global conjectures, there has been a lot of remarkable recent progress: in 2013, Navarro and Tiep [452] proved, for $p = 2$, the height-zero conjecture for all blocks of maximal defect in any finite group. An important step in this is to apply the Gluck–Wolf theorem for $p = 2$. In the situation of [452], it is true that their quotient G/N is always p-soluble, so they could apply the Gluck–Wolf theorem, but for p odd this is not true. That same year though, Navarro and Tiep proved the general Gluck–Wolf theorem [453], removing the p-solubility condition above. This represented one of the main obstacles to proving the height-zero conjecture in full generality.

The following year, Navarro and Späth [449] gave a reduction to quasisimple groups of the height-zero conjecture. Well, not quite. The original Berger–Knörr reduction reduced the 'if' direction for all groups to the 'if' direction for quasisimple groups, a straight reduction theorem. The Navarro–Späth reduction theorem reduces the 'only if' direction to a much more difficult statement for quasisimple groups.

Theorem 4.1.4 (Navarro–Späth [449]) *The 'only if' direction of Brauer's height-zero conjecture is true for all finite groups at the prime p if*

(i) *the 'only if' direction of Brauer's height-zero conjecture is true for all quasisimple groups at the prime p, and*
(ii) *all finite simple groups satisfy the inductive Alperin–McKay condition for the prime p.*

The first of these conditions is self-explanatory, but the second is a difficult generalization of the Alperin–McKay conjecture formulated to reduce that conjecture to simple groups, see Sect. 4.2. Indeed, Kessar and Malle, as well as proving the 'if' direction of the height-zero conjecture for quasisimple groups, a few years later produced a proof of the 'only if' direction for quasisimple groups [351]. Thus if the inductive Alperin–McKay condition is shown to hold for all quasisimple groups, then as well as the Alperin–McKay conjecture, Brauer's height-zero conjecture will also be solved.

If we cannot quite yet prove the 'only if' direction of the height-zero conjecture, what can we prove? In 1996, Robinson conjectured that one may at least able to bound the character heights, even if they can be non-zero. In particular, his guess in [486] was that if $\chi \in \mathrm{Irr}(G)$ lies in a p-block with defect group D, and $\mathrm{ht}(\chi)$ denotes the height of χ, then

$$p^{\mathrm{ht}(\chi)} \leq |D : Z(D)|,$$

with equality if and only if D is abelian.

An alternative description is the following: define the *defect* $\mathrm{def}(\chi)$ of a character $\chi \in \mathrm{Irr}(G)$ to be the integer d such that the p-parts $p^d \cdot \chi(1)$ is equal to $|P|$, where P is a Sylow p-subgroup of G. If χ lies in a block with abelian defect group D, then $|D| = p^{\mathrm{def}(\chi)}$, by the known direction of Brauer's height-zero conjecture. If D is non-abelian though, then $p^{\mathrm{def}(\chi)}$ should be at least $|Z(D)|$, with equality if and only if D is abelian. (These two statements are equivalent.)

This is an extension of the 'if' direction of the height-zero conjecture to non-abelian defect groups, and even this extension has been proved: for odd primes p this is now a theorem of Feng, Li, Liu, Malle and Zhang [220]. Unfortunately, the case $p = 2$ still remains open.

4.2 The McKay Conjecture

This conjecture in its original form, as given by McKay [426, 427], is simple to state: if G is a finite simple group with Sylow 2-subgroup P, then the numbers of irreducible characters of odd degree in G and $N_G(P)$ are the same. A more general version was first mentioned in passing in [312], and later explicitly in [2].

Conjecture 4.2.1 (McKay Conjecture) Let G be a finite group and write $\mathrm{Irr}_{p'}(G)$ for the set of irreducible characters of G of degree prime to p. If P denotes a Sylow p-subgroup of G, then

$$|\mathrm{Irr}_{p'}(G)| = |\mathrm{Irr}_{p'}(N_G(P))|.$$

(Of course, we may assume that p divides $|G|$, as otherwise the result is true, but unsurprising!) This version was proved for groups of odd order by Isaacs in 1973 [312], and then for all soluble groups by Wolf in 1978 [568]. For p-soluble groups, Dade and Okuyama–Wajima independently proved the conjecture. The Okuyama–Wajima proof may be found in [455, 456], but Dade appears to have not published his proof.

A few years later, work on this conjecture, and Alperin's weight conjecture in the next section, was performed on a number of sporadic groups, culminating in [565], where Wilson gives a proof of the McKay conjecture for all sporadic simple groups and all primes. But at this stage there was no way to reduce the conjecture to simple groups, so there was no impetus to really attack the conjecture in this way.

However, once there was a reduction to simple groups, an amazing explosion in results has brought us to the brink of solving this problem. We will talk about this soon, but we for now consider the conjecture and some refinements.

Isaacs and Navarro [315] considered the McKay conjecture for finite groups, refining the equality in the conjecture. The easiest of the two generalizations of the conjecture that they suggest is to refine the statement $\chi(1) \not\equiv 0 \bmod p$ to $\chi(1) \equiv \pm i \bmod p$. In other words, they suggest that the number of irreducible characters of degree congruent to $\pm i$ modulo p for G and $N_G(P)$ are the same, for each $0 < i < p$. For $p = 2, 3$ this offers nothing new, but for $p \geq 5$ there is some extra content here, and it gets increasingly more powerful as p grows.

The second, more difficult, refinement of the McKay conjecture proposed in [315] is the action of Galois automorphisms. If σ is any Galois automorphism of $\bar{\mathbb{Q}}/\mathbb{Q}$ then σ permutes the irreducible (ordinary) characters of G. In particular, we may let σ be an element of the Galois group of the cyclotomic field $\mathbb{Q}_{|G|}$, and consider its actions on the characters of G and of $N_G(P)$. It turns out to be too much to hope for that these actions are the same as permutation actions. Isaacs and Navarro give the example of $G = \mathrm{GL}_2(3)$ and $p = 3$, where the p'-degree characters of $N_G(P)$ are rational valued, but the same is not true of those of G.

What they suggest is that one restricts σ to having p-power order and fixing all p'-roots of unity in $\mathbb{Q}_{|G|}$. In this case σ should fix the same number of points on $\mathrm{Irr}_{p'}(G)$ and $\mathrm{Irr}_{p'}(N_G(P))$.

Navarro [448] cleaned this up into what is sometimes known as the Navarro conjecture, but also as the Galois refinement of the McKay conjecture. The condition on σ is relaxed to the statement that, for some integer e, $\sigma \in \mathrm{Gal}(\mathbb{Q}_{|G|}/\mathbb{Q})$ sends every p'-root of unity ζ to ζ^{p^e}. In fact, letting $\mathcal{H}$ denote the subgroup of the Galois group consisting of such automorphisms, he conjectures that the permutation actions of $\mathcal{H}$ on the sets $\mathrm{Irr}_{p'}(G)$ and $\mathrm{Irr}_{p'}(N_G(P))$ are the same.

One may also combine the two, and ask that $\mathcal{H}$ has the same action on the irreducible characters of degree congruent to $\pm i$ modulo p for the two groups, but the Galois refinement is already incredibly difficult to verify, as it often requires knowledge of the whole character table to compute the exact actions of σ on G and $N_G(P)$, whereas for the congruence refinement we still just need the degrees.

(One of the consequences of Navarro's refinement for $p = 2$ is that, if σ is the Galois automorphism that fixes all 2-power roots and squares all odd roots of unity, then σ fixes all members of $\mathrm{Irr}_{p'}(G)$ if and only if a Sylow 2-subgroup P of G satisfies $P = N_G(P)$. This consequence at least is now a recent theorem of Schaeffer Fry and Taylor [505–507].)

We can also refine this conjecture in a completely different way: block by block. Of course, as with the height-zero conjecture, the McKay conjecture looks like a question of character heights. Irreducible characters of p'-degree must lie in blocks of maximal defect (from the definition of defect, Definition 2.3.1), and must have height zero as well. Thus the McKay conjecture states that the number of height-zero characters in blocks of maximal defect in G and $N_G(P)$ is the same. We can refine the McKay conjecture slightly further, and say that if B is a block of G with defect group P, and B' is its Brauer correspondent in $N_G(P)$, then the number of characters of height zero in B and B' are equal.

Now we have stated it like this, we have an obvious extension to blocks whose defect group is not P. This appears in [4, Conjecture E].

Conjecture 4.2.2 (Alperin–McKay Conjecture) Let G be a finite group. If B is a p-block of G with Brauer correspondent B', then the numbers of ordinary characters of height zero in B and B' are the same.

We can throw the congruence and Galois refinements at the Alperin–McKay conjecture as well, but we have to be a bit more clever.

For the congruence refinement, one considers the p'-parts of the degrees of characters of height zero in blocks of finite groups. Let B be a p-block of G with defect group D, and let B' be its Brauer correspondent. The Alperin–McKay conjecture states that the numbers of characters of height zero in B and B' are the same: the congruence refinement of Isaacs–Navarro is that the number of height-zero characters of B' whose p'-part of the degree is $\pm i$ (modulo p) is equal to the number of height-zero characters of B whose p'-part of the degree is $\pm in$ (modulo p), where $n = |G : N_G(D)|_{p'}$. Of course, if $D = P$ is a Sylow p-subgroup, then $n \equiv 1 \bmod p$ (by Sylow's theorem!) and we get the original congruence refinement of the McKay conjecture back.

For Galois automorphisms the situation is easier: we just require that the Galois automorphism σ fixes the same number of height-zero characters in B and B'. In general, Galois actions permute the blocks though: in the original Isaacs–Navarro statement, σ fixed all p'-roots of unity, so it fixes all Brauer characters, and hence the p-blocks of G, so acts on the ordinary characters in each block. In the more general Navarro conjecture, this is no longer the case, and we instead make the following statement.

Conjecture 4.2.3 (Navarro Refinement of Alperin–McKay Conjecture) Let G be a finite group, and let D be a p-subgroup of G. Let $\bar{B}$ denote the sum of the p-blocks of G with defect group D, and let $\bar{B}'$ denote the sum of the p-blocks of $N_G(D)$ with defect group D. There is a bijection between the height-zero characters in $\bar{B}$ and $\bar{B}'$ such that the subgroup $\mathcal{H}$ of $\mathrm{Gal}(\mathbb{Q}_{|G|}/\mathbb{Q})$ acts permutation isomorphically on the two sets. Furthermore, this bijection respects the Brauer correspondence.

Turull has further strengthened this refinement in [548], to include more invariants of the irreducible characters in the bijection, in particular the 'local field of definition' and 'local Schur index'.

Having discussed the McKay conjecture and the many refinements of it, we should now bring the reader up to date on the progress made.

Up until 2007, the progress was that the Alperin–McKay conjecture was true for p-soluble groups by Okuyama–Wajima [455, 456] and Dade, and it was known for various simple groups, but because there was no reduction theorem, knowing it for simple groups was only evidence of the truth of the conjecture, not a building block of the proof.

2007 saw two papers that constituted the first breakthrough. First, in [411], Malle determined $\mathrm{Irr}_{p'}(G)$ for G a quasisimple group of Lie type, and found a bijection between that and $\mathrm{Irr}_{p'}(N)$, where N is a subgroup containing the normalizer of a Sylow p-subgroup of G (but not necessarily equal to it). The other paper from 2007 was a reduction theorem for the original McKay conjecture. In [314], Isaacs, Malle and Navarro reduced the McKay conjecture for all finite groups down to a much more complicated statement about simple groups. The original statement from [314] has a long list of conditions, that were simplified somewhat by Späth in [525], but are still quite complicated.

Any simple group that satisfies the conditions is called 'McKay-good' for the prime p. We won't describe exactly what those conditions are, but we will give a flavour of what they look like. Let G be a non-abelian finite simple group and p be a prime, and let X be the universal central extension of G, so that $X/Z(X) \cong G$; we let $P \in \mathrm{Syl}_p(X)$, and let A denote the subgroup of $\mathrm{Aut}(X)$ that stabilizes P. We require an A-stable subgroup N such that $N_X(P) \leq N < X$, and an A-equivariant bijection Ω_N^X between the p'-degree characters $\mathrm{Irr}_{p'}(X)$ of X and $\mathrm{Irr}_{p'}(N)$ of N such that if $\chi \in \mathrm{Irr}_{p'}(G)$ then elements of $Z(X)$, which must act as scalar matrices in any irreducible representation, act with the same scalar on both χ and $\Omega_N^X(\chi)$. (Another way to say this is that $\langle \chi\downarrow_{Z(X)}, \Omega_N^X(\chi)\downarrow_{Z(X)}\rangle$ is non-zero.) Furthermore, Ω_N^X must satisfy a couple of technical properties about extending characters to overgroups, which are different (but are always equivalent) in the various papers on the subject.

For all but the most ardent reader, an A-equivariant bijection that respects the action of the centre of X and has some extension properties is a pretty good approximation to what the condition is. So which simple groups are known to be McKay-good? The answer is, almost all of them.

The first groups proved to be McKay-good were in the original article [314], and were $\mathrm{PSL}_2(q)$, ${}^2B_2(q)$ and ${}^2G_2(q)$, together with J_1, for all primes p. This includes all simple groups with abelian Sylow 2-subgroups, and so the McKay conjecture was therefore proved for all finite groups with abelian Sylow 2-subgroups. If G is a finite simple group with no outer automorphisms and no central extensions (there are some such sporadic groups, and a few but not many groups of Lie type), then McKay-good is the same as just checking the McKay conjecture. A year later, Malle proved in [412] that the remaining sporadic simple groups and the alternating groups are McKay-good, together with those groups of Lie type whose universal central extension is not also a group of Lie type (for example, $\Omega_8^+(2)$).

Thus we are left with the groups of Lie type. Späth proved in 2012 [525] that groups of Lie type in characteristic p are McKay-good for p, building on work of Brunat and Himstedt [74, 75, 77], and using a bijection given by Maslowski [423].

We therefore consider the groups of Lie type in characteristic $r \neq p$, which need to be proved for the prime p. In [89], Cabanes and Späth proved that many of the exceptional groups of Lie type are McKay-good, namely ${}^3D_4$, G_2, F_4, ${}^2F_4$ and E_8. (These are the ones where the simple group is equal to its universal central extension.)

For the classical groups PSL and PSU, again Cabanes and Späth proved the result in [91], and in a much shorter paper in the same year also dealt with the case of symplectic groups [92]. (Taylor also proved independently an important part of the result in [92] in [536].)

In very recent work, which appeared just before this book went to press [93], Cabanes and Späth have introduced a more structural approach to understanding the stabilizers of characters, which is an important step in these verifications, and using this have proved that all groups apart from the even-dimensional orthogonal groups $\Omega_{2n}^{\pm}(q)$ are McKay-good.

Along the way, Malle and Späth proved the original McKay conjecture, the case $p = 2$, in [416].

What about the refinements? There is a reduction of the congruence refinement (just adding the congruence refinement to the bijection Ω_N^X above) to simple groups, and there is a reduction of the Alperin–McKay conjecture to simple groups, in a similar fashion to the reduction of the McKay conjecture. Both of these were given by Späth in [526]. As for Navarro's refinement, as I was writing this book a reduction of it to simple groups was unveiled by Navarro, Späth and Vallejo [450].

The congruence refinement of McKay-good can be referred to as being IN-good: Späth in the same article [526] proved that the Ree and Suzuki groups ${}^2G_2$ and ${}^2B_2$ are IN-good for any prime p, that alternating groups are IN-good, and that simple groups of Lie type in characteristic p are IN-good for p. She also proved the result for a few sporadic groups, with the rest being checked by Breuer in unpublished work.

We are again left with the groups of Lie type in characteristic $r \neq p$. The Cabanes–Späth paper [91] that proved that PSL and PSU were McKay-good also proved that they are IN-good. The exceptional groups dealt with in [89] are also IN-good, because the bijection satisfies $\Omega_N^X(\chi)(1) \equiv \pm\chi(1) \bmod p$.

For the inductive Alperin–McKay condition, things are a lot less far advanced. In some cases, most notably PSL and PSU by Cabanes and Späth [90], the inductive AM condition for blocks with maximal defect has been verified. Denoncin [158] verified the condition for alternating groups in characteristic 2. Koshitani and Späth prove [372, 373] the inductive AM condition whenever B is a block of a quasisimple group with cyclic defect groups. Apart from that, Malle proved that the Suzuki and Ree groups satisfy the inductive AM condition [413], but not much more than this has been proved.

4.3 Alperin's Weight Conjecture

The McKay conjecture counts certain irreducible ordinary characters of a finite group, by counting certain characters of $N_G(P)$ for P a Sylow p-subgroup of G. If you want to count $k(G)$, it suffices to count $l(C_G(x))$ for p-elements x by Exercise 2.4. If we had a formula for $l(G)$ in terms of subgroups $N_G(Q)$ for some non-trivial p-subgroups Q, then we can insert it in this formula and be able to express $k(G)$ as a sum of local invariants, i.e., invariants like $k(H)$ and $l(H)$ for p-local subgroups H of G.

A *weight* of kG is a pair (Q, M), where Q is a (possibly trivial) p-subgroup of G and M is a projective simple $N_G(Q)/Q$-module, i.e., a block of defect zero in $N_G(Q)/Q$, or a simple $kN_G(Q)$-module with vertex exactly Q. The group G acts by conjugation on the set of weights by $(Q, M)^x = (Q^x, M^x)$, so we may count the number of conjugacy classes of weights.

Conjecture 4.3.1 (Alperin's Weight Conjecture [7]) If G is a finite group, then $l(G)$ is equal to the number of conjugacy classes of weights of kG.

One sees immediately a few extra pieces of information: since $O_p(H)$ lies in the kernel of every simple kH-module by Theorem 2.4.1, the only p-subgroups Q that possess weights are those for which $O_p(N_G(Q)) = Q$. In particular, this means that $Z(Q)$ is a Sylow p-subgroup of $C_G(Q)$ (such subgroups are called *p-centric*) and $O_p(\mathrm{Aut}_G(Q)) = \mathrm{Inn}(Q)$ (such subgroups are called *radical*).

We can also restrict our attention to a specific block. To do this, we must distribute the weights among the blocks of kG. The easiest way to do this is as follows. Let b be the idempotent of a p-block B of G. If (Q, M) is a weight, then we associate (Q, M) to B if $M \cdot \mathrm{Br}_Q(b) \neq 0$. This condition is equivalent to $\mathrm{Br}_Q(b) \cdot e \neq 0$, where e is the idempotent of the p-block of $N_G(Q)$ containing M. (See [394, Definition 6.10.1] for a slightly different, but equivalent description, or [7] for the original condition.) One sees that this association is well defined, as the conjugate of a weight associated to B is still associated to B. The block-by-block form of the weight conjecture is therefore as follows.

Conjecture 4.3.2 (Alperin's Weight Conjecture, Block Version) Let B be a p-block of kG. Then $l(B)$ is equal to the number of weights of B.

The first big development on Alperin's weight conjecture was a reformulation, due to Knörr and Robinson, that expressed $l(B)$ not as a sum of weights, but as an alternating sum of an invariant over the simplices of a particular simplicial complex. The set we want to consider here is the partially ordered set of all non-trivial p-subgroups of G, and the simplicial complex $\mathscr{P}$ is all chains of subgroups from this set. The n-simplices of $\mathscr{P}$ are all chains

$$Q_1 < Q_2 < \cdots < Q_{n+1}.$$

If C is the chain above define $|C|$ to be the length n. Clearly G acts by conjugation on $\mathscr{P}$, and the stabilizer of the chain C, written G_C, is the intersection of the $N_G(Q_i)$. Note that G_C lies between $C_G(Q_{n+1})$ and $N_G(Q_{n+1})$.

Let B be a sum of blocks of kG (this allows us to deal with the block-by-block and original version of Alperin's weight conjecture simultaneously), and let b denote the central idempotent of B. The image of b under the Brauer map $\mathrm{Br}_{Q_{n+1}}$ is either 0 or a sum of block idempotents of kG_C, so let B_C denote the sum of blocks

$$(kG_C)\mathrm{Br}_{Q_{n+1}}(b),$$

of kG_C. In [366], Knörr and Robinson show that B_C depends only on G_C, and not on C. They prove that (the block-by-block version of) Alperin's weight conjecture is equivalent to the statement that, for any block B of positive defect,

$$\sum_C (-1)^{|C|} l(B_C) = 0,$$

where the sum runs over all chains in $\mathscr{P}$, up to G-conjugation. This is normally known as the *Knörr–Robinson reformulation* of Alperin's weight conjecture. The set $\mathscr{P}$ may be replaced by a number of different complexes, for example, the subcomplex whose subgroups consist solely of elementary abelian subgroups, or of p-centric radical subgroups.

Alperin's weight conjecture has been reformulated a number of times, but we won't discuss all of the different versions: many of them are technical and at the moment, while the different versions provide insight, they don't appear to help with actually proving the conjecture. What does seem to help is that there is a reduction theorem for Alperin's weight conjecture, due to Navarro and Tiep [451], which like the reduction theorem for the Alperin–McKay conjecture, produces a complicated set of conditions. If all finite simple groups are 'AWC-good', then all finite groups satisfy the Alperin weight conjecture. Two years later, in [527] Späth produced the inductive 'blockwise' Alperin weight (BAW) condition, reducing the block-by-block version of Alperin's weight conjecture to simple groups.

In [527], Späth proved that groups of Lie type in defining characteristic satisfy the inductive BAW condition, and a few years later Koshitani and Späth proved the inductive BAW condition for blocks with cyclic defect groups in [372]. For alternating groups, Malle proved the inductive BAW condition in [413], and also for the Suzuki and Ree groups for all primes p.

Cabanes and Späth [89] proved the inductive BAW condition for $p \geq 5$, and maximal defect blocks of all groups ${}^3D_4(q)$, and $G_2(q)$ for $q > 4$, and also for $\mathrm{Sp}_{2n}(2^a)$ for those primes $p > n$ such that the multiplicative order of 2^a modulo p is odd, called linear primes (see Definition 9.3.3). Schulte pushed it further in [511] to all blocks of ${}^3D_4(q)$ and $G_2(q)$, and all primes, but still excluding $G_2(q)$ for $q = 2, 3, 4$. For $q = 3, 4$ there is a central extension of $G_2(q)$, so generic methods of groups of Lie type do not work, but Breuer proved the inductive BAW condition for these individual groups, in as-yet unpublished work. In 2018 Conghui Li and Zhang gave a proof of the inductive BAW condition for those groups $\mathrm{PSL}_n(q)$ and $\mathrm{PSU}_n(q)$ whose outer automorphism group is cyclic [389]. More recently, Feng, Zhenye Li and Zhang have announced a proof of the inductive BAW condition for certain blocks of classical groups, including $\mathrm{PSp}_{2n}(q)$ whenever p divides $q^d - 1$ and p, q and d are all odd [221].

Of course, since there is a reduction of the block version of Alperin's weight conjecture to simple groups, in particular it must be known for p-soluble groups. This is indeed the case, and it is mostly due to Okuyama, or so it appears. The story of Okuyama's proof is quite convoluted, and one may find the details in [26, p. 134], in which a proof of a generalization of Alperin's weight conjecture for blocks of p-soluble groups can also be found.

Things are still considerably further behind the situation for the McKay conjecture, and as with that conjecture we still have no real explanation as to *why* it is true. For some defect groups though, there is a structural explanation for what is going on, and we discuss this in the next section.

4.4 Broué's Abelian Defect Group Conjecture

Broué's conjecture is, as far as I know, the only conjecture in the representation theory of finite groups that seeks to explain numerical coincidences via an equivalence of categories. In the particular case of Broué's conjecture, it is an equivalence of derived categories. The trouble is, it only exists for blocks with abelian defect groups.

Conjecture 4.4.1 (Broué's Abelian Defect Group Conjecture) Let B be a p-block of a finite group G, and suppose that B has abelian defect groups. If B' denotes the Brauer correspondent of B, then B and B' are derived equivalent.

The first guess in this direction that Broué had in [67, Conjecture 3.2] was that if H is a subgroup of G containing a Sylow p-subgroup P of G and H 'controlled fusion' in P with respect to G, i.e., if $g \in N_G(Q)$ for some subgroup $Q \le P$ then there exists some $h \in N_H(Q)$ such that g and h induce the same conjugation action on Q (so $gh^{-1} \in C_G(Q)$), then the principal p-blocks of G and H are derived equivalent. This isn't quite right, and the Suzuki simple groups are counterexamples. This was established by Cliff [117] (see also [488] for a proof by Robinson using perfect isometries, which we will see below).

Just below that first, incorrect, conjecture was [67, Conjecture 3.3], which is the (hopefully) correct conjecture above, now known as Broué's conjecture.

It is easy to see that a derived equivalence between two blocks preserves the number of simple modules (Proposition 3.5.4). This is enough to see that Alperin's weight conjecture for a block B with an abelian defect group D follows from Broué's conjecture: the weights associated to B are exactly the pairs (D, M), where M is a simple module for the Brauer correspondent B' of B in $N_G(D)$ (see [394, Corollary 6.10.10]), so we see that Alperin's weight conjecture for blocks with abelian defect groups is simply $l(B) = l(B')$. It also preserves some other invariants, but the easiest way to see this is with a character-theoretic consequence of a derived equivalence.

The 'shadow' of a derived equivalence at the level of characters is a perfect isometry, and the shadow of a splendid derived equivalence is an 'isotypy'. Let G and H be finite groups, let B be a p-block of G and B' a p-block of H. We consider the lattices $\mathbb{Z}\mathrm{Irr}(B)$ and $\mathbb{Z}\mathrm{Irr}(B')$, and note that these come with inner products on them given by $\langle \chi, \psi \rangle$. Thus we can consider *isometries*: bijective maps

$$I : \mathbb{Z}\mathrm{Irr}(B) \to \mathbb{Z}\mathrm{Irr}(B')$$

that preserve the inner product, i.e., $\langle \chi, \psi \rangle = \langle I(\chi), I(\psi) \rangle$. Of course, if there is an isometry then $|\mathrm{Irr}(B)| = |\mathrm{Irr}(B')|$, and since $\mathrm{Irr}(B)$ forms an orthonormal basis for $\mathbb{Z}\mathrm{Irr}(B)$, if $\chi \in \mathrm{Irr}(B)$ then $I(\chi) = \pm\psi$ for some $\psi \in \mathrm{Irr}(B')$. An isometry I from $\mathbb{Z}\mathrm{Irr}(B)$ to $\mathbb{Z}\mathrm{Irr}(B')$ is *perfect* if the map $\mu_I : G \times H \to \mathbb{C}$ given by

$$\mu_I : (g, h) \mapsto \sum_{\chi \in \mathrm{Irr}(B)} \chi(g) \cdot (I(\chi)(h))$$

has the following two properties, for all $g \in G$ and $h \in H$:

(i) if exactly one of g and h is p-regular then $\mu_I(g, h) = 0$;
(ii) the numbers $\mu_I(g, h)/|C_G(g)|_p$ and $\mu_I(g, h)/|C_H(h)|_p$ are algebraic integers.

Exercise 4.1 asks you to show that μ_I determines I. If I is an isometry then so is I^{-1}, and it is perfect if and only if I is. Similarly, the composition of two isometries is an isometry, and is perfect if the two isometries are (Exercise 4.2). Notice that if B and B' are perfectly isometric, they need not have isomorphic defect groups. An easy example is if B and B' are kP and kQ for two p-groups P and Q with the same character table (so Q_8 and D_8 for instance), and in fact this perfect isometry is the identity, in some sense. It is not known if Morita equivalences preserve defect groups, but the character versions of them definitely need not.

Broué, from whom the definition of a perfect isometry originates, proved in [68] that a perfect isometry, while not preserving a decomposition matrix, does still preserve some relationship. If B and B' have decomposition matrices D and D' respectively, and B and B' are perfectly isometric, then there are matrices X and Y such that

$$XD = D'Y.$$

While the matrix Y is arbitrary in $\mathrm{GL}_{l(B')}(\mathbb{Z})$, the matrix $X \in \mathrm{GL}_{k(B)}(\mathbb{Z})$ is a signed permutation matrix. In particular, this means that B and B' have the same number of characters of each height, so $k(B) = k(B')$ as well. This also means that B has no characters of positive height if and only if B' has none, which is highly relevant for the height-zero conjecture.

This means that Broué's conjecture implies all of the Alperin–McKay, Alperin weight, and Brauer's height-zero conjectures.

But why does a derived equivalence yield a perfect isometry? One has to be a bit careful here, and perform a derived equivalence over the ring O from Sect. 2.2 rather than the field k. If one does this, then there is a bounded complex C of (OG, OH)-bimodules that induces a derived equivalence, and one may form the formal alternating sum

$$\sum_{i\in\mathbb{Z}}(-1)^i C_i.$$

We take the character of this sum, i.e.,

$$\mu = \sum_{i\in\mathbb{Z}}(-1)^i \mathrm{ch}(C_i),$$

and this is the map μ_I for a perfect isometry I.

So Broué's conjecture looks like the 'right' way to prove results about blocks with abelian defect groups. However, it is much harder than the numerical conjectures to prove. One reason is that there is, as of yet, no reduction of it to simple groups, unlike the Alperin–McKay and Alperin weight conjectures. There is a reduction theorem for the case where the block is principal though, but this is more a statement about finite groups than about derived equivalences.

Using the classification of finite simple groups, Fong and Harris [226] proved that if G is a finite group with abelian Sylow p-subgroups then there exist normal subgroups $K \leq H$ of G such that K and G/H are p'-groups, and H/K is a direct product of an abelian p-group and simple groups with abelian Sylow p-subgroups. This is an easy consequence of the fact that if G is a finite simple group with abelian Sylow p-subgroups, then no outer p-automorphism of G centralizes a Sylow p-subgroup of G. In other words, if an almost simple but not simple group has an abelian Sylow p-subgroup, then the Sylow p-subgroup is contained entirely in the simple normal subgroup.

They used the structure of groups with abelian Sylow p-subgroups to reduce the question of the existence of a perfect isometry for the principal block to simple groups, and then along similar lines Marcus gave a reduction for Broué's conjecture [418]. This reduction states that Broué's conjecture is true for the principal p-block of all finite groups if and only if it is true for the principal p-block of all almost simple groups. Slightly stronger, Broué's conjecture is true for all groups if, for each simple group G, there is a derived equivalence that is 'compatible' with the p'-outer automorphisms of G, i.e., the tilting complex extends to a complex for the almost simple groups.

This of course means that Broué's conjecture is true for p-soluble groups. In fact, it turns out that in this case the block and its Brauer correspondent are Morita equivalent [276].

The abelian defect group conjecture has been checked for a number of simple groups, particularly sporadic groups, and also groups with abelian Sylow 2-subgroups [134]. There are a number of simple groups with Sylow 3-subgroup $C_3 \times C_3$ (for example, $\mathrm{PSL}_3(q)$ for $q \equiv 4, 7 \bmod 9$, A_7 and the sporadic Mathieu groups M_{11}, M_{22} and M_{23}), and Broué's conjecture has been checked for each of these families in a series of papers (for example in [378] for $\mathrm{PSL}_3(q)$). Indeed the conjecture is known for the principal 3-block of all finite groups G whose Sylow 3-subgroup has order 9. The paper [369] finishes the proof when the Sylow 3-subgroup is $C_3 \times C_3$. (See the references in [369] for the papers on the simple groups.) Some non-principal blocks of sporadic groups with defect group $C_3 \times C_3$ have also been checked, for example [370]. The other possible Sylow 3-subgroup of order 9 is the cyclic group C_9.

In fact, Broué's conjecture is known for all blocks with cyclic defect groups, and actually more is true. In Sect. 5.2 Brauer tree algebras are introduced, and it turns out that all Brauer tree algebras with the same number of edges and same exceptionality are derived equivalent. This includes all blocks with the same cyclic defect group and that have the same number of simple modules. This result, by Rickard [476] over the field k and Linckelmann [390] over the ring $\mathcal{O}$, formed one of the earliest uses of derived categories in group representation theory.

For finite groups, there are few whole classes of simple groups where we have a proof. Important work of Chuang and Rouquier, which we mention again in Sect. 8.3, solves Broué's conjecture for the symmetric groups (it was pushed down to the alternating groups by Marcus in [419]). The Chuang–Rouquier proof also works for blocks of the group $\mathrm{GL}_n(q)$ for $p \nmid q$, and this proof was modified and extended by Dudas, Varagnolo and Vasserot [179] to unipotent blocks of classical groups at linear primes (see Chap. 9 for definitions, specifically Definitions 9.3.1 and 9.3.3).

One special case of Broué's conjecture is for G a finite group of Lie type. A full description of this would need the notation and ideas from Sects. 9.2 and 9.3, but we will give an overview now. Let $G = G(q)$ be a finite group of Lie type defined over $\mathbb{F}_q$ (for example, $\mathrm{SL}_n(q)$ or $\mathrm{Sp}_{2n}(q)$), where q is a power of $r \neq p$, and let d denote the multiplicative order of q modulo p, so that $p \mid \Phi_d(q)$, the dth cyclotomic polynomial evaluated at q.

The general philosophy of the representation theory of groups of Lie type, after work of Broué, Malle and Michel [71], is that the modular representation theory of G should depend more on d than on p or q. There is a set of characters of G, called 'unipotent characters', which is independent of q and whose degrees are polynomials in q. The distribution of unipotent characters into blocks of G (the blocks containing unipotent characters are called 'unipotent blocks') depends only on d, not p or q, at least if $p \geq 7$. If the defect group D of a unipotent block B is abelian, then the unipotent characters in B are in bijection with the simple B-modules. In fact, the 'unipotent part' of the decomposition matrix, i.e., only the rows corresponding to unipotent characters, should be triangular with respect to a certain ordering on the rows (and columns), and hence there is a canonical bijection between the unipotent characters in B and the simple B-modules.

There is a variety, called the 'Deligne–Lusztig variety' (see Sect. 9.2), which has an action of G on the one side and D on the other. Thus the complex of cohomology (over $\mathcal{O}$) of this variety inherits this action. Conjecturally, this action may be extended to an action of G and $N_G(D)$. This complex should induce a derived equivalence between B and its Brauer correspondent B'.

Thus unlike the general case, where Broué's conjecture simply posits the existence of a derived equivalence, for groups of Lie type there is a *geometric form*, which gives a candidate complex (at least up to this extension statement, which isn't explicitly given). This conjecture was made by Broué and Malle in [70]. Although it looks like a promising method of attack for Broué's conjecture, little has been done directly with it.

The new idea is to use perverse equivalences, which we saw in Sect. 3.5. The equivalence induced by the Deligne–Lusztig variety should be perverse, which means we need a bijection between the simple B- and B'-modules, a perversity function $\pi(-)$, and a stable equivalence to lift. The first of these is given in [70], and comes from a cyclotomic Hecke algebra. The second is given by the Deligne–Lusztig variety itself: the cohomology of it over $\mathcal{O}$, tensored by K, its field of fractions, should have the property that each unipotent character lies in a unique degree. This degree should yield the perversity function.

Helpfully, there is now a conjecture on what this degree is, due to the author [128]: one does not have to prove that this degree is correct, and merely use the conjectural formula to yield a candidate perverse equivalence. This method was used to produce conjectural decomposition matrices for unipotent blocks, which led to a proof that these conjectural decomposition matrices were correct for blocks with cyclic defect group [131]. (It is much easier to prove a theorem when you know its statement.)

The last step is the stable equivalence. At the moment, in full generality, this has not been written down, although it is known if the defect group D has rank 2 [134]. This has been used to prove Broué's conjecture for several groups of Lie type, so it has practical as well as theoretical applications. The theoretical applications include, for a start, the fact that the decomposition matrix should be lower unitriangular [134]. It also explains the fact that the decomposition matrix of B does not depend on p and q, but only on d. It also explains the fact that this latter statement is not true for some small primes p, as the perverse equivalence in large characteristics is generic but can be a little different for small p.

It also suggests that there are Morita equivalences between unipotent blocks at different primes, i.e., that if q and q' are two prime powers that have order d modulo two primes p and p', and 'the same' unipotent block B and B' for the groups $G(q)$ and $G(q')$ respectively, with the same abelian defect group D, then B and B' should be Morita equivalent. This has a deep connection with Donovan's conjecture, which is described in the next section.

4.5 Donovan's Conjecture

Donovan's conjecture appeared in Sect. 3.3. It stated that the number of Morita equivalence classes of blocks with defect group D is finite, for every possible p-group D.

There has been a flurry of recent activity in this area, particularly focusing on the case where D is an abelian 2-group. Before we get into exactly what has been proved though, we can first talk about what is *not* known. The list contains a number of rather basic and embarrassing gaps in our understanding. The first is that if B_1 and B_2 are two blocks that are Morita equivalent, and B_i has defect group D_i, is $D_1 \cong D_2$? Nobody knows, and it is not even known that if D_1 is abelian then D_2 is.

What if B_1 is simply kD_1, the group algebra itself? The question of whether, if $kD_1 \cong kD_2$, then $D_1 \cong D_2$, is another thorny issue in modular representation theory, called the *modular isomorphism problem*. If we replace k by the ring $\mathcal{O}$ that we discussed in Sect. 2.2 then the modular isomorphism problem has a positive answer [492], and if we replace k by the ring $\mathbb{Z}$ then the modular isomorphism problem has a negative answer [280], but for $k = \mathbb{F}_p$ the jury is still out. (There have been partial results in this direction, but it still looks too hard, and this is really a statement about the fact that there are lots of p-groups that all look basically the same.)

A more general question than the modular isomorphism problem is that if B_1 and B_2 are p-blocks that are Morita equivalent over k, are they Morita equivalent over $\mathcal{O}$? So far this is the case in all examples, but it seems difficult at the moment to prove such a statement. It is not even known that there can be only finitely many such Morita classes of blocks over $\mathcal{O}$. In other words, it is not known that Donovan's conjecture over k implies Donovan's conjecture over $\mathcal{O}$.

Having given a couple of ways in which our knowledge falls far short of where we might hope it should be, we can also talk about what we do know of this problem. As with many of these conjectures, Donovan's conjecture is known for certain defect groups D, and for certain classes of groups, with the problem being finite groups of Lie type $G(q)$ in characteristic $p \nmid q$.

If D is cyclic then Donovan's conjecture is true, and this follows because the Morita class of a block is encoded in a tree (see Chap. 5) with at most p vertices, and of course there are only finitely many such trees. If $p = 2$ and D is dihedral or semidihedral then Donovan's conjecture is also known over k by results of Erdmann (see Sect. 6.2). If the defect group is Q_8 then Donovan's conjecture is known over k by Erdmann's work, and even over $\mathcal{O}$ by [185, 300]. If the defect group is Q_{2^n} for $n > 3$, then the number of simple modules $l(B)$ is at most 3, and if $l(B) \neq 2$ then Donovan's conjecture is known over k by Erdmann's results [196, 197] and over $\mathcal{O}$ by the work of Eisele [185], but if $l(B) = 2$ then even over k the result is currently out of reach, despite using the powerful tools of Erdmann's papers. Sticking with $p = 2$, if D is abelian then work, primarily by Eaton and Livesey, has very recently yielded a solution over k for D an abelian 2-group [182], and together with Eisele they pushed the result to one over the ring $\mathcal{O}$.

For most p-groups D, the only block with defect group D is a nilpotent block (see Sect. 6.3) and, up to Morita equivalence, nilpotent blocks with a given defect group are unique. This is a statement about most p-groups being 'boring': a finite group G is *p-nilpotent* if there exists a normal subgroup N of G such that $p \nmid |N|$ and $|G : N|$ is a power of p, i.e., $G \cong N \rtimes P$ for $P \in \mathrm{Syl}_p(G)$. A p-group P is *p-nilpotent forcing* if, whenever G is a finite group with Sylow p-subgroup P, G is p-nilpotent.

A theorem of Martin [422] states that 'almost all' p-groups P have no outer automorphisms of order prime to p, a necessary condition for P to be p-nilpotent forcing. It isn't sufficient, of course, as the dihedral 2-groups show (which are the Sylow 2-subgroups of $\mathrm{PSL}_2(q)$ for q odd). Henn and Priddy show [279] that most p-groups are p-nilpotent forcing. The translation from finite groups to p-blocks of finite groups goes via fusion systems, in particular a result of Stancu [530] (see also [127, Theorem 5.29]),[1] which shows that most p-groups only have the trivial fusion system on them (the fusion system version of p-nilpotent forcing), and this means that all p-blocks with that defect group are nilpotent.

Thus Donovan's conjecture is true for most p-groups, but this is more a statement that there are lots of p-groups that are 'boring'. If P is a Sylow p-subgroup of a group that is not p-nilpotent, then P is 'interesting', and for this much smaller class very little is known about Donovan's conjecture in general.

The weak Donovan conjecture was given in Sect. 3.1, and states that there is a function $f : \mathbb{N} \to \mathbb{N}$ such that if B is a block with defect group D, then the entries of the Cartan matrix of B are bounded by $f(|D|)$. Of course, if there are only finitely many Morita equivalence classes of blocks with defect group of order $|D|$ then we take $f(|D|)$ to be the maximum of these entries. The gap between the weak Donovan and Donovan conjectures—that there are only finitely many Cartan matrices, and that there are only finitely many Morita classes—can be exactly specified.

Recall that a Morita equivalence carries a k-linear structure, i.e., that the two blocks have equivalent module categories as k-linear categories. Thus we can apply

[1] I recently learned that this result was proved by Thévenaz in the early 1990s, not in the language of fusion systems, but it appears he did not publish it.

the map $\sigma : \lambda \mapsto \lambda^p$ on the field and obtain from a block B a new k-algebra B^σ. Maybe B is Morita equivalent to B^σ, maybe not.

Definition 4.5.1 The *Morita–Frobenius number* of a block B, denoted $\mathrm{mf}(B)$, is the smallest positive integer n such that B^{σ^n} is Morita equivalent to B.

If G is a finite group then this map σ permutes the blocks of kG (the easiest way to see this is if $b = \sum \alpha_g g$ is the block idempotent, then the block idempotent of B^σ is $\sum \alpha_g^p g$). Thus for any given block B, $\mathrm{mf}(B)$ is finite. In addition, this shows that the *Frobenius number*, the smallest positive integer n such that B and B^{σ^n} are isomorphic, rather than just Morita equivalent, is also finite.

The obvious first question is: is the Morita–Frobenius number of a block ever not equal to 1? For principal blocks, because $B^\sigma = B$ we have that $\mathrm{mf}(B) = 1$. In fact, it seems difficult to construct examples where $\mathrm{mf}(B) \neq 1$. First one needs a group G with a lot of p-blocks with the same defect group, and then the map σ needs to permute those blocks. Then, and this is the difficult part, you need to show that B and B^σ are not Morita equivalent. The point is, their module categories *are* equivalent, just not as k-linear categories.

In 2007, Benson and Kessar gave an example of a block B with $\mathrm{mf}(B) = 2$ in [36]. This example cannot be used in any obvious way at the moment to produce examples with larger Morita–Frobenius numbers, which leads us to three possibilities:

(i) there exists an integer n such that $\mathrm{mf}(B) \leq n$ for all primes p and all defect groups D;
(ii) for a given p-group D, there is an integer $n = n(D)$ such that $\mathrm{mf}(B) \leq n = n(D)$ for all p-blocks with defect group D, but there is no integer N such that $n(D) \leq N$ for all D;
(iii) there is an infinite sequence of p-blocks B_i such that $\mathrm{mf}(B_i) \geq i$ and the B_i all have the same defect group D.

Notice that if the third case holds then there must be infinitely many Morita equivalence classes represented by the B_i, so Donovan's conjecture is false. Thus we should hope that either the first or second case holds.

Conjecture 4.5.2 (Morita–Frobenius Conjecture) Let D be a finite p-group. There exists an integer $n = n(D)$ such that, for all p-blocks B with defect group D, $\mathrm{mf}(B) \leq n$.

In 2004, Kessar proved the following theorem [346], which links all of these notions.

Theorem 4.5.3 (Kessar) *Let D be a finite p-group. If Donovan's conjecture is true for D over k, then the weak Donovan conjecture and the Morita–Frobenius conjecture are true for D, both over k. Conversely, if the weak Donovan conjecture and Morita–Frobenius conjecture are true for D over k then Donovan's conjecture is true for D over k.*

Eaton, Eisele and Livesey have very recently proved a version of this result over the ring $\mathcal{O}$ as well [181].

If I were forced to choose, I would guess that $\mathrm{mf}(B)$ is bounded in terms of D, but not globally for all blocks, i.e., the second option above. One might compare the situation for Morita–Frobenius numbers to the situation for 1-cohomology of modules for simple groups, which was initially thought to be globally bounded by a small number because all examples had a small number, and now is believed to be unbounded, as we talked about in Sect. 3.4. I would be very wary of predicting that Morita–Frobenius numbers are globally bounded.

Düvel proved [180] a reduction for the weak Donovan conjecture to quasisimple groups, but Donovan's conjecture itself seems so far to resist such a reduction. The problem is that there is no obvious way to relate the Morita–Frobenius number of a block of an arbitrary, potentially quite complicated group G, to the composition factors of G. The kernel of the issue is that if G has a normal subgroup H such that $|G : H| = p$, then it is difficult to bound the Morita–Frobenius number of a p-block B of G in terms of those of p-blocks of H: this type of information is crucial for reduction theorems, and it is not clear how to proceed otherwise. However, if D is abelian, then there are techniques available, and Eaton and Livesey managed to do it [182], but needed the *Frobenius number* (we replace 'Morita equivalent' by 'isomorphic as a k-algebra' in the definition) rather than the Morita–Frobenius number, and needed it over $\mathcal{O}$. Since [182], with Eisele they have proved the reduction to quasisimple groups of the Donovan conjecture over $\mathcal{O}$, again if the defect group is abelian [181]. This needs a technical, but crucial step: one may assume that G is generated by conjugates of the defect group D in proving Donovan's conjecture. (In other words, Donovan's conjecture is true for groups if and only if it is true for groups generated by conjugates of the defect groups.) This was known over k by work of Külshammer [376], but recently Eisele provided a proof over $\mathcal{O}$ [186], that enables the reduction to work.

Having talked about reducing it to quasisimple groups, we should now talk about what is known about the weak Donovan, Donovan, and Morita–Frobenius conjectures for quasisimple groups. For symmetric groups, Donovan's conjecture was proved by Scopes, and this proof descends to the alternating groups by work of Hiss [289] (see also [345]). Kessar proved Donovan's conjecture for the double covers of the alternating and symmetric groups in [343].

For Lie type groups in characteristic p, as we will see in Theorem 9.1.9, the only possible defect groups are the Sylow p-subgroup and the trivial group. There are only finitely many groups of Lie type in characteristic p with a given Sylow p-subgroup (as the order of the Sylow p-subgroup grows with q as well as the rank of the group) and so Donovan's conjecture automatically holds for them.

Thus we end up with groups of Lie type $G(q)$ with $p \nmid q$. Jost [337] proved Donovan's conjecture for a subclass of blocks of $\mathrm{GL}_n(q)$ called unipotent blocks (see Sect. 9.3), and this is enough to prove the statement for all blocks for $\mathrm{GL}_n(q)$ using a deep result of Bonnafé and Rouquier, Theorem 9.4.7.

For the other classical groups, in [290] and [291], Hiss and Kessar used the ideas from Scopes's work in [516] to prove that Donovan's conjecture holds for some collections of unipotent blocks again, but there are some issues in this case. For some primes p called 'linear' primes (see Definition 9.3.3), they could show that if one fixes q and lets the rank of the group vary (for example, $\mathrm{Sp}_{2n}(q)$ for various n) then Donovan's conjecture holds, but could not handle different q, and for 'unitary' primes, they could show the opposite, that they could not let n vary completely, but could allow q to vary. For $\mathrm{GL}_n(q)$, all primes $p \nmid q$ are in some sense both linear and unitary, so one can vary both parameters. For linear primes, the work of Dudas, Vasserot and Varagnolo [179] that was mentioned in the context of Broué's conjecture now fills in this gap, so Donovan's conjecture holds for unipotent blocks and linear primes, but it doesn't help for unitary primes. Another issue is that the Bonnafé–Rouquier theorem for groups other than $\mathrm{GL}_n(q)$ doesn't get us all the way to proving Donovan's conjecture for all blocks, and still leaves a collection of blocks at the end, called 'quasi-isolated' blocks (see Sect. 9.4).

For exceptional groups, things are even worse in some sense. We don't really know very much at all about relating say $G_2(q)$ to $G_2(q')$. We will talk about this more later in this section.

So Donovan's conjecture looks difficult for Lie type groups in characteristic different from p. What about the weak version of Donovan's conjecture and the Morita–Frobenius conjecture? Weak Donovan is in better shape, because we can compute decomposition numbers better than we can compute Morita equivalences. Recently, strong geometric methods developed mostly by Dudas have been able to obtain previously unheard-of detail about decomposition numbers of groups of Lie type. For more on this topic, see Sect. 9.3, but note that these results have in particular proved the weak version of Donovan's conjecture for large swathes of groups.

For the Morita–Frobenius conjecture, things are looking even better.

Theorem 4.5.4 (Farrell–Kessar [206]) *Let B be a p-block of a finite quasisimple group G.*

(i) *If G is sporadic, alternating, Lie type in characteristic p, or a linear, unitary, symplectic, or odd-dimensional orthogonal group, then* $\mathrm{mf}(B) = 1$.
(ii) *If G is not a simple group $E_8(q)$ then* $\mathrm{mf}(B) \leq 2$.
(iii) $\mathrm{mf}(B) \leq 4$ *in all cases.*

Furthermore, these bounds hold for k and for $\mathcal{O}$.

It is important that these bounds hold over $\mathcal{O}$ rather than just k because it is becoming increasingly clear from the Eaton–Livesey work in [182] that we need more than just the Morita–Frobenius conjecture, but the boundedness of the Frobenius numbers over $\mathcal{O}$. Hopefully, in a future reduction of Donovan's conjecture to quasisimple groups, the needed statement about Morita–Frobenius numbers will not be too strong, so that it can be deduced from the work of Farrell and Kessar in [206].

We end this section by discussing Puig equivalences and Puig's finiteness conjecture. The precise definition of a Puig equivalence is technical, and we will not give it right now (see Sect. 6.3), but more or less it is a Morita equivalence between two blocks B and B' of finite groups that also preserves defect groups, and vertices and sources of the simple B- and B'-modules. Of course, Puig equivalent blocks are Morita equivalent, but the converse is not true. One may ask whether there are only finitely many Puig equivalence classes of blocks of a given defect group, and this is the content of Puig's finiteness conjecture, from 1982.

Conjecture 4.5.5 (Puig's Finiteness Conjecture) There are only finitely many Puig equivalence classes of blocks with a given defect group D.

(This conjecture is often called simply the Puig conjecture, but then so is the conjecture on endopermutation sources in Sect. 6.3.) This was never published by Puig; see [394, Conjecture 6.4.2], for example.

Since it keeps the notion of a defect group, it doesn't suffer the potential deficiencies of Morita equivalences between blocks that we talked about before. The downside is that it can be fairly technical to check, even if you already know that the blocks are Morita equivalent.

Many of the Morita equivalences talked about above have been upgraded to Puig equivalences, but many more have not. Puig's finiteness conjecture is known to hold for blocks with cyclic defect group, by the work of Linckelmann [391]: this is already non-trivial, as the Brauer tree from Chap. 5 only determines the block up to Morita equivalence, not Puig equivalence. It was also shown to hold for Klein four defect groups, in the paper [132] by Eaton, Kessar, Linckelmann and myself, improving work of Erdmann [192, 194], who proved Donovan's conjecture for the Klein four group. Outside of these groups, Puig's finiteness conjecture is only known in 'boring' cases, i.e., the p-nilpotent forcing cases mentioned earlier. Due to the structure of the p-group D in question, all blocks must be Puig equivalent to kD, the nilpotent blocks that we haven't mentioned yet, but will in Sect. 6.3.

As for the Morita equivalences for specific groups, for symmetric groups the Morita equivalences were upgraded to Puig equivalences by Puig himself [471], and the same paper proved the conjecture for p-soluble groups. For alternating groups and the double covers of alternating and symmetric groups, Puig's conjecture was proved by Kessar in [345] and [343] respectively. Just as Puig upgraded Scopes's Morita equivalences for symmetric groups to Puig equivalences, Kessar upgraded Jost's Morita equivalences for unipotent blocks of $\mathrm{GL}_n(q)$ to Puig equivalences in [344]. The work of Hiss and Kessar on blocks of classical groups mentioned before [290, 291] also extends to Puig equivalences, but not the work of Dudas–Varagnolo–Vasserot.

In Chap. 9 we will talk about the representation theory of groups of Lie type $G = G(q)$ when $p \nmid q$, and how it depends less on q than you might imagine. One of the big wide open problems is to try to relate the representation theory of $G(q)$ at the prime p to the representation theory of $G(q')$ at the prime p. For example, the principal 2-blocks of $\mathrm{PSL}_2(q)$ and $\mathrm{PSL}_2(q')$ are Puig equivalent if q and q' are both congruent to 3 mod 8, and also if q and q' are congruent to -3 mod 8 [132]. If $q \equiv \pm 1$ mod 8 then the precise power of 2 dividing $|\mathrm{PSL}_2(q)|$ depends on q, so we have to be a bit more careful: if q and q' are both odd prime powers, the same power of 2 divides $|\mathrm{PSL}_2(q)|$ and $|\mathrm{PSL}_2(q')|$, and $q - q'$ is divisible by 4, then the principal 2-blocks of $\mathrm{PSL}_2(q)$ and $\mathrm{PSL}_2(q')$ are Puig equivalent. (The condition on $q - q'$ is so that the powers of 2 dividing $q - 1$ and $q' - 1$ are the same: we will see in Chap. 9 that the multiplicative order of q modulo 4 is what really guides the representation theory, and this condition states that the orders of q and q' modulo 4 are the same.)

Although it was known from the work of Erdmann [191] that the simple modules in the principal 2-blocks have the same Green correspondents, vertices and sources as q varies in this way, it appears an actual Puig equivalence between them was found only recently by Koshitani and Lassueur [371].

There are similar statements for other groups of Lie type, but unfortunately they are usually only conjectural in nature. The statement that varying q (as long as it satisfies some congruence conditions modulo p^n for certain n) results in Puig equivalent p-blocks of groups of Lie type is one of the most useful statements in the representation theory of groups of Lie type, and also one of the most challenging to prove.

4.6 Feit's Conjecture

Feit's conjecture, which we saw in Sect. 3.2, states that, given a p-group Q, there are only finitely many isomorphism types of indecomposable kQ-module that can act as the source of simple kG-modules with vertex Q, as G ranges over all finite groups. Unsurprisingly, this is again unsolved in full generality.

If $Q = 1$ then the source must be the trivial module (and the module is projective), and so this case is clear. If Q is cyclic then there are only finitely many indecomposable kQ-modules, so again this case is true.

There is not much more that is known in general; it isn't even known in the case of the Klein four group, for example. But if we aren't quite as general then we can give a few more results: if M is a simple kG-module, then M lies in a block with defect group D, and M has a vertex Q. Up to conjugacy, Q is contained in D, so one may be more restrictive about Feit's conjecture, and consider it for the pair (D, Q), rather than just Q.

The first question you might want to ask is whether, given any Q, there are any constraints on the potential D that can appear. In other words, are there constraints on the vertices of simple modules in blocks with defect group D?

We have already seen a couple of such constraints: Theorem 3.2.12 stated that if Q is cyclic then $Q = D$, but the case Q cyclic is already fine. Theorem 3.2.11 showed that $C_D(Q) \leq Q$: such a subgroup is called *self-centralizing*. This can be a very powerful tool for certain isomorphism types of Q; for example, if Q is

$C_p \times C_p$ then this means that D is a maximal-rank p-group, so for $p = 2$ this means that D is either dihedral or semidihedral. There is a general classification of blocks with dihedral or semidihedral defect groups by Erdmann (see Sect. 6.2) which would help with this, but as far as I know nobody has tried to pin down the sources of the simple modules in such blocks. If $D = Q$ is Klein four, then this follows from a much tighter classification of blocks with Klein four defect group due to Eaton, Kessar, Linckelmann and me [132] which, as mentioned in Sect. 4.5, built upon Erdmann's work in [192, 194] but managed to obtain the sources of the simple modules as well. The sources have dimension at most 2, and are definable over $\mathbb{F}_4$.

There are two things to see from this specific case that can be considered more generally. The first is that for Q Klein four, the index $|D : Q|$ may be arbitrarily large (as there are simple modules in the principal 2-block of $\mathrm{PSL}_2(q)$ with $q \equiv 1 \bmod 4$ that have Klein four vertices [191]), but is this true for other groups, and other primes? The second is that to state that there are only finitely many sources for simple modules, I had to give a bound on both the dimension and the field over which the module is defined. Of course, both of these statements must hold if there are only finitely many modules, but can they be decoupled, and perhaps at least one be proved?

We can talk about the index $|D : Q|$ first. A subgroup Q has the *vertex-bounded defect property* if there exists an integer n, dependent only on Q, such that if B is a block of a finite group with defect group D, and there is a simple B-module with vertex isomorphic to Q, then $|D : Q| \leq n$.

Question 4.6.1 (Puig) If p is odd, does every p-group have the vertex-bounded defect property?

There are no known counterexamples to this, but there are no non-cyclic p-groups for which this is known in full generality. The combination of this and Puig's finiteness conjecture (Conjecture 4.5.5) implies Feit's conjecture (naturally, only for odd primes), as was proved by Danz and Müller in [154].

We can ask whether the vertex-bounded defect property holds for a class of groups other than all of them, for example p-soluble groups or symmetric groups. Zhang [574] has reduced this question of Puig to quasisimple groups, so it makes sense to consider these. Of course, since these are questions of finiteness, one can always exclude finitely many groups from contention by just increasing every bound; thus sporadic groups are not of interest.

If G is a symmetric group and p is odd then in [152] Danz and Külshammer proved that $|D| \leq |Q|!$, and the same bound was proved [154] by Danz and Müller for the double covers of symmetric groups; this also holds for the double covers of alternating groups. As we mentioned in the previous section, Puig's conjecture is already known for symmetric groups by the work of Scopes [516] and then Puig [471] (for $p = 2$, the alternating groups were done by Kessar in [345]), and Kessar [343] for the double covers, so this proves Feit's conjecture for these groups. Thus we are left with groups of Lie type.

One can be completely naïve and pose an obvious question: if Q is a p-group, is there an integer $n = n(Q)$ such that, whenever G is a finite group and D is a defect group of a p-block of G, and Q is a self-centralizing subgroup of D, the index $|D : Q|$ is at most n? Since vertices are self-centralizing subgroups by Theorem 3.2.11, this would imply the vertex-bounded defect property for Q. It is far too hopeful in general, but what about using it to prove that certain simple groups have the vertex-bounded defect property at least?

We concentrate on groups of Lie type for p odd. Luka [400] proved this exact statement for the symplectic and orthogonal groups, so we actually do not need to worry about what the vertices are for these groups, as *any* self-centralizing subgroup will do. For linear groups, he proved that $\mathrm{GL}_n(q)$ and $\mathrm{GU}_n(q)$ also have the vertex-bounded defect property, but that moving to $\mathrm{SL}_n(q)$ and $\mathrm{PGL}_n(q)$ doesn't always work if $p \mid (q-1)$ (and $\mathrm{SU}_n(q)$ and $\mathrm{PGU}_n(q)$ for $p \mid (q+1)$). He also proved that $G_2(q)$ and ${}^3D_4(q)$ do not have this property for $p = 3$, so any proof of the vertex-bounded defect property for these groups must rely on a better understanding of the possible vertices than simply that they are self-centralizing.

As two quick examples, we first consider $G = G_2(q)$ for $p = 3$ and $3 \nmid q$. In this case there is a self-centralizing $C_3 \times C_3$ subgroup, even though the order of a Sylow 3-subgroup of $G_2(q)$ can become arbitrarily large. On the other hand, a Sylow 7-subgroup P of $E_8(q)$ is either abelian or has the form $(C_{7^n} \wr C_7) \times C_{7^n}$ for some integer n. As $Z(P)$ has order $C_{7^{2n}}$ in this group, we obtain the bound $|P| \leq |Q|^4$ for any self-centralizing subgroup Q of P.

So it is clear that we cannot proceed completely naïvely, although I think that we did well getting symplectic and orthogonal groups just from this method. A general proof of the vertex-bounded defect property would be important, of course. However, for it to be useful in proving Feit's conjecture, it would need to be coupled with strong results relating the representation theories of $G(q)$ and $G(q')$ for two prime powers q and q', as mentioned in the previous section. More than a simple Morita equivalence between blocks of groups of Lie type at different prime powers, it would also need to include statements about the vertices, sources and Green correspondents of the simple modules for groups of Lie type. This is widely believed (myself included), but outside of certain cases for low ranks and some cases that come as a consequence of the proof of Broué's conjecture for some groups (Conjecture 4.4.1), it seems out of reach with our current understanding of these groups and their representation theory.

There is another way to attack this, as we mentioned: by bounding the dimension of the source of a simple module with vertex Q and bounding the field of definition of such a module. This approach does have promise: Dade proved in [147] that there are only finitely many kQ-modules of a given dimension d that can be sources of simple modules for a finite group G, in essence, proving that the field of definition is bounded (although he actually proved more than that). Even stronger, for groups of Lie type, the simple groups that are left to worry about, we can ask that the dimension of the Green correspondent can be bounded in terms of $|Q|$ and the rank of the group, which should be true if the Puig equivalences between different prime powers that are conjectured do actually exist.

As with Donovan's conjecture, it all seems to come down to being able to relate the representation theory of (for example) $\mathrm{GL}_n(q)$ with $\mathrm{GL}_n(q')$ for two different prime powers q and q', and some $p \nmid q, q'$. As we said in the previous section, being able to do this anything deeper than merely numerically will yield a complete revolution in the representation theory of these groups.

4.7 Brauer's $k(B)$-Conjecture

Brauer's $k(B)$-conjecture, Conjecture 2.3.3, places a sharp bound on the number of ordinary irreducible characters in B. It states that $k(B) \leq |D|$ if B is a p-block with defect group D. As we have already mentioned, the best bound known that holds for all blocks B is the Brauer–Feit bound from [59].

Theorem 4.7.1 (Brauer–Feit Theorem) *If B is a p-block of a finite group, with defect group D, then $k(B) \leq p^{2d}/4 + 1$.*

This improves Brauer's bound of $p^{d(d+1)/2}$ in [51], where I believe he first posited the $k(B)$-conjecture. Brauer was able to prove the $k(B)$-conjecture for $|D| \leq p^2$, but above that it seemed too hard. If D is abelian though, Sambale has proved that $k(B) < p^{(3d-1)/2}$ in [502].

Let us first note that the bound is sharp if it is true, as G an abelian p-group satisfies this. More interestingly the group $C_p \rtimes C_{p-1}$, the normalizer of a Sylow p-subgroup of S_p has a defect group C_p and exactly p irreducible characters, $p-1$ of degree 1 and one of degree $p-1$. This is true for all blocks with defect group C_p and $p-1$ simple modules, by Theorem 5.1.2 in the next chapter. More generally, there exists an element x of order $p^d - 1$ in $\mathrm{GL}_d(p)$, so if D is an elementary abelian p-group of order p^d, then we may form the group $D \rtimes \langle x \rangle$, and this group has exactly $|D|$ many irreducible ordinary characters again. (In the two cases of an abelian p-group and the group $D \rtimes \langle x \rangle$, $l(B) = 1$ and $l(B) = |D| - 1$ respectively, while $k(B) = |D|$ in both cases.)

The idea is that these examples in some sense have 'the most' conjugacy classes, i.e., the largest number $k(G)$. Since they have one block, $k(G) = k(B)$, and so one might feel that $k(B)$ cannot become larger than $|D|$. A lot of progress has been made on this problem, but the general case still looks elusive.

We start with the case where G is p-soluble. By Fong reduction, which we will see in Sect. 7.4, we can reduce to the case where the finite group has the form $V \rtimes G$, where G is a p'-group and V is a faithful, irreducible $\mathbb{F}_p G$-module. (The groups $C_p^d \rtimes \langle x \rangle$ are examples of these.) The *$k(GV)$-conjecture* is the specialization of the $k(B)$-conjecture to this case, namely that the number of conjugacy classes of $V \rtimes G$ is at most $|V|$.

Although Knörr proved the $k(GV)$-conjecture for certain groups G called 'supersoluble' [363], and Gluck used his methods in [250] to prove the case where $|G|$ is odd, the first major breakthrough in this conjecture came with the work [489] of Robinson and Thompson in 1996. They showed that the $k(GV)$-conjecture was true for all primes greater than 5^{30}, using the idea of a *real vector*, a vector $v \in V$ such that $V{\downarrow}_{C_G(v)}$ has a faithful self-dual submodule.

One may reduce the structure of G still further, and it splits into two cases: the 'extraspecial', also known as 'symplectic', case, and the 'quasisimple' case.

The quasisimple case was attacked by Goodwin, who in two papers [256, 257] proved the result for $p > 53$. Robinson [487] dealt with the extraspecial case for $p > 211$. Papers of Gluck and Magaard [251], Köhler and Pahlings [368], and Riese [481, 482] eventually proved that the $k(GV)$-conjecture is true using these real vectors, except for $p \in \{3, 5, 7, 11, 13, 19, 31\}$. These remaining cases were finally dispatched by Gluck, Magaard, Riese and Schmid in [252, 253, 483], so we have the following result.

Theorem 4.7.2 *If G is p-soluble then the $k(B)$-conjecture is true.*

There is no reduction of the $k(B)$-conjecture to a statement about simple groups at the moment. Part of the issue is that such a reduction theorem would seem to encapsulate the $k(GV)$-conjecture. There is been some recent work of Rizo on 'θ-blocks' [485], producing a 'relative' version of the $k(B)$-conjecture that involves normal subgroups.

The remaining work on this conjecture has been based on specific defects or defect groups, and on specific classes of groups. There have been results on various quasisimple groups, but given that there is no known reduction to simple groups, they are currently not as useful as the known results for the McKay conjecture, the Alperin weight conjecture, and so on.

Malle recently proved [414] that for $p \geq 5$ a minimal counterexample to the $k(B)$-conjecture is not a quasisimple group. This doesn't show that quasisimple groups satisfy the $k(B)$-conjecture, because the proof relates blocks of quasisimple groups to blocks of groups of smaller order, but these groups are not quasisimple. For some quasisimple groups, more is known, and the $k(B)$-conjecture itself can be verified for any group whose character table is known. For example, this means that sporadic groups satisfy the $k(B)$-conjecture. The $k(B)$-conjecture is also known for alternating groups [460], and for groups of Lie type in characteristic p the result appears in [414]. This paper also proves the result for 'unipotent' blocks of groups of Lie type in other characteristics (see Sect. 9.3).

As we said earlier, Brauer was able to prove the $k(B)$-conjecture for blocks of defect at most 2, but couldn't get any further. It took until 2017 for the $k(B)$-conjecture for blocks of defect 3 to be proved, by Sambale [504]. In fact, he proved a more general result, which was that if D is an abelian p-group that has no elementary abelian direct summand of order p^4 then the $k(B)$-conjecture holds for blocks of defect group D (for $p = 2$ he obtains the same result but with 2^8 instead of p^4 in [503]). Together with the proof of the $k(B)$-conjecture for non-abelian groups of order p^3 in [503], this gives the proof for all blocks of defect 3, and 2-blocks of defect at most 7.

For $p = 2$, Sambale manages to get much better results than for p odd. For example, he proves all of the numerical conjectures we discuss in this book for 2-

blocks of defect at most 4 in [501, Theorem 13.6]. The book [501] collects lots of Sambale's results that were in the literature by 2014, and contains a lot of useful information on how to go about understanding $k(B)$ in terms of the local structure of the block. I don't have the space to give fusion systems any treatment here: the interested reader is recommended to read the excellent [21]. I also wrote a book on the subject [127].

Exercises

Exercise 4.1 In the definition of perfect isometry, prove that I is determined by μ_I, i.e., given the map μ_I and the knowledge that it arises from a perfect isometry, prove that I is determined uniquely.

Exercise 4.2 Let $I : \mathbb{Z}\mathrm{Irr}(B) \to \mathbb{Z}\mathrm{Irr}(b)$ and $I_1 : \mathbb{Z}\mathrm{Irr}(b) \to \mathbb{Z}\mathrm{Irr}(b_1)$ be two isometries. Show that I^{-1} and $I \circ I_1$ are isometries. In addition, show that if I and I_1 are perfect, then so are I^{-1} and $I \circ I_1$.

Exercise 4.3 Using the representation theory of symmetric groups from Chap. 8, prove Brauer's height zero conjecture for blocks B of symmetric groups. To do this, first show that the bijection on irreducible characters induced by a Scopes move preserves character heights. Thus we may assume that B is the principal block of S_{pw}, where w is the weight of B. Then prove that all characters have height zero if $w < p$, and find a character of positive height if $w \geq p$.

Exercise 4.4 Again, using the representation theory of symmetric groups, prove Brauer's $k(B)$-conjecture for symmetric groups.

Exercise 4.5 We will reduce Brauer's $k(B)$-conjecture for p-soluble groups to the $k(GV)$-conjecture. Let G be a p-soluble group.

(i) Suppose that Q is a normal p-subgroup of G such that $C_G(Q) \leq Q$. Prove that G possesses a unique p-block.
(ii) Prove that if $O_{p'}(G) = 1$ then $k(G) \leq |P|$, where P is a Sylow p-subgroup of G, assuming the $k(GV)$-conjecture holds. (Note that this will use the theory of p-soluble groups, so unless you know a little of this, do not attempt this part.)
(iii) Using Exercise 7.7 below, prove that the $k(B)$-conjecture holds for G (of course, assuming as in (ii) that the $k(GV)$-conjecture holds).

Chapter 5
Blocks with Cyclic Defect Groups

One of the crowning achievements of the representation theory of finite groups is the more or less complete understanding of blocks with cyclic defect groups. In Chap. 2 we saw that blocks with trivial defect group are simply matrix algebras. The next simplest p-group is a cyclic p-group, and sure enough, blocks with cyclic defect groups are the next simplest to understand. They are no longer matrix algebras, but we can write down a family of algebras given by generators and relations, and blocks with cyclic defect groups are always Morita equivalent to one of them.

In this chapter we will describe the combinatorial theory of blocks with cyclic defect groups, and then talk about the recent near-classification of all possible blocks with cyclic defect groups up to Morita equivalence. Along the way we give a description of a family of algebras, called Brauer tree algebras, that include all blocks with cyclic defect groups among them (although almost all of them are not themselves blocks of finite groups) and show how to construct all indecomposable modules for a block with cyclic defect groups.

5.1 The Brauer Tree

The Brauer tree is a combinatorial object, a finite tree, that encodes the Morita equivalence class of a block B with cyclic defect groups. The vertices correspond to irreducible ordinary characters, the edges to the irreducible Brauer characters, or equivalently the simple B-modules, and the planar embedding of the tree encodes information about the structure of the projective modules, i.e., Ext^1 between simple B-modules.

We begin with an example.

Example 5.1.1 Let G be the group $C_{p^n} \rtimes C_e$, where $e \mid (p-1)$ and C_e acts faithfully on C_{p^n}. (There is a single such group up to isomorphism.) If $e = 2$ then this is the dihedral group D_{2p^n}, and in general it is a Frobenius group. If $n = 1$ and $e = p - 1$

D. A. Craven, *Representation Theory of Finite Groups: a Guidebook*, Universitext,
https://doi.org/10.1007/978-3-030-21792-1_5

then G is the normalizer of a Sylow p-subgroup of the symmetric group S_p. Write P for the Sylow p-subgroup C_{p^n}, and E for the acting group C_e.

The conjugacy classes are as follows: there are e conjugacy classes (including the identity) that intersect E, and these are p-regular. This takes care of all non-trivial elements not in the normal subgroup P, which are split into $(p^n - 1)/e$ classes by the action of E. From the character side, we see e irreducible ordinary characters χ_i of degree 1 and $(p^n - 1)/e$ irreducible ordinary characters θ_i of degree e.

We can see e irreducible Brauer characters as well, simply the 1-dimensional characters from E, so the decomposition numbers for the χ_i are easy. For the θ_i, we can easily show from the Mackey formula, inducing 1-dimensional characters from P up to G and restricting down to E, that θ_i reduces modulo p to the sum of all of the irreducible Brauer characters. Thus the decomposition matrix is as follows:

$$\begin{pmatrix} 1 & & 0 \\ & \ddots & \\ 0 & & 1 \\ 1 & \cdots & 1 \\ & \vdots & \\ 1 & \cdots & 1 \end{pmatrix}.$$

This example is representative of the structure of any block with cyclic defect groups. It has the following properties:

(i) There is an arrangement of the rows and columns of the decomposition matrix that makes it lower unitriangular.
(ii) All of the decomposition numbers are either 0 or 1.
(iii) There is at most one row of the decomposition matrix that is repeated. There is always a repeated row unless the defect groups have order p and there are $e = p - 1$ irreducible Brauer characters.
(iv) Remove any repeats of this repeated row of the decomposition matrix, so that each row occurs exactly once. Choose any two columns of this new matrix. There is at most one row where both columns have a 1 rather than a 0 (i.e., the dot product of the two rows is at most 1).

These conditions hold for all decomposition matrices for blocks with cyclic defect groups. For blocks of defect 1, Brauer proved these statements in [48], and extending these to all blocks with cyclic defect groups was the work of Dade in [139], building on results of Thompson [543]. If B is a block with defect group D and idempotent b, write $N_G(D, b)$ for the set of all $g \in N_G(D)$ such that $b^g = b$.

Theorem 5.1.2 *Let B be a p-block of a finite group G with defect group D and block idempotent b, and suppose that D is cyclic. Write $e = |N_G(D, b) : C_G(D)|$. Let $t = (|D| - 1)/e$, called the* exceptionality *of B.*

(i) *There are exactly e irreducible Brauer characters in B.*

(ii) *There are exactly $e + t$ irreducible ordinary characters in B.*
(iii) *Suppose that $t > 1$. There are exactly t rows of the decomposition matrix that are identical to one another, labelled by* exceptional characters. *All other rows are distinct from one another.*
(iv) *If $t > 1$, then each irreducible Brauer character is a constituent of exactly two non-exceptional irreducible ordinary characters, or one non-exceptional irreducible ordinary character and all exceptional characters. If $t = 1$ then each irreducible Brauer character is a constituent of exactly two irreducible ordinary characters.*
(v) *All decomposition numbers are either* 0 *or* 1.
(vi) *If $t > 1$, there is a unique bijection between the irreducible Brauer characters and the non-exceptional irreducible ordinary characters with respect to which the decomposition matrix is lower unitriangular. If $t = 1$, the same statement holds after declaring any one character to be 'exceptional'.*
(vii) *The sum of the irreducible ordinary characters (including exceptional) with a fixed irreducible Brauer character as a constituent has degree divisible by the order of a Sylow p-subgroup of G.*

This theorem gives very strong information about the form of the decomposition matrix for a block with cyclic defect groups, together with precise information about the number $k(B)$ of irreducible ordinary characters and $l(B)$ of irreducible Brauer characters that belong to B.[1]

The structure of the decomposition matrix may be constrained still further. Consider a graph with vertices labelled by the ordinary irreducible characters, with the proviso that there is a single vertex for all exceptional characters. Since every irreducible Brauer character is a constituent of exactly two non-exceptional characters, or one non-exceptional character and all exceptional characters, we can associate to each Brauer character an edge in this graph. (If $e = 1$ then the Brauer tree is a single line with one edge, one exceptional vertex of exceptionality $|D| - 1$ and one non-exceptional vertex according to the definition above. However, this case is special: we see that the exceptional and non-exceptional characters have the same reduction modulo p, so the 'non-exceptional' character is, in fact, exceptional. Thus there is no natural choice for the non-exceptional character on our Brauer tree when $e = 1$.)

The final substantial constraint on blocks with cyclic defect groups is this: the graph just constructed is a tree, called the *Brauer tree* of the block B. It completely encodes the decomposition matrix, but it does much more than that. Before we explore its effect on the structure of B-modules, we give a few examples of Brauer trees.

[1]Much more is known about the irreducible ordinary and modular characters in a block with cyclic defect groups. For example, there is lots of technical information about fields of values, what happens over the ring $\mathcal{O}$ rather than the field k, character values on elements, and other things besides. One source of information is [217, Chapter VII]. We will avoid most of the technicalities, but if you happen to need them for some reason, you will find some of them there.

Example 5.1.3 If G is the group $C_{p^n} \rtimes C_e$ that we had above, then the Brauer tree of kG (it has only one block) is a star, with exceptional vertex (if there is one) in the centre of the star. We fill the exceptional node in here, whereas some other authors place a ring around it. We draw the Brauer tree on the plane as below.

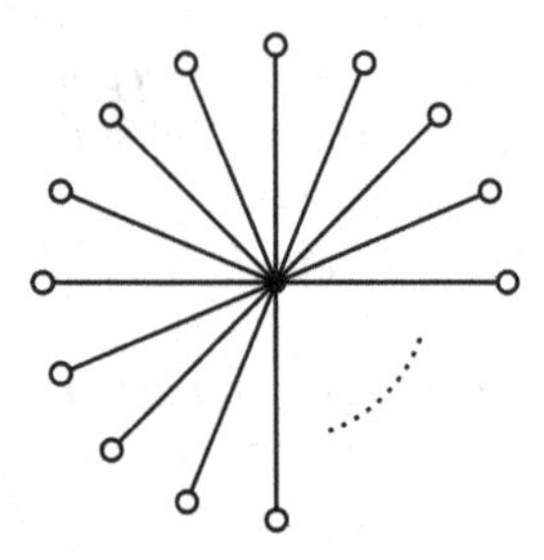

For example, if $G = C_5 \rtimes C_4$, a Frobenius group of order 20, then the Brauer tree is a star with four vertices and no exceptional node.

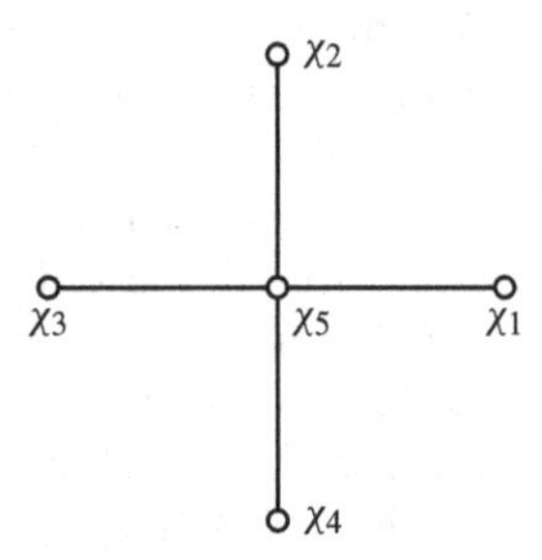

Write x for a generator for the C_4 quotient. In the picture above, χ_1 is the trivial character, χ_2 and χ_4 are degree 1 characters that send x to $\pm$i and $\chi_3 = \chi_2^{\otimes 2}$ sends x to -1. The character χ_5 has degree 4.

On the other hand, if $G = C_{25} \rtimes C_4$ or $C_{13} \rtimes C_4$, then the Brauer tree is exactly the same, except this time there is an exceptional node in the middle, of exceptionality 6 and 3 respectively.

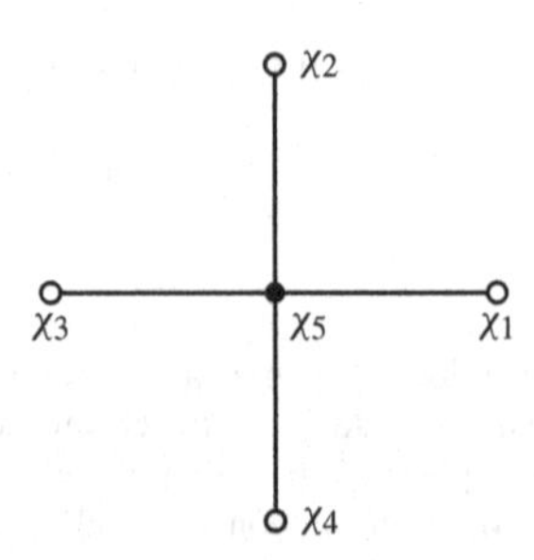

The characters χ_1 to χ_4 are defined in the same way as before, but χ_5 is now exceptional (but still of degree 4), so labels multiple characters. We can use χ_{exc} for this rather than a number, to signify that it is not a single character.

Every block of every soluble, in fact every p-soluble, group has this form (see Sect. 5.3.1 below, although it follows from the Fong–Swan theorem, Theorem 7.1.6), for some number of edges and some exceptionality. If we want any other types of Brauer tree we have to move into the realm of non-abelian simple groups.

Example 5.1.4 The easiest examples of insoluble groups with cyclic Sylow p-subgroups are the symmetric groups S_n for $p \leq n < 2p$. These have Brauer tree a line with p vertices, $p - 1$ edges, and no exceptional node.

For example, for S_5 and $p = 5$, the Brauer tree has the form

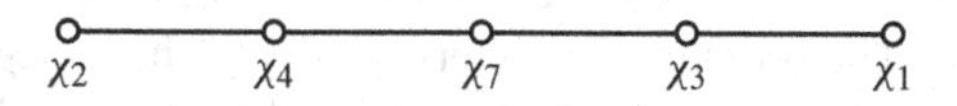

where the labelling of the characters is as in Example 2.2.8, whereas for $p = 3$ they have the form

In fact, every block of every symmetric or alternating group with cyclic defect groups has Brauer tree a line. If B is a block of S_n with cyclic defect groups, then there are always $p - 1$ irreducible Brauer characters in B, and the order of a defect group is p, so there is no exceptional node. We can even label the vertices and edges with explicit ordinary and Brauer characters, see Theorem 8.3.13.

However, for the alternating groups this is not always the case. For example, we saw the decomposition matrix of A_5 for $p = 5$ in Example 2.2.7, and this yields the tree below.

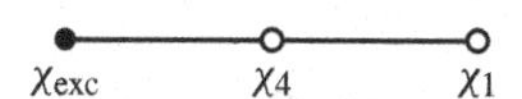

Here χ_{exc} labels both χ_2 and χ_3. The principal 5-block of A_6 is also like this, with two edges, but the principal 5-block of A_7 looks like that of S_7, with four edges. If the two end nodes of the tree of S_n restrict to the same character of A_n (or equivalently the one character, tensored by the sign character, is equal to the other character), then the tree for A_n is 'folded over', i.e., has an exceptional vertex. The precise meaning of 'folded over' will be given in Sect. 5.3.

(I have called this tree a line here, because that's what it is. Brauer in [49], and numerous authors afterwards, for example Feit [218] in an important paper that we will come back to later in this chapter, call it an open polygon.)

Example 5.1.5 Another famous example of a group with a cyclic Sylow p-subgroup is $\mathrm{SL}_2(p)$. For $\mathrm{SL}_2(p)$ and p odd, there are three p-blocks: The first is the principal block, which has the same ordinary and Brauer characters, and decomposition matrix, as the principal block of $\mathrm{PSL}_2(p)$ by Theorem 2.4.2. The second is a block of defect zero, containing the 'Steinberg module', which has dimension equal to the order of the Sylow p-subgroup of $\mathrm{SL}_2(p)$, in this case p. The third block is a faithful block (i.e., a block consisting of faithful representations) that contains the natural 2-dimensional representation of $\mathrm{SL}_2(p)$ over a field of characteristic p, sending matrices to themselves, and has maximal defect. The first and third blocks again have Brauer tree a line, this time with exceptionality 2, just as for the alternating group A_p, so we have to locate the exceptional node: it lies at one end of the tree, again, just as in A_p.

We will use this as a fundamental example in the next section, so for now we give the case $\mathrm{SL}_2(7)$. The ordinary characters of $\mathrm{SL}_2(p)$ have degrees 1, $(p \pm 1)/2$ (two of each of these if p is odd), p (the Steinberg character), and several of degree $p \pm 1$, depending on p. Thus for $p = 7$ we have ordinary characters of dimensions $1, 2, 3, 6, 8$, and the Steinberg of degree 7. The modular characters have dimension $1, 2, \ldots, p$.

Here we have labelled the Brauer characters by their dimension, as each number occurs once, so this is a unique labelling.

In general, the principal p-block of $\mathrm{PSL}_2(p)$ for p odd has the structure

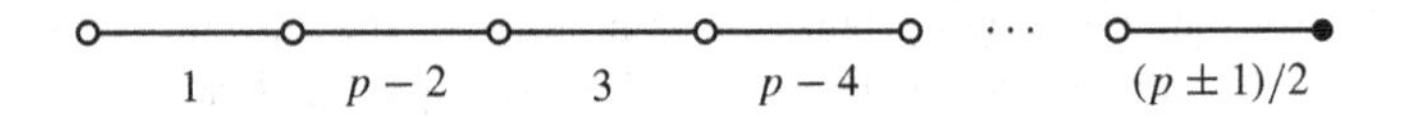

and the faithful p-block has the structure

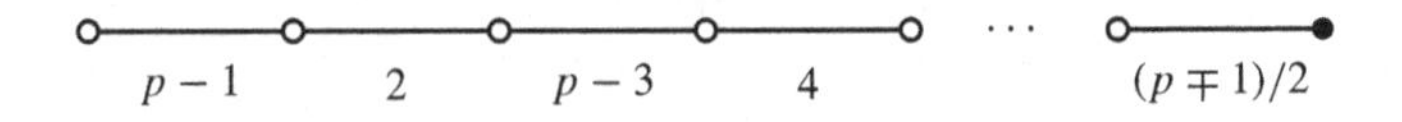

It is no coincidence that the trivial character is always at the end of a branch of the tree: notice that an ordinary character has degree equal to the sum of the Brauer characters that label edges that are incident to it, so since the trivial character has degree 1, it cannot be incident to more than one edge (as dimensions are positive integers!).

In order to write more concise sentences, I will pretend that the vertices themselves *are* the irreducible ordinary characters, and the edges themselves are the irreducible Brauer characters or simple modules. Thus we can talk about two ordinary characters being adjacent, or a Brauer character being incident to an ordinary character. This is mainly to stop having to write 'the ordinary character labelling the vertex' so often. Thus the trivial character is always at one end of the Brauer tree.

The trivial character is real, so lies in a real block B (see Exercise 2.8). The real characters of any block B form a line in the Brauer tree, called the *real stem* of the tree. Duality of ordinary and Brauer characters becomes a reflection in the real stem. If a block B is not real, then it and its dual block B^* have the same Brauer tree, and the same planar-embedded Brauer tree, except one is a reflection of the other (Exercise 5.1).

A real block must always have a real stem: Exercise 5.2 proves this when there is no exceptional node, and of course if there is an exceptional node then it lies on the real stem. However, this does not mean that there are any real characters in the block, since the exceptional characters might be dual to one another.

John Murray provided me with an example of such a block, and he himself was alerted to the existence of such blocks by Blau and Gow: for $p = 5$ and $G = 6 \cdot S_6$, a 6-fold cover of S_6. He also suggested the smaller example, a group of order 120, with presentation

$$\langle x, y \mid x^{30} = 1, y^4 = x^{15}, x^y = x^{17} \rangle,$$

which has a real 5-block (of cyclic defect) with no real characters.

So far we have only given two examples of Brauer trees: the star and the line. In some sense 'most' trees are of this form, or something very similar called a 'windmill', that we will see in Sect. 5.3. However, they can get a bit complicated, as the next example shows.

Example 5.1.6 Let G denote the Thompson sporadic simple group, and let $p = 19$. The Sylow 19-subgroups of G are cyclic of order 19, so the principal 19-block of G possesses a Brauer tree. It was given up to two possibilities in [294], and determined completely in [123]. There are eighteen edges, and therefore no exceptional node. The tree is given in Fig. 5.1. (Here, the ordinary characters are labelled as in [122].) The real characters are χ_1, χ_{25}, χ_{44}, χ_{48}, χ_{43}, χ_{26} and χ_2. The conjugate character to χ_{22} is χ_{23}, for example, so complex conjugation of characters, i.e., duality, does indeed induce a reflection in the real stem.

Example 5.1.7 Let $G = G_2(q)$ and let $p > 3$ be a divisor of $q^2 + q + 1$. In these cases, the Brauer tree of the principal p-block of G is given in Fig. 5.2.The notation for the characters is taken from [107, Section 13.9]. The characters $G_2[\theta]$ and $G_2[\theta^2]$ are dual, and the others lie on the real stem.

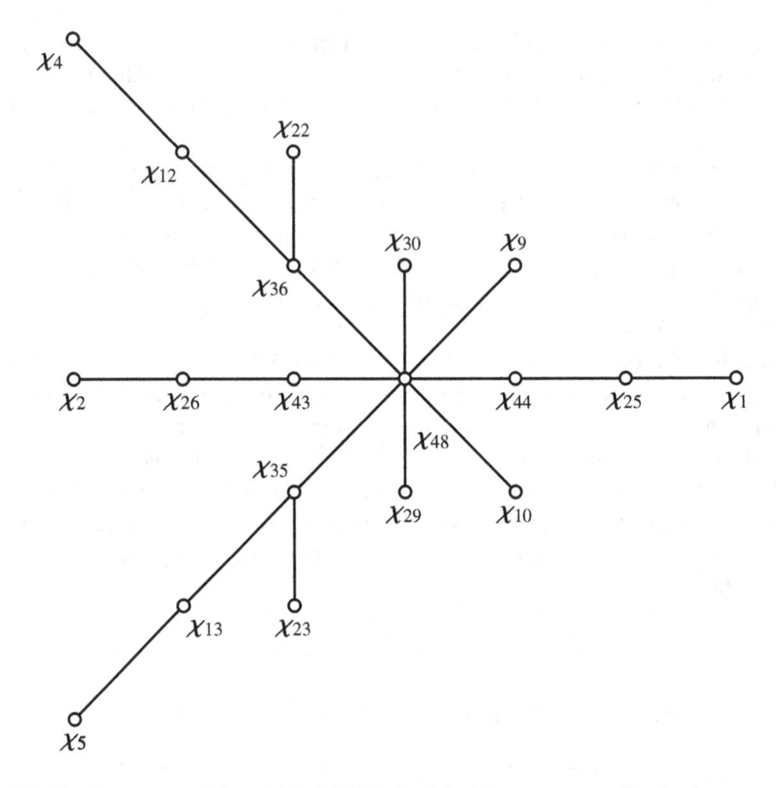

Fig. 5.1 The Brauer tree of the principal 19-block of the Thompson sporadic simple group

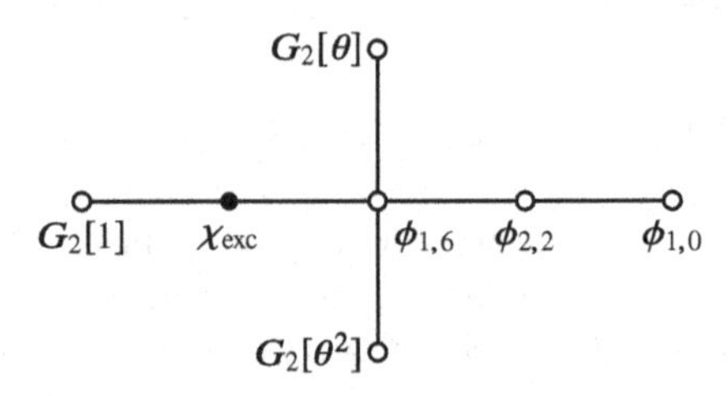

Fig. 5.2 The Brauer tree of the principal p-block of the simple group $G_2(q)$ for $p > 3$ a divisor of $q^2 + q + 1$

As the Brauer tree is a tree, hence a bipartite graph, it splits the vertices, the ordinary irreducible characters, into two disjoint sets. By Theorem 5.1.2, the sum of two adjacent characters in the tree has degree divisible by the order of a Sylow p-subgroup P, but since Brauer's height-zero conjecture is true for blocks with cyclic defect groups (see Sect. 4.1), each character degree is only divisible by the index $|P : D|$. If $e = 1$ then the Brauer tree is a single edge with all ordinary characters having the same, irreducible, reduction modulo p, and this is the case if $p = 2$ (as $e \mid (p-1)$). If p is odd and $e > 1$, then the two sets of characters in the bipartite graph have degrees congruent to $\pm i$ mod $|P|$ for some i. In [294] the members of these two sets were labelled $\times$ and $\circ$. If B is the principal block then the degree congruences are ± 1 mod p, so we could label these as $(+)$- and $(-)$-type. For other blocks, the degree congruences could be something like ± 10 mod 25, and to define which should be $(+)$ and which should be $(-)$ is a bit more convoluted. As the Brauer tree of the Brauer correspondent of a block B is a star, all of the non-exceptional characters ϕ in the correspondent have the same degree modulo $|P|$. All of the non-exceptional characters χ of B have, up to sign, the same degree modulo $|P|$ as those ϕ of the Brauer correspondent, and so we can write $(+)$ if $\chi(1)$ and $\phi(1)$ are congruent modulo p, and $(-)$ if $-\chi(1)$ and $\phi(1)$ are congruent modulo $|P|$.

If the Brauer tree is a line, then there is only one way to draw it, but in the last example I chose a very specific way to draw it on the plane, as opposed to one of myriad other ways to do so, and with the example of $C_5 \rtimes C_4$ before, I was very specific about which of the four degree 1 characters was which. That is because the Brauer tree can encode not just numerical information about the decomposition matrix of a block, but also the structure of the projective indecomposable modules, and in general the Morita equivalence class of the block.

To do this we will introduce the class of Brauer tree algebras, which includes all blocks with cyclic defect groups, but many other algebras as well. Every block with cyclic defect groups is Morita equivalent to a Brauer tree algebra, and there is one Brauer tree algebra for every Brauer tree *with planar embedding*, together with a choice of exceptional vertex. Strictly speaking, we are only interested in the cyclic ordering of the edges around each vertex, which is sometimes referred to as a 'ribbon graph', but it is usually referred to as a planar-embedded Brauer tree in representation-theoretic circles.

The next section will achieve this goal.

5.2 Brauer Tree Algebras

In this section we construct the Brauer tree algebras that will help us better understand the module structure of blocks with cyclic defect groups. Brauer tree algebras first appeared in the work of Janusz [335], in his attempt to describe the indecomposable modules for such blocks. Kupisch [379, 380] independently computed the structure of the projective indecomposable modules, and then all indecomposable modules, for blocks with cyclic defect groups. These ideas were recast by Gabriel in [237], and the foundations of the theory were laid proper by Gabriel and Riedtmann in [238], who worked with Brauer tree algebras in general.

Brauer tree algebras are often described by a characterization, rather than a construction. To construct Brauer tree algebras directly is best done using quiver algebras with relations, also known as path algebras with relations, and this is what is done in [238]. We will introduce as much of the theory as needed to define the algebras. Path algebras are a little outside of our main interests in this book, but the beauty and versatility of the object are enough for us to make the detour.

We first describe what the planar embedding means in terms of extensions between simple modules, so let B be a p-block of G with cyclic defect groups: if M and N are simple B-modules labelling edges of the Brauer tree of B, then $\mathrm{Ext}^1_{kG}(M,N) = 0$ if M and N do not have a vertex in common. If they have a vertex v in common, then $\mathrm{Ext}^1_{kG}(M,N) = 0$ if M is not the edge immediately preceding N in the cyclic ordering around v, and $\mathrm{Ext}^1_{kG}(M,N) = k$ if it is.

For example, if we take the tree from Example 5.1.6 again, let M_1 denote the edge between χ_1 and χ_{25}. It has an extension with only one other module: that between χ_{25} and χ_{44}. However, if M_2 denotes the simple B-module labelling the edge between χ_{44} and χ_{48}, then $\mathrm{Ext}^1_{kG}(M_2,N) \neq 0$ only when N is the simple module for the edge between χ_9 and χ_{48}, and $\mathrm{Ext}^1_{kG}(N',M_2) \neq 0$ only when N' is the module between χ_{10} and χ_{48}.

It is utterly remarkable that one may imbue the Brauer tree with this extra information—a planar embedding—and that enriches the information sufficiently to encode Ext^1. In fact, it does more than that: from this you can read off the projective indecomposable modules, and indeed two blocks with the same planar-embedded Brauer trees, the same exceptional node and the same exceptionality, are Morita equivalent.

Suppose that M is a simple B-module, incident to the vertices v and w. To begin with, suppose that neither v nor w is the exceptional node of the tree. In this case, the projective cover $\mathcal{P}(M)$ of M is easy to define. The socle and top of $\mathcal{P}(M)$ are both M, as noted in Sect. 3.1, so we are interested in the *heart* of $\mathcal{P}(M)$, i.e., the quotient $\mathrm{rad}(M)/\mathrm{soc}(M)$. This is the direct sum of two uniserial modules M_v and M_w, one for v and one for w as the notation suggests. The composition factors of the module M_v are all simple B-modules other than M that are incident to v, each occurring exactly once. They are arranged in the only way that is possible given our conditions on Ext^1 earlier: the top of M_v is the first module that appears after M in the cyclic ordering around v, then the second radical layer is the next module in the ordering, and so on, until we reach the module that is just before M in the ordering around v, which is the socle of M_v. The construction of M_w is identical, except with w instead of v. If v has valency 1 (i.e., M is the only edge incident to v) then M_v is the zero module, so that $\mathcal{P}(M)$ is uniserial.

For example, if the tree is

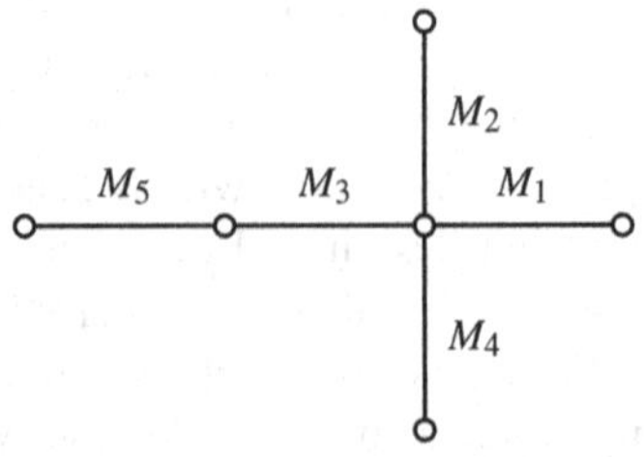

M_2

M_5 M_3 M_1

M_4

then the projective modules are

$$\begin{array}{c} M_1 \\ M_2 \\ M_3 \\ M_4 \\ M_1 \end{array}, \quad \begin{array}{c} M_2 \\ M_3 \\ M_4 \\ M_1 \\ M_2 \end{array}, \quad \begin{array}{ccc} & M_3 & \\ M_4 & & \\ M_1 & \oplus & M_5 \\ M_2 & & \\ & M_3 & \end{array}, \quad \begin{array}{c} M_4 \\ M_1 \\ M_2 \\ M_3 \\ M_4 \end{array}, \quad \begin{array}{c} M_5 \\ M_3 \\ M_5 \end{array}.$$

If (without loss of generality) v is the exceptional node, then M_w stays the same, and M_v looks similar to what it was previously, except you wind around the exceptional node t times, where t is the exceptionality. For example, if the tree is the same as above with exceptionality 2 at the node incident to M_3 and M_5 then the projective covers of M_1, M_2 and M_4 stay the same, but those of M_3 and M_5 are affected, and are

$$\begin{array}{ccc} & M_3 & \\ M_4 & & M_5 \\ M_1 & \oplus & M_3 \\ M_2 & & M_5 \\ & M_3 & \end{array}, \quad \begin{array}{c} M_5 \\ M_3 \\ M_5 \\ M_3 \\ M_5 \end{array}.$$

Notice that the first tree we gave must be hypothetical, and cannot occur as the Brauer tree of a block. The number $l(B)$ of simple modules is five, and the exceptionality t is 1, so that $|D|$, which is given by $|D| - 1 = te$, must be 6, which is not a prime power. So we can define the projective modules for a block even if it doesn't really exist. (In the second case, $t = 2$ and so $|D| = 11$, which of course is a prime power.)

So what did we construct the projective modules of in the first case, if it isn't a block of a finite group? The answer is a Brauer tree algebra. The definition of a Brauer tree algebra is fairly easy.

Definition 5.2.1 A finite-dimensional k-algebra A is a *Brauer tree algebra* if there exist

(i) a finite planar-embedded tree T,
(ii) a designated vertex of T, called the exceptional node,
(iii) a positive integer t, called the exceptionality, and
(iv) a bijection f between the edges of T and the simple A-modules,

such that the projective indecomposable A-modules have the same structure as given by the algorithm above applied to T.

Of course, this is a characterization rather than a construction. The theorem that we want is of course the following.

Theorem 5.2.2 (Janusz [335], Kupisch [379]) *If B is a block with cyclic defect groups then B is a Brauer tree algebra with respect to its Brauer tree.*

The obvious question is, given a tree, a designated vertex and an exceptionality, is there a Brauer tree algebra with those data? The answer is 'yes', and furthermore it is unique up to Morita equivalence. To provide one, we will define the path algebra, and then give the path algebra that realizes the data of a Brauer tree algebra.

Let Q be a finite quiver, that is, a directed graph in which loops and multiple edges are allowed. The *path algebra* kQ of Q over a field k is a vector space with basis all paths of finite length in Q, including of length 0 (i.e., a vertex). The product $\alpha\beta$ of two paths α and β is defined to be 0 if the end of α is not the start of β, and the composition of α and β if they may be composed. Thus kQ is finite-dimensional if and only if there are no circuits in Q. In general, path algebras have some interesting properties, but in order for them to be useful for us, we need to impose relations, for example that some path is equal to 0, or that two paths are equal. We then take the quotient by this ideal I of relations to produce a path algebra with relations kQ/I. The quiver Q is closely related to the *Ext-quiver* of an algebra A, which has vertices the simple modules (up to isomorphism) and the number of directed edges from M to N is equal to the dimension of $\mathrm{Ext}_A^1(M, N)$. In order to produce the Ext-quiver of an algebra, a generating set for I should be finite and consist of relations whose terms are paths of length at least 2. (This is simply so we cannot express an edge in terms of the other edges, thus creating a degeneracy in the quiver.) In order to make the definition easier though, we will include edges in the Brauer tree algebra that are *not* in the Ext-quiver. These should be removed to produce the Ext-quiver.

An algebra is *basic* if all simple modules for it are 1-dimensional.

Theorem 5.2.3 *Let A be a finite-dimensional k-algebra.*

(i) *There exists a basic algebra B that is Morita equivalent to A.*
(ii) *If A is basic and B is Morita equivalent to A and also basic, then A and B are isomorphic algebras.*
(iii) *Every basic algebra is isomorphic to a path algebra with relations.*

These results are fundamental to representations of algebras, and may be found in, for example, [200, I.2.6, I.2.7 and I.5.7]. The last of these is attributed to Gabriel. Writing $A = kQ/I$, the simple A-modules are in bijection with the vertices of Q.

Having defined path algebras, we now want to describe the path algebra for a Brauer tree algebra B. The path algebra itself is easy: it is simply the Ext-quiver for B, so as we said above the vertices are labelled by the isomorphism classes of simple modules, and such that there are $\dim(\mathrm{Ext}_{kG}^1(M, N))$ arrows from M to N. Since Ext^1 is non-zero along the cyclic ordering around each vertex, Q is a union of directed cycles, in bijection with the vertices of the Brauer tree. If there is an exceptional node, the corresponding cycle will be called exceptional.

We also add a loop at each simple module that is incident to a vertex of valency 1. This does not belong in the Ext-quiver, but does help with the description of the relations of the path algebra. There are always two arrows entering and leaving each simple module.

Example 5.2.4 We consider the tree we had before, which is as follows.

M_2

M_5 M_3 M_1

M_4

The vertices of the corresponding quiver Q are the M_i, there are arrows $M_1 \to M_2$, $M_2 \to M_3$, $M_3 \to M_4$ and $M_4 \to M_1$, and there are arrows in both directions between M_3 and M_5, as in this picture.

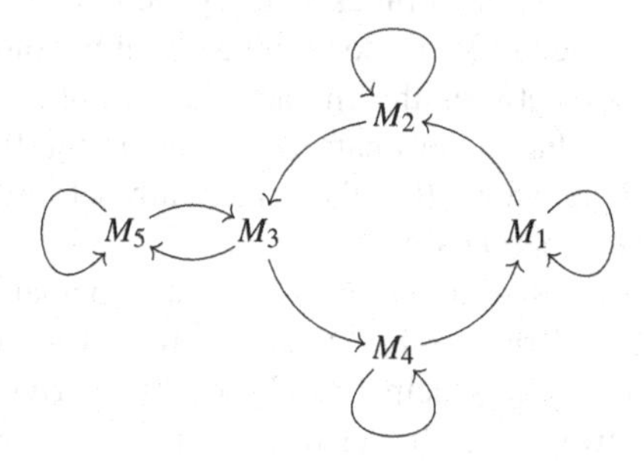

The hard part in constructing quivers with relations is usually determining the relations that need to be imposed on the Ext-quiver to produce the path algebra with relations that is Morita equivalent to B.

In the case of blocks with cyclic defect groups, and Brauer tree algebras in general, because of the restrictive nature of the projective modules it is actually possible to write down the relations that we need to quotient out by to produce the correct algebra.

There are two types of relations: if $\alpha : u \to v$ and $\beta : v \to w$ are arrows that belong to different cycles of Q then $\alpha\beta = 0$. The second type of relation is that if α and β are cycles that start (and therefore end) at the same vertex, then $\alpha = \beta$. The only modification of this is if α is the exceptional cycle, in which case $\alpha^t = \beta$, where t is the exceptionality of the cycle.

The loops are really important: they are there to say you cannot go round a cycle more than once, or around the exceptional cycle more than the exceptionality number of times. To see this, by the second relation, a cycle α starting at a vertex v is equal to any other cycle starting at v, and since the loops are there, every vertex has exactly two cycles emanating from it. Thus the composition of α with any other edge β starting at v must be 0, by replacing α by the cycle that does not contain β and applying the first relation. (If α were the exceptional cycle, this would need to be α^t for it to work, but that is what we want.)

We give an example of this now.

Example 5.2.5 If the Brauer tree is a star with e edges and with exceptional node in the centre, with exceptionality t, then it is a single cycle of length e, with a loop coming off each vertex. The important relation is that any path around the cycle of length at least $et + 1$ is 0.

If $G = C_{p^n} \rtimes C_e$, then the Brauer tree of the group algebra kG is a star with e edges and exceptionality $t = (p^n - 1)/e$, so that $et + 1 = p^n$. Thus this quiver has the condition that any path around the cycle of length p^n is 0.

In this case the algebra kQ/I is actually isomorphic to kG by Theorem 5.2.3, since all simple kG-modules are of dimension 1 and so kG is basic.

Before we describe the indecomposable modules, we will discuss *Green's walk*. The Green correspondence was introduced in Sect. 3.2, but so far we haven't used it very much. It is actually fairly easy to describe the Green correspondents of the simple B-modules when B is a block with cyclic defect groups, at least after Green wrote [267]. You need to know something about the block before you start, essentially the comparison between the character degree of a single character in B and the degrees of the irreducible ordinary characters in the Brauer correspondent of B. We give the example where B is the principal block first, and then describe how this differs from the general case.

Suppose that there are e simple modules in B, a principal block (so the trivial module is a B-module). By Brauer's third main theorem (Theorem 2.4.4) the Brauer correspondent B' of B is also a principal block, so it has a trivial module, which we label 1. The block B' always has Brauer tree a star, so we label the edges of the Brauer tree of B' by $1, 2, \dots, e$, around the exceptional node. The projective cover of the trivial module has structure

$$\begin{matrix} 1 \\ 2 \\ 3 \\ \vdots \\ e \\ 1. \end{matrix}$$

With respect to this labelling on the simple B'-modules, we can describe the Green correspondents of the simple B-modules. We start by describing the socles and tops of the Green correspondents, recalling that by Example 3.1.8, all indecomposable B'-modules are uniserial. We will label the edges of the Brauer tree of B, by 'walking' around the edges. Each edge will be labelled twice, so there are $2e$ labels in total. The labels are ordered $1, \delta(1), 2, \delta(2), \ldots, e, \delta(e)$, so that 1 and $\delta(1)$ label the edge corresponding to the trivial module.

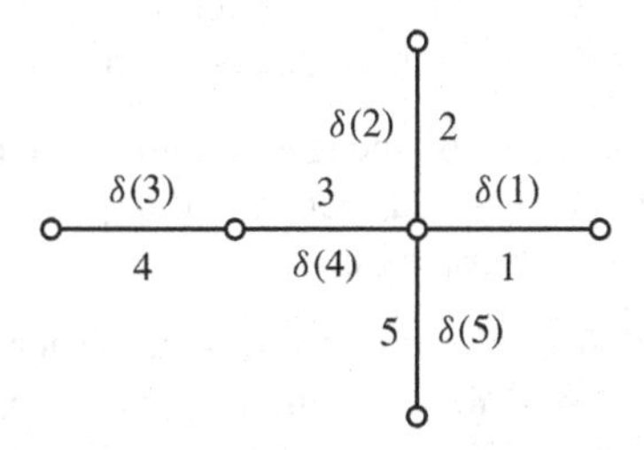

This determines a permutation δ, via $\delta(i) = j$ if $\delta(i)$ and j label the same edge of the tree. The Green correspondent of the simple B-module corresponding to the edge labelled by i (and $\delta(j)$ for some j) has socle $\delta(i)$ and top i (this is [267, Theorem 1]).

If the exceptionality t of B (and hence B') is 1, then an indecomposable B'-module is uniquely determined by its socle and top, but in general there are t non-isomorphic such modules, each of a different dimension. Thus we need to describe the number of composition factors of the Green correspondent. The important piece of information is that the Green correspondents of simple modules have either at most e composition factors, or at least $|D| - e$ composition factors, which brings us down to two possible modules.

To determine which it is, we need to know which it is for one of the modules in the block; if a given simple module M has Green correspondent with at most e composition factors, then all simple modules adjacent to M have Green correspondents with at least $|D| - e$ factors, and vice versa. This means that if we know how many factors are in one of the Green correspondents, then we know all of them. In our case, one of the modules is the trivial module, which of course restricts to the trivial module, hence has trivial Green correspondent. In an arbitrary block, it is not quite so easy to make this determination.

To do so, we note that for any edge incident to a 'leaf' on the tree, i.e., a vertex of valency 1, the corresponding simple module M has Green correspondent either a simple module S or $\Omega(S)$ (where $\Omega(-)$ is the Heller translate) for S simple. This is true for *any* simple module in that position, if the corresponding node is not exceptional (and there must be such a module). Of course, for the Brauer correspondent block, all simple modules S have the same dimension (modulo $|P|$) since the Brauer tree is a star (and therefore all edges are of the same type). Thus we may compare the congruences of M and S modulo $|P|$. If they are the same, then M has simple Green correspondent, and otherwise it is $\Omega(S)$ for some simple module S.

We did this at the end of Sect. 5.1, to assign (+) and (−) to each vertex of the tree of B. So (+)-type vertices of valency 1 have incident simple modules with simple Green correspondents.

In the first case, we match the labellings and continue exactly as above. In the second case, we do the same labelling as above, but then take Heller translates of every module obtained in the way above to find the Green correspondents of the simple modules (and label the simple modules for the Brauer correspondent so that this one simple module has the correct Green correspondent).

We are now in a position to discuss the classification of indecomposable modules for a block with cyclic defect groups, as performed in [335]. We give two examples of this, and then describe the general case.

Example 5.2.6 Suppose that the Brauer tree is a star. Then every indecomposable module is a submodule (or equivalently a quotient) of a projective indecomposable module, and hence is uniserial. The description of such modules is determined by the structure of the projectives.

Example 5.2.7 Suppose that the Brauer tree is a line, with no exceptional vertex, with simple modules $M_1, \dots, M_e$. The projective indecomposable modules are

$$\begin{matrix} M_1 \\ M_2 \\ M_1 \end{matrix}, \qquad \begin{matrix} M_i \\ M_{i-1} \oplus M_{i+1} \\ M_i \end{matrix}, \qquad \begin{matrix} M_e \\ M_{e-1} \\ M_e \end{matrix}$$

for $2 \leq i \leq e-1$. By taking quotients of these, we find modules of the form

$$\begin{matrix} M_i \\ M_{i-1} \oplus M_{i+1} \end{matrix} \qquad \begin{matrix} M_{i+2} \\ M_{i+1} \oplus M_{i+3} \end{matrix}.$$

Take the sum of these two, and quotient out by a diagonal submodule isomorphic to M_{i+1}. This yields a module with structure

$$\begin{matrix} M_i \oplus M_{i+2} \\ M_{i-1} \oplus M_{i+1} \oplus M_{i+3} \end{matrix}.$$

Similarly, we can take a diagonal submodule

$$\begin{matrix} M_{i-1} \oplus M_{i+1} \oplus M_{i+3} \\ M_i \oplus M_{i+2} \end{matrix} \subseteq \begin{matrix} M_{i-1} \oplus M_{i+1} \\ M_i \end{matrix} \oplus \begin{matrix} M_{i+1} \oplus M_{i+3} \\ M_{i+2} \end{matrix}.$$

In general, we draw the diagram

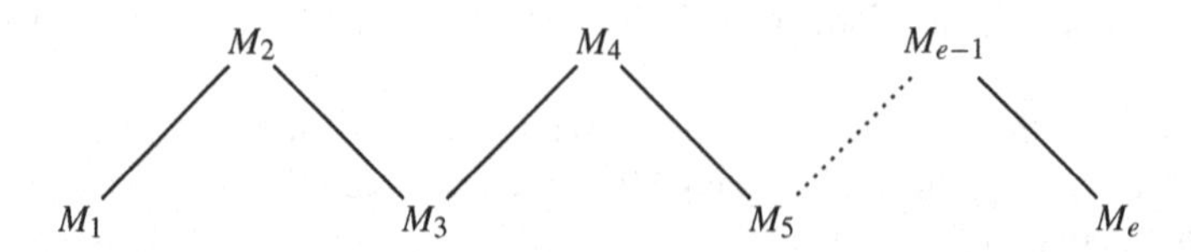

(In this diagram e is odd. If e is even then M_e would be above M_{e-1}.) For every connected subset of this diagram, there is an indecomposable module with socle the bottom layer and top the top layer of this diagram, by iterating the procedure above. To make the rest of the indecomposable modules, take duals of these. There are $e(e-1)/2$ connected subsets of this diagram of size at least 2, so this yields $e(e-1)$ non-simple, non-projective, indecomposable modules. There are e simple modules, and e projective indecomposable modules, yielding $e(e+1)$ indecomposable modules in total.

So what about the general case, for an arbitrary Brauer tree algebra? Here we take sequences of uniserial modules, but the sequences need to be special, because we are going to take a submodule of a direct sum and then quotient out by a diagonal submodule. Thus the socles of the $2i$th and $(2i+1)$th modules are isomorphic, and the tops of the $(2i+1)$th and $(2i+2)$th modules are isomorphic, or possibly the other way round.

Formally, let $S_1, \dots, S_r$ denote a sequence of simple B-modules, such that S_i and S_{i+1} are both incident to the same vertex for each $1 \leq i < r$. Let I be either all odd integers or all even integers in the set $\{1, \dots, r\}$. Let M_i denote a uniserial module with socle S_i and top S_{i+1} if $i \in I$, and socle S_{i+1} and top S_i if $i \notin I$. We form two submodules of $\bigoplus M_i$: the first, N_1, is the sum of the diagonal submodules of the socles of M_i; the second, N_2, is the kernel of the analogous map that sends $\bigoplus M_i$ to the sum of the diagonal quotients of the tops of the M_i. Formally, N_1 is the sum of diagonal submodules of $\mathrm{soc}(M_i \oplus M_{i+1}) \cong S_{i+1}^{\oplus 2}$ for $1 \leq i < r$ and $i \notin I$, and N_2 is the intersection of kernels of maps $\bigoplus M_j \to S_{i+1}$ for $1 \leq i < r$ and $i \in I$ that do not contain either M_i or M_{i+1}.

The modules N_2/N_1 for the various possibilities for the sequences of the M_i furnish us with a complete set of indecomposable modules, but also with many decomposable modules. We finally need to identify those sequences of the S_i that yield precisely the indecomposable modules. There are two types: the first type is where all of the S_i are distinct; the second is where there exist $1 \leq a < b \leq r$ such that $b-a$ is odd, $S_{a+i} \cong S_{b-i}$ for all $0 \leq i \leq (b-a)/2$, with $S_{(a+b-1)/2}$ being incident to the exceptional node and $S_{(a+b-3)/2}$ not incident to it, and all of the S_i are distinct except for the isomorphisms given above. A diagram of the second type is given in Fig. 5.3.

We should also briefly discuss real blocks, which were mentioned in the previous section and Exercise 2.8. There we said that if B is real then the Brauer tree can be written on the plane in such a way that the real characters form a line, called the real stem, and complex conjugation, the duality on $\mathrm{Irr}(B)$, was given by a reflection on the plane. (Exercise 5.1.)

Things make more sense if we consider our Brauer tree embedded in $\mathbb{C}$. Now the real characters should obviously be placed on the real line, the exceptional node should go at 0, and the complex conjugate of a character appearing at position $x+\mathrm{i}y$ should appear at position $x-\mathrm{i}y$. Doing this, we obtain not only a reflection of the Brauer tree, but also a reflection of the *planar-embedded* Brauer tree. This is because $\mathrm{Ext}^1_{kG}(M, N) \cong \mathrm{Ext}^1_{kG}(N^*, M^*)$.

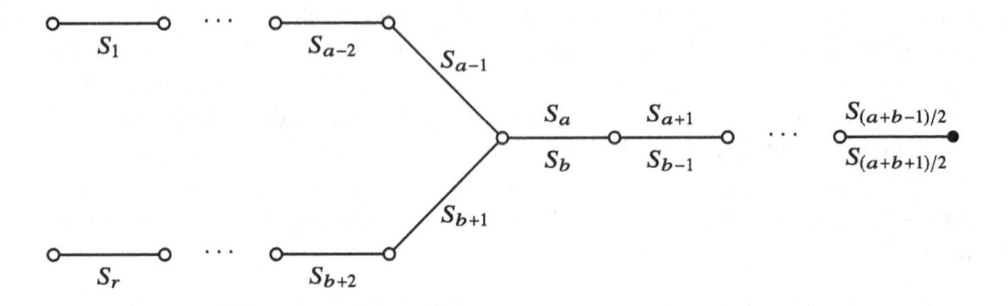

Fig. 5.3 Paths in the Brauer tree yielding indecomposable modules

For example, consider the tree from Example 5.1.6. The reflection in the real line swaps χ_{30} and χ_{29}, and also χ_9 and χ_{10}. Since there is an extension between some of those simple modules and not others, given by the cyclic ordering, one cannot reflect some of the characters in the tree and not others.

Part of the subtlety of computing a Brauer tree is in working out which of two dual characters lies on which side of the real stem. If there is only one pair then it does not matter, as complex conjugation means that which goes where is simply a choice of which 4th root of unity is called i. However, once you have more than one pair of non-real characters in B, then you cannot normally choose where to place all pairs simultaneously. For example, one may flip χ_9 and χ_{10} if one desires, but then one is forced to swap χ_{30} and χ_{29}, and χ_{36} and χ_{35}, and so on.

This also explains why we place χ_1, which is supposed to label the trivial character, on the far right. The exceptional vertex should go at the origin of $\mathbb{C}$ in the planar embedding, and the obvious place for χ_1, the trivial character, to go, is at the position of the real number 1, so to the right of the exceptional node. For the Brauer tree of $C_{p^n} \rtimes C_e$, arrange ζ a primitive eth root of unity and x a generator of C_e so that for the 1-dimensional simple module M such that $\mathrm{Ext}^1_{kG}(k, M) \neq 0$ (i.e., next to k in the cycling ordering around the exceptional node), the action of x on M is as multiplication by ζ. With this arrangement, the ordinary character value $\chi_j(x)$ is ζ^j, where χ_j is the character in the $\mathbb{C}$-embedded Brauer tree at the position $(1, 2\pi j/e)$ in polar co-ordinates. (With this labelling, χ_0 is the trivial character. We do this in order not to have to write a clunky exponent in the character value $\chi_j(x)$.)

5.3 Classification of Brauer Trees

The classification of Brauer trees of all finite groups was a distant prospect when Brauer's first paper [49] on the problem appeared in 1942. It took two things to make it a possibility: the first was the classification of the finite simple groups, and the second was a 1984 paper [218] by Feit. This paper reduced the problem of determining the possible Brauer trees to determining all Brauer trees of finite quasisimple groups. The second thing is the detailed theory from Chap. 9, which can then be brought to bear on it. Over the decades that followed an almost complete picture has emerged of what the Brauer trees of quasisimple groups are, which we will detail in this section. We begin with a few words about Feit's reduction to quasisimple groups.

5.3.1 From All Groups to Simple Groups

Feit's work relies heavily on Clifford theory, which we discuss in Chap. 7. The first thing we need to define is an unfolding of a tree. To make the definition cleaner, we embed Brauer trees in $\mathbb{C}$ rather than $\mathbb{R}^2$, as we did in the previous section, because we use polar co-ordinates.

Informally, to 'unfold' a tree at a vertex v, you perform the following steps: first, place v at the origin of $\mathbb{C}$. Second, redraw the tree so that all edges appear in the first $2\pi/n$ radians around the origin. Third, place $n-1$ more copies of the redrawn tree, each with their vertex v at the origin, and each rotated by $2\pi/n$ radians around the origin from the last copy. Then take the union of all of these trees. The formal definition is as follows.

Definition 5.3.1 Let T be a tree embedded in $\mathbb{C}$, with a fixed vertex v that lies at 0. The *n-unfolding* of T is the tree T^n obtained from T by the following procedure:

(i) Let T' denote an isomorphic copy of T, embedded in $\mathbb{C}$ by compressing T so that it fits in the first $2\pi/n$ radians of $\mathbb{C}$, i.e., for each point of T with argument α, the corresponding point of T' has argument α/n. Note that v still lies at 0.
(ii) Take n copies of T', labelled $T_0,\dots,T_{n-1}$, obtained from T' by multiplying by $\mathrm{e}^{2\pi\mathrm{i}/n}$ for $i=0,\dots,n-1$.
(iii) Let T^n be the union of the T_i.

An *unfolding* is an n-unfolding for some n. The *folding class* of a tree is the equivalence class of a tree under the equivalence relation generated by unfolding.

Note that a tree (with designated vertex) is determined by its folding class and the number of edges in it, by Exercise 5.3.

We will consider unfoldings of Brauer trees, where the fixed vertex is of course the exceptional node, if it exists. For example, the 2-unfolding of the tree from Example 5.2.4, with fixed vertex between M_3 and M_5, is

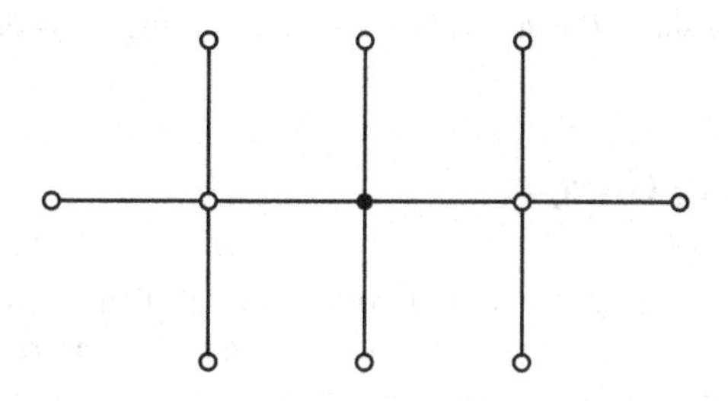

As a non-example, the tree in Example 5.1.6 is *not* the 2-unfolding of a tree with nine edges. Everything works apart from the edges incident to χ_{22} and χ_{23}, which are reflections rather than rotations of one another, and thus the potential 2-unfolding does not preserve the planar embedding.

The original definition of unfolding, as it appears in [218], does not mention the planar embedding, and as far as I can tell that was added only in the recent paper by Dudas, Rouquier and myself [131], and the as-yet unpublished work of Kessar and myself [133]. For example, the trees given in [308, Figure 4] and [294, Figure 1.4] are fine as trees but wrong as planar-embedded trees. Of course, the planar embedding is necessary if one wants to know the Morita equivalence class of the block, but often people are only interested in the decomposition matrices.

The main theorem of [218] may be summarized as follows.

Theorem 5.3.2 (Feit [218]) *Every Brauer tree has the same folding class as that of a block of a quasisimple group or a cyclic group.*

Note that unfolding can work both ways: if we have a group G and a normal subgroup H, and a block B whose defect group D is contained in H, then the Brauer tree of B is the same as that of a block B' of H, but only up to unfolding, and this can be a folded tree, an unfolded tree, or a folding of an unfolding. In other words the number of edges in the trees for B and B' can satisfy any relation, as long as they are both multiples of the number of edges in the smallest tree in the folding class of the tree.

The proof of this theorem, the reduction to quasisimple groups, follows the 'standard' use of Clifford theory. Let G be an arbitrary finite group and let H be a normal subgroup of G such that G/H is a (possibly cyclic) simple group. Let B be a block of kG with cyclic defect groups. There are three possibilities for $D \cap H$:

(i) If $D \leq H$ then let B_0 be a block of kH such that, for some χ an ordinary character of B, there is a constituent of $\chi\downarrow_H$ in B_0. (Any such B_0 will do.) The Brauer trees of B and B_0 lie in the same folding class.
(ii) If $D \not\leq H$ but $D \cap H \neq 1$ then the Brauer tree of B is always a star.
(iii) If $D \cap H = 1$ then the Brauer tree of B is in the same folding class as that of a block B_0 of $\bar{G}$, where $\bar{G}$ is a finite group such that $\bar{G}/Z(\bar{G}) \cong G/H$.

Thus we need to look at quasisimple groups. In [218] Feit also proved that the Brauer trees of quasisimple groups whose simple quotient is alternating or a classical group are always in the folding classes of lines.

Definition 5.3.3 A *windmill* is a tree that is in the folding class of a line.

5.3.2 Alternating Groups

The alternating groups are pretty easy to deal with: all Brauer trees are windmills, as we said before. However, because of the combinatorial nature of symmetric and alternating groups, we can say a lot more. In fact, we can be completely explicit about the exact trees that occur, and even which ordinary characters label the vertices of the planar-embedded tree.

If $G = S_n$, then since every conjugacy class of G is real, so is every ordinary irreducible character. Hence every Brauer tree of G is its own real stem, so it is a line. Also, we will see from the Nakayama conjecture (Theorem 8.3.1) that there are always $p - 1$ irreducible ordinary characters in a block with cyclic defect groups, and that the defect groups are always C_p, so that there is no exceptional node. In particular, when restricting to a block of $G' = A_n$, we cannot unfold the Brauer tree any more, because it already has the maximum number of edges.

Thus a Brauer tree of G' is either isomorphic to the Brauer tree of G (e.g., S_9 and A_9, principal blocks, $p = 7$), or is a 2-folding of the line, so a line with $(p-1)/2$ edges and exceptional node at the end (as for $\mathrm{SL}_2(p)$) (e.g., S_7 and A_7, principal blocks, $p = 7$). The labelling of the vertices for the tree for G is given by Theorem 8.3.13.

We also should consider the quasisimple groups $2 \cdot A_n$. Here again there is a combinatorial parametrization of the irreducible ordinary characters (see Sect. 8.4), and in [439] Müller determined the labelled, planar-embedded Brauer trees for both the quasisimple groups $2 \cdot A_n$ and the groups $2 \cdot S_n$ (this last group is not well defined, as there are two such groups, see Sect. 8.4). In all cases they are windmills, but now there are many different isomorphism types of tree, rather than just one in the case of S_n. Indeed, if p is odd, then there are $\lfloor (p+3)/4 \rfloor$ different Brauer trees as one ranges over all n and considers all group $2 \cdot S_n$.

5.3.3 Sporadic Groups

The twenty-six sporadic simple groups yield many of the more interesting and unique trees, including the tree in Example 5.1.6 earlier in the chapter. Many of the trees are collated in [294], but not all of them were computed at the time, and some still have not been determined. Sometimes the tree is determined but not the labelling of the vertices by ordinary characters, usually because there are several characters of the same degree that all look very similar, i.e., they are algebraically conjugate. (Two characters χ_1 and χ_2 are *algebraically conjugate* if there is a Galois automorphism σ of $\bar{\mathbb{Q}}/\mathbb{Q}$ such that $\chi_1^\sigma = \chi_2$.)

As well as giving around 450 pages of trees, the first few chapters of [294] give a very readable overview of methods used in determining Brauer trees, particularly of sporadic simple groups. Unfortunately, they often only work up to algebraic conjugacy, as this is usually a subtle question as to *exactly* which of the characters that form an irreducible character over $\mathbb{Q}$ lie in which position in the tree. There are two types of problem: the first is that there are two ordinary characters that appear somewhere in the tree, they are algebraically conjugate, but we cannot decide which goes where. The other problem is where the tree is determined completely (i.e., the decomposition matrix), but not the planar-embedded tree, and the algebraic conjugacy problem is to do with the precise ordering of the edges around a given vertex.

Methods since the publication of [294] have progressed to the point where these subtleties have mostly been removed for the first type, where determining the decomposition matrix is enough. Unfortunately, the module-theoretic information given by the planar-embedded tree is often too difficult to deduce for now.

The first issue is $12 \cdot M_{22}$, a quasisimple group with centre of order 12. Because there are four primitive 12th roots of unity, there are four algebraically conjugate blocks that are faithful, that come in two dual pairs. The first paper on this was by Humphreys [308], which determined the trees, with a couple of errors in the case $p = 11$, which were corrected by Benson [31]. He was very careful and dealt with the algebraic conjugacy problems completely. For $p = 5, 7$ the problems remained, although nowadays one can ask a computer to construct a given instance of the group and find out every piece of information you want. As far as I can tell this has never appeared in print for the two smaller primes though. (The trees are available in [294, Section 6.4], with the ambiguities.)

The problems with the trees for the Janko group J_3 were resolved by Wilson in [564], and those for the Lyons simple group by Müller, Neunhöffer, Röhr and Wilson in [440]. Two Diploma theses [461, 493] in 2000 by Ottensmann and Röhr resolved the algebraic conjugacy problems for the Rudvalis and O'Nan groups.

The Monster simple group was accidentally missing for $p = 29$ from [294], and this was partially rectified by Naehrig in [445], although it was not possible to completely determine the trees.

There are, however, a few other cases where the algebraic conjugacy problems left over in [294] have been resolved, but appear to have never been published. They, along with decomposition matrices for other primes, for various sporadic simple groups, have been made available in online form at the *Modular Atlas* website [64]; this is a website that lists the known decomposition matrices for 'interesting' groups, normally groups associated to the sporadic simple groups and other small simple groups. The website provides the resolution for the problems with the Rudvalis group for $p = 7$, the Fischer group Fi_{23} for $p = 23$, and the Baby Monster group for $p = 17$.

There remain a few trees for five groups where not everything is known at the time of writing, and only for two of them where the unlabelled tree is not known. We summarize what isn't known in Table 5.1. (Here, as we mentioned, the first type of algebraic conjugacy is where the decomposition matrix is not known, and the second type is where the decomposition matrix is known but the planar embedding is not known.)

5.3.4 Groups of Lie Type

Let G be a finite group of Lie type. Then G is defined over some finite field $\mathbb{F}_q$, and we have two possibilities: $p \mid q$ or $p \nmid q$. The representation theories of G when $p \mid q$ and $p \nmid q$ are completely different (see Chap. 9).

Group	Prime	Block number	Problem
Th	13	1	Algebraic conjugacy (second type)
Th	31	1	Algebraic conjugacy (second type)
J_4	23, 29, 31, 43	1	Algebraic conjugacy (first type)
$3.Fi_{24}'$	13	4, 5, 6	Algebraic conjugacy (first and second type)
$3.Fi_{24}'$	17	2, 3	Algebraic conjugacy (first type)
$3.Fi_{24}'$	29	2, 3	Algebraic conjugacy (first and second types)
$2.BM$	13	6	Tree not known
BM	19	2	Tree not known
$2.BM$	31	2	Tree not known
BM	47	1	Tree not known
M	19	3	Algebraic conjugacy (second type)
M	29, 41, 47, 59, 71	1	Tree not known
M	29	2, 3	Trees not known
M	47	2	Algebraic conjugacy (second type)

Table 5.1 The remaining unknown trees for sporadic groups

If $p \mid q$, then there are only two possible defect groups for blocks of G: Sylow p-subgroups and the trivial group (Theorem 9.1.9). Thus if G contains a block with a cyclic, non-trivial, defect group, then G has cyclic Sylow p-subgroups, and this means that $G = \mathrm{SL}_2(p)$. We already saw that the Brauer trees of G are lines in Example 5.1.5, so we may assume that $p \mid |G|$ but that $p \nmid q$.

In this case, we let d denote the multiplicative order of q modulo p, i.e., the smallest positive integer such that $q^d \equiv 1 \bmod p$. The representation theory of G depends more on d than on p and q.

We explain more in Chap. 9 of course, but broadly speaking there is a set of ordinary irreducible characters called 'unipotent', that does not depend on q (or d, of course), so the set of labels is the same for $\mathrm{GL}_9(2)$ as $\mathrm{GL}_9(19)$. The p-blocks that contain unipotent characters are called unipotent blocks. The distribution of unipotent characters among the p-blocks only depends on d (at least if p is not too small, see Sect. 9.3), and for unipotent blocks with cyclic defect groups, all Brauer trees are known.

For classical groups, the precise description of the Brauer trees is given in a series of papers [227, 229, 231] by Fong and Srinivasan, and as Feit proved in [218], they are windmills. In fact, they are lines, at least for unipotent blocks. For the exceptional groups, work on the small-rank groups was done in a few papers. Hiss dealt with the Ree groups in [288], and Shamash in [521–523] produced the trees for $G_2(q)$. Hiss–Lübeck [292] analysed the groups $F_4(q)$ and ${}^2E_6(q)$, and together with Malle [293] attacked $E_6(q)$. These papers left open a few possibilities, either for planar embeddings or for certain congruences of q. The cases of $E_7(q)$ and $E_8(q)$, together with clearing up the odd cases that were left out, and setting up a general framework for understanding the Brauer trees for unipotent blocks of exceptional groups, was given by Dudas, Rouquier and me in [131], building on three papers of Dudas and Dudas–Rouquier [172, 174, 178].

Of course, there are all of the blocks that are not unipotent. An important paper of Bonnafé and Rouquier [44], recently extended to quasisimple groups by those authors and Dat [42] (see Theorems 9.4.7 and 9.4.9) proves that every block of a finite group of Lie type is Morita equivalent to a 'quasi-isolated' block. This is a generalization of the notion of a unipotent block that we consider in Sect. 9.4. By passing from G to a larger group (e.g., from $\mathrm{SL}_n(q)$ to $\mathrm{GL}_n(q)$), at the expense of now only working up to unfolding, we may assume that G comes from a group with 'connected centre', and in this case we may work with 'isolated' blocks, rather than quasi-isolated blocks. Isolated blocks are less restrictive than unipotent blocks, but more than quasi-isolated blocks. Moreover, it is expected that every isolated block is Morita equivalent to a unipotent block, but this is definitely not true for quasi-isolated blocks.

Kessar and I have been working on extending the Bonnafé–Rouquier theorem to cover this last case, at least for blocks with cyclic defect groups. In [133] this result is proved for all groups other than $E_8(q)$, and for $E_8(q)$ apart from $d = 12, 18$ (where d is the order of q modulo p, as stated before), and even then it is proved for all but one p-block for each of these primes. These last couple of cases, at the time of writing, defy the techniques of [133], but there is enough theory present that this last gap should be filled, and all Brauer trees of all blocks of all groups of Lie type will be known.

Exercises

Exercise 5.1 If B is a real block, prove that duality induces a reflection of the planar-embedded Brauer tree of B. If B is not a real block, show that the planar-embedded Brauer trees of B and B^* are reflections of one another.

Exercise 5.2 Let p be odd, and let B be a real p-block of a finite group. Prove that if B has no exceptional vertex then B possesses a real ordinary character. (You will need the remarks after Exercise 2.8.)

Exercise 5.3 Let Γ be a finite tree embedded in $\mathbb{C}$. Suppose that Γ is an n-unfolding of some tree for $n > 1$, and let v be the vertex at the origin of $\mathbb{C}$. Show that v is the vertex lying at the centre of any path of maximal length in Γ. Deduce that there is a unique tree Π with a minimal number of edges, and a fixed vertex v of Π, such that every tree in the folding class of Γ is an n-unfolding of Π around the vertex v, for some $n \geq 1$.

(In particular, this shows that, given any finite tree Γ, there is at most one vertex v around which you may fold Γ.)

Exercise 5.4 Let B be a block with cyclic defect groups. Prove that if $\mathrm{Ext}_B^1(M, N)$ is known for all simple B-modules M and N, then the planar-embedded Brauer tree is determined (but not, of course, the exceptional node).

Exercise 5.5 Prove Brauer's height zero conjecture for blocks with cyclic defect groups.

Chapter 6
Blocks with Non-cyclic Defect Groups

The theory of blocks with cyclic defect groups is more or less complete: the Brauer tree is a powerful combinatorial object, Feit's reduction to quasisimple groups offers us an effective way to compute the Brauer trees of an arbitrary finite group from those of the composition factors, and the classification of the Brauer trees for finite simple groups is more or less complete, with a few trees missing for the Baby Monster and Monster groups, and a few trees for $E_8(q)$ for specific primes p.

When moving beyond Brauer trees, to potentially an arbitrary defect group, one almost immediately hits the wildness problem (Sect. 3.3). If $p = 2$ and you stay within the confines of dihedral, semidihedral and quaternion defect groups life is a little more civilized, and we will remain within these relatively peaceful pastures for the first two sections of this chapter. Beyond that, even for $C_3 \times C_3$ defect group we are in wild territory, but there is still Broué's conjecture (Sect. 4.4) to hopefully give some kind of structure to the category.

Crossing this last boundary, from tame or abelian, takes us into the general case. Here there is little that we can do structurally, but there have been a few numerical results about, say, the number of ordinary and modular characters in a block with a given defect group.

One particular case stands out: the theory of nilpotent blocks. A nilpotent block with defect group D is one whose source algebra is isomorphic to kD; these sometimes appear during Clifford theory, which we will see in the next chapter, and also occur fairly regularly in examples. A good structure theory of nilpotent blocks is available, and so we can include this.

What happens in the general case? Here we are (conjecturally) hemmed in by our local-global conjectures, and there are some highly speculative and nebulous ideas as to how to understand this case, but right now there is basically no structure theory available.

D. A. Craven, *Representation Theory of Finite Groups: a Guidebook*, Universitext,
https://doi.org/10.1007/978-3-030-21792-1_6

6.1 Klein Four Defect Groups

Once one moves beyond blocks with cyclic defect groups, and eventually one must do so, the next simplest case should be the Klein four group. In fact, we have a complete understanding of blocks with Klein four defect group, starting with a construction of the indecomposable modules for the Klein four group itself. These were more or less given by Basev [28] and Heller–Reiner [278], and given very explicitly by Conlon in [121], but we use a mixture of the notation and terminology of Ringel [484] and Conlon [121]. We are deliberately setting this up in greater generality than is needed for V_4 so that we can transport this to general dihedral 2-groups in the next section.

Let $\mathscr{W}$ denote all words in the alphabet $a^{\pm}$ and $b^{\pm}$, such that every $a^{\pm}$ is followed by a $b^{\pm}$ and vice versa. Let $\mathscr{W}_1$ be the subset of $\mathscr{W}$ of words w such that there is no instance of ab, ba, $a^{-1}b^{-1}$ or $b^{-1}a^{-1}$ in w. Thus $\mathscr{W}_1$ consists of words $ab^{-1}ab^{-1}\dots$ and $a^{-1}ba^{-1}b\dots$, together with $ba^{-1}ba^{-1}\dots$ and $b^{-1}ab^{-1}a\dots$, so in particular there are four strings of a given length. If w is a string with $n-1$ symbols, we construct a *string module* as follows: let α and β be two matrices in $\mathrm{GL}_n(\mathbb{F}_2)$, and start with α and β both the identity matrix. We will set some off-diagonal entries to be 1:

(i) if the ith symbol in w is a, set $\alpha_{i+1,i}=1$, and if it is a^{-1} set $\alpha_{i,i+1}=1$;
(ii) if the ith symbol in w is b, set $\beta_{i+1,i}=1$, and if it is b^{-1} set $\beta_{i,i+1}=1$.

As an example, if $w=ab^{-1}ab^{-1}a$, then α and β are the 6×6 matrices

$$\alpha=\begin{pmatrix}1&0&0&0&0&0\\1&1&0&0&0&0\\0&0&1&0&0&0\\0&0&1&1&0&0\\0&0&0&0&1&0\\0&0&0&0&1&1\end{pmatrix},\quad \beta=\begin{pmatrix}1&0&0&0&0&0\\0&1&1&0&0&0\\0&0&1&0&0&0\\0&0&0&1&1&0\\0&0&0&0&1&0\\0&0&0&0&0&1\end{pmatrix}.$$

The matrices α and β commute and both have order 2, so yield a representation $V_4\to\mathrm{GL}_n(2)$. It turns out that, if $M(w)$ denotes the string module with string w, then $M(w)$ is always indecomposable, and $M(w)$ and $M(w')$ are isomorphic if and only if $w'=w$ or $w'=w^{-1}$, where w^{-1} is obtained from w by swapping a and a^{-1}, swapping b and b^{-1}, and reversing the word. Thus there are, up to isomorphism, two string modules of a given dimension.

Let $n\geq 1$ be an integer. Thought of as a right kG-module, if $A_n=M(w)$ for w of length $2n$ starting with a^{-1}, and $B_n=M(w)$ for w of length $2n$ starting with a, then $A_n^*=B_n$, and $\Omega^n(k)=A_n$ for $n\geq 0$.

If w is a word of length $2n-1$ starting with a write $C_n(0)$ for $M(w)$, and if w is a word of length $2n-1$ starting with b write $C_n(\infty)$ for $M(w)$. In this case we have $\Omega(C_n(0))=C_n(0)$ and $\Omega(C_n(\infty))=C_n(\infty)$.

We now construct the *band modules*. This time the set $\mathscr{W}_1'$ is all words w that are not powers of shorter words, and such that every power of them is also a word in $\mathscr{W}_1$. That is just the words of length 2 in our case. We will not make a distinction between w and w^{-1}, nor between w and the word obtained from w by moving the last letter to the front. Thus $\mathscr{W}_1'$ actually just consists of one word, ab^{-1}. We construct two matrices, α' and β', from this word, of the form

$$\alpha' = \begin{pmatrix} I & 0 \\ \phi & I \end{pmatrix}, \qquad \beta' = \begin{pmatrix} I & 0 \\ I & I \end{pmatrix},$$

for ϕ some linear transformation. (When we consider all dihedral 2-groups there will be other words.)

We now let V be a finite-dimensional k-vector space, where k is a field of characteristic 2. The I above stands for the identity transformation on V, and ϕ will be an automorphism of V. Notice that α', β', and $\alpha'\beta'$ all have order 2, so this a representation of V_4 for any map $\phi \in \mathrm{Hom}_k(V, V)$. In order to pick up only indecomposable modules, we require that ϕ is an indecomposable linear transformation. This is equivalent to ϕ having rational canonical form a single block, which is associated with a power π^n of an irreducible polynomial π over k. Write $M(w, \pi^n)$ for the band module, and in Conlon's notation $C_n(\pi)$. We have $\Omega(C_n(\pi)) = C_n(\pi)$.

Theorem 6.1.1 *The string modules $M(w)$ for $w \in \mathscr{W}_1$, the band modules $M(w, \phi)$ for $w \in \mathscr{W}_1'$ and ϕ an indecomposable automorphism of a k-vector space, and the free module kV_4, form a complete set of indecomposable kV_4-modules up to isomorphism.*

In [121], Conlon computed the complete tensor product structure on the indecomposable modules for V_4. We already know that the string module $\Omega^i(k)$, the Heller translate of the trivial module, is endotrivial, and we know how to take tensor products of such modules modulo projectives:

$$\Omega^i(k) \otimes \Omega^j(k) = \Omega^{i+j}(k).$$

Thus by counting dimensions we can get an exact formula.

We now want to understand the tensor product of $\Omega^i(k)$ and the modules $C_n(\pi)$. Since $\Omega(C_n(\pi)) = C_n(\pi)$, we get that

$$\Omega^i(k) \otimes C_n(\pi) = C_n(\pi)$$

modulo projectives. The tensor product of $C_n(\pi)$ and $C_{n'}(\pi')$ is projective unless $\pi = \pi'$, and then it is $C_m(\pi)^{\oplus 2}$ plus projectives if $\pi = \pi'$, where $m = \min(n, n')$. (This is almost true, but $C_1(\pi_2)^{\otimes 2} = C_2(\pi)$ if $\pi \neq X, X + 1$.) This completely describes the tensor product structure for indecomposable modules for the Klein four group.

In [121], Conlon also considers the tensor product operation for modules for kA_4, the group algebra of the alternating group. This has a normal Sylow 2-subgroup, and so there is a single 2-block, as we noted in Sect. 2.4, and it has three simple modules, coming from the simple modules for the cyclic quotient. (By Theorem 2.4.1 the simple kA_4-modules are those of $k(A_4/V_4) = kC_3$.) The other ordinary character is 3-dimensional, and the 2-decomposition matrix has the form

$$\begin{pmatrix} 1 & 0 & 0 \\ 0 & 1 & 0 \\ 0 & 0 & 1 \\ 1 & 1 & 1 \end{pmatrix}.$$

This has a completely different structure to kV_4, which has 2-decomposition matrix

$$\begin{pmatrix} 1 \\ 1 \\ 1 \\ 1 \end{pmatrix}.$$

There is one other obvious group with Klein four Sylow 2-subgroup, which is A_5. As there is an ordinary irreducible character of degree 4 for A_5, it has a block of defect zero, and there are exactly two 2-blocks. The irreducible ordinary characters have degrees 1, 3, 3, 5 and 4. Thinking of A_5 as $\mathrm{SL}_2(4)$, we see a 2-dimensional simple module, and it is defined over $\mathbb{F}_4$ rather than $\mathbb{F}_2$, so one may twist the action by a Galois automorphism to obtain another simple module over $\mathbb{F}_4$. This yields three simple modules, of dimensions 1, 2 and 2, and we have our projective 4-dimensional module of course, which comes from a block of defect zero (as there is a character of degree 4 for G). There are four 2-regular classes in A_5 (with representatives 1, $(1, 2, 3)$, $(1, 2, 3, 4, 5)$ and $(1, 2, 3, 5, 4)$) so we have found all simple modules. The 2-decomposition matrix is fairly easy to compute, and it is

$$\begin{pmatrix} 1 & 0 & 0 \\ 1 & 1 & 0 \\ 1 & 0 & 1 \\ 1 & 1 & 1 \end{pmatrix}.$$

This has a different structure to kA_4 and kV_4, and so the principal block of kA_5 is another Morita equivalence class of blocks with Klein four defect groups.

In fact, there are only three Morita equivalence types of block with Klein four defect group, a fact proved by Erdmann.

Theorem 6.1.2 (Erdmann [192, 194]) *Let B be a 2-block of a finite group G. If B has Klein four defect groups, then B is Morita equivalent to one of kV_4, kA_4 and the principal block of kA_5.*

Let B be a block with Klein four defect group D. In all cases, the simple B-modules have vertex the whole of D, either by Knörr's theorem, Theorem 3.2.11, or Erdmann's theorem, Theorem 3.2.12. In [194], Erdmann proves that there is a simple module S in B whose Green correspondent in the Brauer correspondent block of $N_G(D)$ is simple up to a Heller translate, i.e., $\Omega^i(f(S))$ is simple for some $i \in \mathbb{Z}$, where f denotes the Green correspondent.

In fact, she proves that if B is Morita equivalent to kA_4 then, denoting the three simple B-modules by S, X and Y, there is an integer $i \in \mathbb{Z}$ such that $\Omega^i(f(S))$, $\Omega^i(f(X))$ and $\Omega^i(f(Y))$ are the three simple modules for the Brauer correspondent B' of B. In the other case, she proves that if B is Morita equivalent to the principal block of kA_5 then, again denoting the simple B-modules by S, X and Y, there is an integer $i \in \mathbb{Z}$ such that $\Omega^i(f(S))$ is simple, and $\Omega^i(f(X))$ and $\Omega^i(f(Y))$ are both indecomposable but with only two composition factors, and $\Omega^3(X) = X$ and $\Omega^3(Y) = Y$. Erdmann doesn't go as far as to conjecture that i is always 0 for all finite groups, but notes that $i = 0$ in all known examples.

In fact, i does always equal 0, but we had to wait just under 30 years for a proof of this, using the classification of finite simple groups.

Theorem 6.1.3 (Craven–Eaton–Kessar–Linckelmann [132]) *If B is a p-block with Klein four defect group D, there is a simple B-module whose Green correspondent in $N_G(D)$ is simple.*

The proof of this actually proves a statement about the source algebra of the blocks, but we won't see this until Sect. 6.3, so we will not pursue it any further, at least for now.

6.2 Tame Defect Groups

The blocks with tame representation type have defect groups that are dihedral, semidihedral or quaternion. Much more is known about blocks with these defect groups than for the general case, and we will attempt to give a flavour of this understanding now.

We begin with dihedral defect groups, and with the group algebra kD_{4q}, where q is a power of 2. Let $\mathscr{W}$ be as in the previous section, and this time let $\mathscr{W}_q$ denote the subset of $\mathscr{W}$ of words that have no occurrence of $(ab)^q$, $(ba)^q$, $(a^{-1}b^{-1})^q$ or $(b^{-1}a^{-1})^q$. Notice that $\mathscr{W}_1$ is as we defined it to be in the previous section. A *string module* $M(w)$ for a word $w \in \mathscr{W}_q$ is exactly as in the previous section, and it is a (not necessarily faithful) representation of D_{4q}.

For *band modules*, the statements about the map ϕ transfer over, and we just need to be a bit more careful about defining $\mathscr{W}_q'$, and how to go from $w \in \mathscr{W}_q'$ to the block matrices for α' and β'. Set $\mathscr{W}_q'$ to be all words w in $\mathscr{W}_q$ such that all powers of w lie in $\mathscr{W}_q$, and w is not itself a proper power of a word. As before, we only take words in $\mathscr{W}'_q$ up to cycling letters and inverting, so we may assume that w starts with a. If w has length n, then α' and β' are $n \times n$-matrices with I down the diagonal, the $(2, 1)$-position in α' is ϕ, and we then run through the letters of w, just as with string modules, so for $2 \leq i \leq n-1$, we do the following:

(i) if the ith symbol in w is a, set $\alpha'_{i+1,i} = I$, and if it is a^{-1} set $\alpha'_{i,i+1} = I$;
(ii) if the ith symbol in w is b, set $\beta'_{i+1,i} = I$, and if it is b^{-1} set $\beta'_{i,i+1} = I$.

If the last symbol is b (remember that n is even) then place I in the $(n, 1)$ position of β', and if it is b^{-1} place I in the $(1, n)$ position of β'. As an example, if $w = aba^{-1}b$ then

$$\alpha' = \begin{pmatrix} I & 0 & 0 & 0 \\ \phi & I & 0 & 0 \\ 0 & 0 & I & I \\ 0 & 0 & 0 & I \end{pmatrix}, \quad \beta' = \begin{pmatrix} I & 0 & 0 & 0 \\ 0 & I & 0 & 0 \\ 0 & I & I & 0 \\ I & 0 & 0 & I \end{pmatrix}.$$

Again, write $M(w, \phi)$ for this representation.

It turns out that these are all of the indecomposable representations of dihedral 2-groups.

Theorem 6.2.1 (Bondarenko [40], Ringel [484]) *For q a power of 2, the set $M(w)$ for $w \in \mathscr{W}_q$, $M(w, \phi)$ for $w \in \mathscr{W}'_q$ and ϕ an indecomposable k-vector space automorphism, and the free module kD_{4q}, form a complete set of indecomposable representations of D_{4q} over the field k.*

Unlike the Klein four case, not everything is known about the tensor products of string and band modules. Archer [19] proved that, modulo band modules, the set of string modules of odd dimension form an abelian group under $\otimes$, which is torsion-free and not finitely generated, but there is no algorithm which, given two words $w_1, w_2 \in \mathscr{W}_q$, gives the string module summand of $M(w_1) \otimes M(w_2)$. For products of even-dimensional string modules, things are slightly more complicated; some work in this direction by the author is in [126], but little is known.

What about general blocks with dihedral defect group? Brauer proved [57] that if the principal 2-block B has dihedral defect group of order 2^n, then $k(B) = 3+2^{n-2}$ and $l(B) \leq 3$. He generalized these results in [58] to include non-principal blocks, proving again that $k(B) = 3 + 2^{n-2}$, with four characters of height 0 and the rest of height 1, and $l(B) \leq 3$ in general. Furthermore, he showed that one elementary divisor of the Cartan matrix is 2^n and the rest are 1.

Shortly after Brauer's second paper, Erdmann went back to the principal-block case, using the classification of groups with dihedral Sylow 2-subgroups obtained by Gorenstein and Walter [258], to determine in [191] the vertices and sources of the simple modules, together with the structure of the projective indecomposable modules. Along with this paper, Landrock [382], and Erdmann–Michler [201] also considered blocks with dihedral defect group from a numerical point of view.

A decade later in 1987, Erdmann returned to the subject, but this time to non-principal blocks as well. Her paper [195] focused on the case $l(B) = 3$: if $l(B) = 1$ then B is a nilpotent block, which we will look at in Sect. 6.3, and in [468], Puig proved that nilpotent blocks are isomorphic to matrix algebras over kD, where D is the defect group of B; if $l(B) = 2$, Donovan [170] determined the Cartan matrix of B up to three possibilities; $l(B) = 3$ was considered the hardest case.

She actually classified 'algebras of dihedral type' up to Morita equivalence, a certain collection of algebras with specified Auslander–Reiten quivers. We won't see Auslander–Reiten theory in this book, but to talk about tame blocks in reasonable detail we would need it.

If you have read about path algebras from Chap. 5, I can give a few more details. Since $l(B) \leq 3$, there are a limited number of possibilities for the path algebra, although the relations have not yet been determined. A case-by-case analysis works based on which of the simple modules are periodic. Assumptions about periodicity, the general structure of projectives for blocks with dihedral defect group, plus the restrictions given by the Auslander–Reiten theory, allow the precise relations to be written down, leading to a complete classification of the possible path algebras with relations.

Then, from the list that is produced, it has to be decided which can be blocks of finite groups. Some obviously can, because we have examples. Some cannot, because no decomposition matrix can be produced from the Cartan matrix, for example, or because some other property of blocks does not hold, such as $k(B)$ being the dimension of $Z(B)$, or the determinant of the Cartan matrix being a power of p (both of which are true for blocks). Then there are usually a few other cases that pass all of the obvious tests, but there are no known examples.

Theorem 6.2.2 (Erdmann [195]) *Let B be a 2-block with dihedral defect groups of order 2^n for $n \geq 3$, and suppose that $l(B) = 3$. One of the following holds:*

(i) *B has two periodic simple modules and B is Morita equivalent to the principal block of $\mathrm{PSL}_2(q)$ for some odd prime power q such that 2^n is the exact power of 2 dividing $q - 1$, with decomposition matrix*

$$\begin{pmatrix} 1 & 0 & 0 \\ 1 & 1 & 0 \\ 1 & 0 & 1 \\ 1 & 1 & 1 \\ 2 & 1 & 1 \end{pmatrix}$$

with the last row repeated $2^{n-2} - 1$ times;

(ii) *B has no periodic simple modules and B is Morita equivalent to the principal block of $\mathrm{PSL}_2(q)$ for some odd prime power q such that 2^n is the exact power* *of 2 dividing $q + 1$, with decomposition matrix*

$$\begin{pmatrix} 1 & 0 & 0 \\ 0 & 1 & 0 \\ 0 & 0 & 1 \\ 1 & 1 & 1 \\ 0 & 1 & 1 \end{pmatrix}$$

with the last row repeated $2^{n-2} - 1$ times;

(iii) *B has one periodic simple module and B has decomposition matrix*

$$\begin{pmatrix} 1 & 0 & 0 \\ 1 & 1 & 0 \\ 1 & 0 & 1 \\ 1 & 1 & 1 \\ 0 & 1 & 0 \end{pmatrix},$$

with the last row repeated $2^{n-2} - 1$ times. If $n = 3$ then B is Morita equivalent to the principal block of A_7. No example is known of such a block for $n \geq 4$.

(The result in [195] is incorrect for (iii) above, see [200, X.4(a)].)

This follows a trend of Erdmann's methods being powerful enough to *almost* completely classify the Morita classes of blocks, but there are one or two cases where it is not clear whether the algebra appears as a block.

From [200], we see that for one and two simple modules we get the following.

Theorem 6.2.3 *Let B be a 2-block with dihedral defect groups of order 2^n for $n \geq 3$, and suppose that $l(B) < 3$. One of the following holds:*

(i) *$l(B) = 1$ and B is Morita equivalent to kD_{2^n};*

(ii) *$l(B) = 2$ and B is Morita equivalent to the principal block of $\mathrm{PGL}_2(q)$ for some odd prime power q such that 2^{n-1} is the exact power of 2 dividing $q - 1$, with decomposition matrix*

$$\begin{pmatrix} 1 & 0 \\ 1 & 0 \\ 1 & 1 \\ 1 & 1 \\ 2 & 1 \end{pmatrix}$$

with the last row repeated $2^{n-2} - 1$ times;

(iii) *$l(B) = 2$ and B is Morita equivalent to the principal block of $\mathrm{PGL}_2(q)$ for some odd prime power q such that 2^{n-1} is the exact power of 2 dividing $q + 1$,* *with decomposition matrix*

$$\begin{pmatrix} 1 & 0 \\ 1 & 0 \\ 1 & 1 \\ 1 & 1 \\ 0 & 1 \end{pmatrix}$$

with the last row repeated $2^{n-2} - 1$ *times.*

(iv) $l(B) = 2$ *and* B *is Morita equivalent to an algebra with the same decomposition matrix as one of the previous two cases, and with the same quiver, but with a certain parameter set to* 1 *instead of* 0 *in the relations for the path algebra.*

In the classification of the basic algebras in [195], there is a particular relation $\alpha^2 = c(\alpha\beta\gamma)$, with $c = 0, 1$. In other words, it is not clear whether, for a particular arrow in the quiver, it squares to either 0 or the composition of three arrows in the quiver, and this is where (iv) above arises. Eisele, in his Ph.D. thesis [183], proved that (iv) does not occur (this result appears in [184]).

Theorem 6.2.4 (Eisele [184]) *Case (iv) in Theorem 6.2.3 cannot occur.*

He did this by showing that only those k-algebras with $c = 0$ can be lifted to $\mathcal{O}$-algebras. Since blocks of finite groups can always be lifted to $\mathcal{O}$-algebras, this means that $c = 0$ for blocks. In addition, he proved that for $c = 0$ there is a unique lift, so that two blocks B_1 and B_2 with dihedral defect group and $l(B_i) = 2$ are Morita equivalent over k if and only if they are over $\mathcal{O}$. In [183], Eisele showed the same unique lifting property for the case $l(B) = 3$.

Thus for B a block with dihedral defect group and $l(B) \leq 2$, there is a perfect classification, both over k and $\mathcal{O}$, and for $l(B) = 3$ there is a near-perfect classification, also over k and $\mathcal{O}$, because one of the three possibilities might only occur for $|D| = 8$.

The way to understand Theorems 6.2.2 and 6.2.3 is:

- if $l(B) = 1$ then B is Morita equivalent to kD_{2^n};
- if $l(B) = 2$ then the Morita equivalence type of B is determined by the Cartan matrix up to a parameter, and by changing basis that parameter may be chosen to be either 0 or 1. Only one of these should occur (and was later shown to be the case, but needed tools outside of Auslander–Reiten theory).
- If $l(B) = 3$ then the Morita equivalence type of B is determined by the Cartan matrix completely, not up to a parameter.

This theme will repeat itself.

Not satisfied with blocks with dihedral defect groups, Erdmann continued to semidihedral defect groups, the second class of 2-groups yielding tame algebras. Again, she gives a more general definition of algebras of 'semidihedral type' and classifies those. In [198, 199], Erdmann used Auslander–Reiten theory again to classify algebras of semidihedral type.

The indecomposable modules of the semidihedral groups were classified by Bondarenko and Drozd in [41], then again in an easier-to-understand way by Crawley-Boevey in [136]. The description is quite similar to that of dihedral 2-groups above. We won't give it here, but refer instead to [136] for the construction.

For general blocks with semidihedral defect groups, Olsson in [458] proved that if B has semidihedral defect groups then again $l(B) \leq 3$, and $k(B)$ is either $2^{n-2}+3$ or $2^{n-3}+4$. Erdmann [193] studied some principal blocks with semidihedral defect groups, using the classification of finite simple groups with semidihedral Sylow 2-subgroups by Alperin, Brauer and Gorenstein [9], and obtained information on Green correspondence, vertices and sources of simple modules, and whether simple modules were periodic.

As with blocks with dihedral defect groups, we split the result up into two theorems to make it more manageable. We start with the case where B has semidihedral defect groups and $l(B) \leq 2$.

Theorem 6.2.5 (Erdmann [198]) *Let B be a 2-block with semidihedral defect groups SD_{2^n} for $n \geq 4$, and suppose that $l(B) \leq 2$. One of the following holds:*

(i) *$l(B) = 1$ and B is Morita equivalent to kSD_{2^n};*
(ii) *$l(B) = 2$ and B is Morita equivalent to the principal block of $\mathrm{GU}_2(q)$ for some odd prime power q such that 2^{n-2} is the exact power of 2 dividing $q-1$, with decomposition matrix*

$$\begin{pmatrix} 1 & 0 \\ 1 & 0 \\ 1 & 1 \\ 1 & 1 \\ 0 & 1 \\ 2 & 1 \end{pmatrix}$$

with the last row repeated $2^{n-2}-1$ times;
(iii) *$l(B) = 2$, B is Morita equivalent to the principal block of a group $\mathrm{PSL}_2(q^2).2$ (this is the field-diagonal automorphism) for some odd prime power q such that 2^{n-1} is the exact power of 2 dividing q^2-1, and the decomposition matrix is as in Theorem 6.2.3(ii).*
(iv) *$l(B) = 2$ and B is Morita equivalent to the principal block of $\mathrm{GL}_2(q)$ for some odd prime power q such that 2^{n-2} is the exact power of 2 dividing $q+1$,*

with decomposition matrix

$$\begin{pmatrix} 1 & 0 \\ 1 & 0 \\ 1 & 1 \\ 1 & 1 \\ 2 & 1 \\ 0 & 1 \end{pmatrix}$$

with the last row repeated $2^{n-2} - 1$ *times;*

(v) $l(B) = 2$ *and* B *has decomposition matrix as in Theorem 6.2.3(iii). If* $n = 4$ *then* B *is Morita equivalent to a non-principal block of the central extension* $3 \cdot M_{10}$*, whereas no example is known of such a block for* $n \geq 5$*;*

(vi) $l(B) = 2$ *and* B *has decomposition matrix*

$$\begin{pmatrix} 1 & 0 \\ 1 & 0 \\ 0 & 1 \\ 0 & 1 \\ 1 & 1 \end{pmatrix}$$

with the last row repeated $2^{n-2} - 1$ *times. No example is known of such a block.*

(vii) B *is Morita equivalent to an algebra with the same decomposition matrix as one of the previous cases, and with the same* Ext*-quiver, but with a different parameter in the relations for the path algebra. No example is known of such a block.*

(Note that in the table in [200], the examples blocks in cases (iv) and (v) are swapped round. They are correct in [198].)

For three simple modules there are seven possible Morita equivalence classes of blocks, only two of which are known to occur, as the principal 2-block of $\mathrm{PSL}_3(q)$ for $q \equiv 3 \bmod 4$ and the principal 2-block of $\mathrm{PSU}_3(q)$ for $q \equiv 1 \bmod 4$. The other simple group with a semidihedral Sylow 2-subgroup, M_{11}, has a principal 2-block that is Morita equivalent to that of $\mathrm{PSL}_3(3)$. This is contrary to the assertion in [199, 200] (itself relying on [510]), but you may check it for yourself in Magma. Simply type in the following commands:

```
load m11; //Now G is M11
G2:=PSL(3,3);
Proj:=ProjectiveIndecomposableModules(G,GF(2));
Proj2:=ProjectiveIndecomposableModules(G2,GF(2));
Remove(~Proj,3); Remove(~Proj2,4);
A:=BasicAlgebraOfBlockAlgebra(Proj);
A2:=BasicAlgebraOfBlockAlgebra(Proj2);
IsIsomorphic(A,A2);
```

Theorem 6.2.6 (Erdmann [199]) *Let B be a 2-block with semidihedral defect groups SD_{2^n} for $n \geq 4$, and suppose that $l(B) = 3$. The decomposition matrix of B is one of the following:*

$$\begin{pmatrix}1&0&0\\1&1&0\\1&0&1\\1&1&1\\0&0&1\\0&1&0\end{pmatrix}, \begin{pmatrix}1&0&0\\1&1&0\\1&0&1\\1&1&1\\0&0&1\\2&1&1\end{pmatrix}, \begin{pmatrix}1&0&0\\1&1&0\\1&0&1\\1&1&1\\2&1&1\\0&1&0\end{pmatrix}, \begin{pmatrix}1&0&0\\0&1&0\\1&0&1\\0&1&1\\1&1&1\\0&0&1\end{pmatrix}, \begin{pmatrix}1&0&0\\0&1&0\\0&0&1\\1&1&0\\1&0&1\\1&1&1\end{pmatrix}, \begin{pmatrix}1&0&0\\0&1&0\\0&0&1\\1&1&0\\1&1&1\\1&0&1\end{pmatrix},$$

with the last row repeated $2^{n-2} - 1$ times. There are two Morita equivalence classes of blocks with the first decomposition matrix, and one for each of the others.

The first two decomposition matrices are realized by the principal blocks of $\mathrm{PSL}_3(q)$ *for* $q \equiv 3 \bmod 4$, *and* $\mathrm{PSU}_3(q)$ *for* $q \equiv 1 \bmod 4$, *respectively. For the other matrices, and for one of the two Morita equivalence classes for the first decomposition matrix, no blocks are known.*

The particular class to which $\mathrm{PSL}_3(q)$ belongs is labelled $\mathrm{SD}(3\mathcal{B})_1$ in [200], and has a single simple module with a non-trivial self-extension (i.e., $\mathrm{Ext}^1_{kG}(M, M) \neq 0$ for a single simple module) whereas for the Morita class that has no known examples, labelled $\mathrm{SD}(3\mathcal{D})$, two simple modules have non-trivial self-extensions.

The final class of tame blocks is those with quaternion defect group. Here Erdmann was not able to even prove the finiteness of the number of Morita classes of blocks when the number of simple modules is two, because unlike in the dihedral and semidihedral cases, the parameters can now have infinitely many possibilities.

If $l(B) = 3$ then we obtain a very similar result to the dihedral case, Theorem 6.2.2.

Theorem 6.2.7 (Erdmann [197]) *Let B be a 2-block with quaternion defect groups Q_{2^n} for $n \geq 4$, and suppose that $l(B) = 3$. The decomposition matrix of B is one of the following:*

$$\begin{pmatrix}1&0&0\\1&1&0\\1&0&1\\1&1&1\\0&1&0\\0&0&1\\2&1&1\end{pmatrix}, \begin{pmatrix}1&0&0\\0&1&0\\0&0&1\\1&1&1\\1&1&0\\1&0&1\\0&1&1\end{pmatrix}, \begin{pmatrix}1&0&0\\1&1&0\\1&0&1\\1&1&1\\2&1&1\\0&0&1\\0&1&0\end{pmatrix},$$

with the last row repeated $2^{n-2} - 1$ times. Any two blocks with quaternion defect groups, three simple modules, and the same decomposition matrices are Morita equivalent.

The first two decomposition matrices are realized by the principal blocks of $\mathrm{SL}_2(q)$ *for* $q \equiv 1 \bmod 4$, *and* $\mathrm{SL}_2(q)$ *for* $q \equiv 3 \bmod 4$, *respectively. If* $n = 4$ *then the third decomposition matrix is realized by the principal block of* $2 \cdot A_7$. *If* $n \geq 5$ *then no example is known.*

If B has defect groups that are quaternion of order 8 and $l(B) = 3$, then there are only two decomposition matrices and Morita classes, the principal blocks of $\mathrm{SL}_2(5)$ and $\mathrm{SL}_2(11)$. Notice that in this case, $2^{n-2} - 1 = 1$, and the first and third matrices from Theorem 6.2.7 are, up to permuting the rows, the same.

Eisele has upgraded this theorem to the ring $\mathcal{O}$, showing that in each case there is a unique $\mathcal{O}$-algebra (up to Morita equivalence) lifting each k-algebra that is a block. In other words, he has shown that if B_1 and B_2 have quaternion defect groups and $l(B_i) = 3$, then B_1 and B_2 are Morita equivalent over k if and only if they are over $\mathcal{O}$ [185].

If $l(B) = 1$ then B is Morita equivalent to kQ_{2^n}, so the last case is $l(B) = 2$. Here, there are now two-parameter families that appear in Erdmann's list, and even up to Morita equivalence there are potentially infinitely many possibilities. At the moment, this problem for $l(B) = 2$ has not been resolved. We can say what the decomposition matrices are though.

Theorem 6.2.8 (Erdmann [196, 200]) *Let* B *be a 2-block with quaternion defect groups* Q_{2^n} *for* $n \geq 4$, *and suppose that* $l(B) = 2$. *The decomposition matrix is one of the following:*

$$\begin{pmatrix} 1 & 0 \\ 1 & 0 \\ 1 & 1 \\ 1 & 1 \\ 0 & 1 \\ 2 & 1 \end{pmatrix}, \quad \begin{pmatrix} 1 & 0 \\ 1 & 0 \\ 1 & 1 \\ 1 & 1 \\ 2 & 1 \\ 0 & 1 \end{pmatrix},$$

with the last row repeated $2^{n-2} - 1$ *times.*

Both of these are realized by principal blocks of groups $\mathrm{SL}_2(q^2).2$. *For any fixed* $n \geq 4$, *there are only two known Morita equivalence classes of blocks of finite groups, but there are potentially infinitely many Morita equivalence classes of blocks with each decomposition matrix.*

If B has quaternion defect groups and $l(B) \neq 2$, then Kessar and Linckelmann proved an intriguing result, which can be thought of as a Brauer–Suzuki theorem for blocks. The *Brauer–Suzuki theorem* [61] states that if G is a finite group with a quaternion Sylow 2-subgroup P and $O_{2'}(G) = 1$, then $Z(G) = Z(P)$. A block-theoretic version of this is as follows: If B is a block with a quaternion defect group P, and B' is the Brauer correspondent of B in $C_G(Z(P))$, then B and B' are Morita equivalent. Under the assumption that $l(B) \neq 2$, this was proved by Kessar and Linckelmann in [348], but for $l(B) = 2$ they could not show this because the information about Morita equivalence classes in Theorem 6.2.8 is not strong enough: there is a perfect isometry (see Sect. 4.4) between the blocks, and they even have the same decomposition matrix, but this is not enough to give the Morita equivalence because B and B' having the same decomposition matrix does not mean that they are Morita equivalent.

6.3 Nilpotent Blocks

The idea of a nilpotent block is a generalization of the idea of a nilpotent group, or at least the idea of a p-nilpotent group. As mentioned in Sect. 4.5, a p-nilpotent group is a finite group G such that $G/O_{p'}(G)$ is a p-group. *Frobenius's normal p-complement theorem* (see, for example, [127, Theorem 1.12]) states that G is p-nilpotent if and only if, for every p-subgroup Q of G, $N_G(Q)/C_G(Q)$ is a p-group.

The generalization to blocks of finite groups involves generalizing the subquotient $N_G(Q)/C_G(Q)$ to $N_G(Q, e)/C_G(Q)$, where e is a central idempotent associated to the p-block B. If D is a defect group of B then e is the block idempotent of the Brauer correspondent, and the group $N_G(D, e)$ was seen in Chap. 5 as controlling the number of simple modules in B ($l(B) = |N_G(D, e)/C_G(D)|$); it is defined simply as

$$N_G(D, b) = \{g \in N_G(D) \mid b^g = b\},$$

where b is the block idempotent of B, or equivalently of the Brauer correspondent of B (the group so defined is the same). You might guess that a block with cyclic defect groups is nilpotent if and only if $l(B) = 1$, and you would be right.

What about if $Q \neq D$?

Definition 6.3.1 Let G be a finite group and let B be a p-block of G, with defect group D and block idempotent b.

(i) A *Brauer pair* is a pair (Q, e) where Q is a p-subgroup of G and e is a block idempotent of $kC_G(Q)$.
(ii) A *b-Brauer pair* (or B-Brauer pair) is a Brauer pair (Q, e) such that $\mathrm{Br}_Q(b)e \neq 0$.

Since G acts by conjugation on the set of p-subgroups, we obtain a conjugation action of G on the Brauer pairs of G. Thus we may define $N_G(Q, e)$ to be the stabilizer of (Q, e) under this action. (The pedantic reader might ask for more brackets in this: we should write $N_G((Q, e))$. We will not.)

The definition of a defect group in Definition 2.3.8 is that there is a b-Brauer pair (Q, e) for every $Q \leq D$. We now give the definition of a nilpotent block.

Definition 6.3.2 Let B be a p-block of a finite group G, with defect group D. We say that B is *nilpotent* if, for every B-Brauer pair (Q, e), we have that $N_G(Q, e)/C_G(Q)$ is a p-group.

If B is the principal block then for any B-Brauer pair (Q, e), e is the principal block idempotent of $kC_G(Q)$. Since $N_G(Q, e) = N_G(Q)$ if e is a principal block idempotent, we see that the principal p-block is nilpotent if and only if G is p-nilpotent. Thus nilpotency for blocks is a genuine extension of p-nilpotency for groups.

There are lots of nilpotent blocks in non-nilpotent groups though. Inside quasisimple groups, An and Eaton proved that all nilpotent blocks have abelian defect groups in [11, 12], and also proved a conjecture of Puig on nilpotent blocks, which is still open for arbitrary finite groups.

Conjecture 6.3.3 (Puig) Let B be a p-block of a finite group G. Then B is nilpotent if and only if, for all p-subgroups Q of G, and all blocks B' of $C_G(Q)$ whose Brauer correspondent is B, we have $l(B') = 1$.

Another conjectural characterization of nilpotent blocks is as follows.

Conjecture 6.3.4 (Malle–Navarro [415]) Let B be a block of a finite group. All characters of B of height zero have the same degree if and only if B is nilpotent.

The 'if' direction of this is a result of Broué and Puig [73, (3.11)], so we have to check the converse. In [415], Conjecture 6.3.4 was verified whenever B has abelian defect groups and all irreducible ordinary B-characters have height zero (i.e., Brauer's height-zero conjecture is true for B). Since this direction of Brauer's height-zero conjecture is now known (Theorem 4.1.2), this second condition is no longer required.

They then prove that blocks of classical groups do not produce counterexamples to Conjecture 6.3.4, but leave out blocks of the quasisimple groups $2 \cdot A_n$ and certain blocks of exceptional groups of Lie type. The former of these was dealt with by Gramain [260], and then the exceptional groups were dealt with by the same paper of Kessar and Malle that finished off the proof of Brauer's height-zero conjecture [349]. Thus this conjecture is also true for all quasisimple groups.

However, neither of these conjectures has been reduced to quasisimple groups. Even for p-soluble groups Conjecture 6.3.4 is not known in general. Malle and Navarro prove that the conjecture would follow from a statement about large orbits on modules, like the $k(GV)$-conjecture from Sect. 4.7. At the moment any reduction seems far off.

Nilpotent blocks were introduced by Broué and Puig in [73]. They proved that nilpotent blocks possess a unique simple module, and if (Q, e) is a B-Brauer pair and B is nilpotent, then $kC_G(Q)e$ is nilpotent. The statement that $l(B) = 1$ is not enough to characterize nilpotent blocks though, as there are non-nilpotent blocks with a single simple module [73, Example 1.3].

Puig [468] later gave a strong theorem about nilpotent blocks, one facet of which is that a nilpotent block with defect group D is always Morita equivalent to the group algebra kD (and hence has one simple module). In order to give the full-strength version of his theorem we need the notion of a source algebra of a block.

Notice that whether a block is nilpotent or not does not depend on whether we are dealing with k or the ring $\mathcal{O}$ from Sect. 2.2. This is simply because the subgroup $N_G(Q, e)$ does not depend on where the idempotent lies. Much of the more structural representation theory takes place over $\mathcal{O}$, and in the literature $\mathcal{O}$ is used almost exclusively, partly because results over $\mathcal{O}$ descend to results over k but not vice versa. We saw a benefit of working over $\mathcal{O}$ in Sect. 4.5, namely that if kD_1 and kD_2 are Morita equivalent, we don't know if $D_1 \cong D_2$, but if $\mathcal{O}D_1$ and $\mathcal{O}D_2$ are Morita equivalent then we do. Thus working over $\mathcal{O}$ seems to provide us with a richer structure that bypasses some thorny issues that we don't yet understand.

Definition 6.3.5 Let G be a finite group and let k be a field of characteristic p. Let B be a block of kG with defect group D.

(i) An *interior G-algebra* is a k-algebra A together with a homomorphism $G \to A^\times$.
(ii) A *source algebra* of B is an interior D-algebra eBe, where (D, e) is a B-Brauer pair, together with the map $x \mapsto xe = exe = ex$ for $x \in D$.

If two blocks have isomorphic source algebras then they are Morita equivalent, as we will see soon. Moreover, because the defect group is built into the definition of a source algebra, it is preserved by a source algebra equivalence (as opposed to Morita equivalences, which as we saw in Sect. 4.5 are currently not known to preserve defect groups). If one prefers $\mathcal{O}$ to k, then one makes exactly the same definition but with $\mathcal{O}$ instead of k.

Note that source algebras are not unique up to isomorphism. If (D, e) and (D, f) are two B-Brauer pairs, then eBe and fBf are isomorphic as k-algebras, but not necessarily as *interior* D-algebras. There must be $x \in N_G(D)$ and $a \in (B^D)^\times$ (the multiplicative group of the D-fixed points of B) such that $f = e^{xa}$, and the map $e\sigma e \to (e\sigma e)^{xa} = f(\sigma^{xa})f$ is an isomorphism of k-algebras. However, the map $D \to (eBe)^\times$ sending $g \in D$ to ge gets sent to $g \mapsto (ge)^{xa} = g^x f$ (a is a D-fixed point), not gf. Thus if we are given a source algebra eBe, with map $D \to (eBe)^\times$, then all other source algebras from that block can be obtained by composing the map $D \to (eBe)^\times$ with an automorphism of D.

Source algebras are interesting because they keep a lot of the invariants of the block. Puig proved [467, Corollary 3.5] that the module categories of a block and its source algebra are equivalent, in fact via the map $M \mapsto Me$. Hence blocks with equivalent source algebras are Morita equivalent.

Indeed, this map $M \mapsto Me$ preserves not just the categorical structure, but also vertices and sources. The map $D \to (eBe)^\times$ yields an inclusion of k-algebras $kD \to eBe$. Thus we may form the tensor product $V \otimes_{kQ} eBe$ for any kQ-module V and subgroup $Q \leq D$. Let M be an indecomposable B-module. If Q is minimal such that Me is a summand of $V \otimes_{kQ} eBe$ for some kQ-module V, then Q is a vertex of M. Furthermore, if V is chosen to be indecomposable, then V is a source of M.

Since source algebras preserve so much of the structure of B, it is unsurprising that proving equivalences of source algebras is even more difficult than proving

Morita equivalences. We described in Sect. 4.5 some instances where equivalences of source algebras, called *Puig equivalences* or *splendid Morita equivalences*, have been found. (Splendid Morita equivalences are called such because they are the analogues of the splendid derived equivalences from Sect. 3.5 for Morita equivalences.) Another such situation is nilpotent blocks.

Theorem 6.3.6 (Puig [468]) *Let B be a nilpotent block of a finite group G with defect group D. For any source algebra eBe of B, there is an endopermutation module V with vertex D such that eBe is of the form* $\mathrm{End}_k(V) \otimes_k kD$.

This looks technical, but what it means is that B is Morita equivalent to kD, in fact B is isomorphic as a k-algebra to a matrix ring $M_n(kD)$ for some integer n. In addition, there is a unique simple B-module, and it has vertex D (by Proposition 3.2.5(ii)) and source V. All of this is true over $\mathcal{O}$ as well.

Conjecture 6.3.7 (Puig Conjecture on Nilpotent Blocks) The endopermutation module in Theorem 6.3.6 is always of finite order, i.e., torsion, in the Dade group.

Puig proved in [469] (see [539, Theorem 30.5]) that the source of any simple kG-module is endopermutation for any p-soluble group G. In 1980, Feit proved that all simple modules for p-soluble groups are algebraic [215] (and for endopermutation modules, torsion and algebraic are the same thing), so Conjecture 6.3.7 is true for p-soluble groups. Any group for which all simple modules are algebraic will satisfy Conjecture 6.3.7 as well, so for example all groups with abelian Sylow 2-subgroup for $p = 2$ by the author [125], and the result holds for blocks with cyclic defect group (since all modules are algebraic) and blocks with Klein four defect group by [132].

Salminen, using results of Kessar [347], proved [499, 500] that if B is a nilpotent block of G with defect group D, and there exists a normal subgroup $H \trianglelefteq G$ such that $G = H \rtimes D$, then Conjecture 6.3.7 holds for B, provided $p \geq 5$. (For $p = 3$ the remaining open case is when H is an 8-dimensional orthogonal group. For $p = 2$ the reduction to finite simple groups in [499, 500] does not work.)

In general, one might ask if every simple module with endopermutation source is algebraic, i.e., if M is a simple kG-module with source S, and S is endopermutation, then S is torsion in the Dade group. The answer to this is 'no', but examples are somewhat rare. First, we should look in simple groups, as it is true for p-soluble groups. There are some examples known in sporadic simple groups: among them, there are two 10-dimensional simple modules for M_{11} in characteristic 3 whose source is $\Omega^{\pm 2}(k)$, which has infinite order in the Dade group. For the quasisimple group $3 \cdot McL$ there are four 126-dimensional simple modules for $p = 5$ whose source is again $\Omega^{\pm 2}(k)$. That these are endotrivial, and not just with endotrivial source, was given in [387], and their sources were determined in [130].

In the next chapter we will discuss Clifford theory, which relates the representation theory of a group to that of a normal subgroup. One of the important parts of this theory is to try to understand the p-blocks of a normal subgroup N of a finite group G in terms of the p-blocks of G, and vice versa.

If e is a primitive central idempotent of N such that $e^g = e$ for a set of coset representatives for G/N, then e is a central idempotent in kG, hence is a sum of block idempotents b_i of kG. We say that b_i *covers* e. Suppose that kNe is a nilpotent block. It is not true that kGb_i is nilpotent, as of course we may take N to be the trivial group! We say that e is *G-stable* if $e^g = e$ for all $g \in G$. (Of course, since $e^g = e$ for all $g \in N$, it suffices to check a transversal to N in G.) We will see in Theorem 7.4.2 that requiring the covered block to be G-stable is not really a restriction, as we may always assume that we are in such a case.

The next result appears in [83, Theorem 2], and seems to be folklore, with no clear originator.

Proposition 6.3.8 *If G/N is a p-group and B_0 is a G-stable nilpotent block of kN, then any block B of kG that covers B_0 is nilpotent. Furthermore, if D_0 is a defect group of B_0, then B possesses a defect group D such that $D_0 \leq D$ and $ND = G$.*

If G/N is a p'-group then we cannot prove a similar result, because for example if N is a p-group and $G = N \rtimes E$ for some p'-group E then the group algebra kG is certainly not nilpotent, and covers kN, which is. A result of Linckelmann and Puig proves that this is more or less the general situation. We only include the case where G/N acts faithfully on a defect group of B_0 here, because the general case is a little more complicated.

Theorem 6.3.9 (Linckelmann–Puig [395]) *Let G be a finite group, let N be a normal subgroup, and suppose that $G/N \cong E$ is a p'-group. Let B_0 be a G-stable nilpotent block of kN with defect group D, and assume that the conjugation map $E \to \mathrm{Aut}(D)$ is injective. The block idempotent b_0 of B_0 is a block idempotent of kG, and*

$$\mathrm{End}_k(V) \otimes_k k_\alpha(P \rtimes E)$$

is a source algebra of $B = kGe$, where

(i) *V is the endopermutation module associated to B_0 (see Theorem 6.3.6), and*
(ii) *$k_\alpha(P \rtimes E)$ is a twisted group algebra (see Sect. 7.2).*

The general case was described in the work of Külshammer and Puig. We again assume that the nilpotent block B_0 of N is G-stable. In [377], Külshammer and Puig prove that if a block B of G covers B_0 then a source algebra of B is determined by the endopermutation module associated to B_0, together with a twisted group algebra $k_\alpha(\widehat{G/N})$, where $\widehat{G/N}$ is a finite group with a quotient group G/N and a defect group of B_0 as a normal subgroup, so it is an extension of the defect group by G/N (remember that the group theory convention on extensions is the other way round to describing extensions of modules). Although a precise description is difficult in this chapter, we will have slightly more to say about extensions of blocks in the next chapter.

6.4 What Happens in General?

Outside of abelian defect, nilpotent blocks, and tame blocks, the general case looks incredibly difficult. Of course, numerically the local-global conjectures from Chap. 4 should still hold, but there is no structural explanation of them, i.e., some categorical equivalence, like Broué's conjecture. This even extends to the case where $N_G(P)$ controls fusion in P: if G has an abelian Sylow p-subgroup P, then an old theorem of Burnside from 1900 shows that any two elements of P that are conjugate in G are also conjugate in $N_G(P)$. This property also holds if the Sylow p-subgroup is a *trivial intersection* subgroup, i.e., $P \cap P^g = 1$ whenever $g \notin N_G(P)$. Groups with trivial intersection Sylow p-subgroups look like a natural laboratory in which we can test theories about how to extend Broué's conjecture to at least some groups with non-abelian Sylow subgroups, but as of yet no such extension has been proposed.

What we are left with is numerical information about blocks, such as $k(B)$ and $l(B)$, potentially together with statements about decomposition and Cartan numbers, the number of socle layers in projective modules, vertices and sources of simple modules, and so on.

There has been a recent explosion in using the theory of fusion systems to get numerical information about $k(B)$ and $l(B)$ in particular. The book [501] by Sambale gives lots of information, much proved by Sambale himself, about these numbers in terms of the defect group of B. Local control of certain numerical invariants such as $k(B) - l(B)$ has been known since the time of Brauer, but the language of fusion systems has clarified the situation and allowed us to progress a lot further in this area.

Another possible clarification and codification of the concept of local control has appeared very recently in the work of Isaacs and Navarro, as yet unpublished, which defines the concept of a 'local invariant' in terms of stabilizers of chains of p-subgroups, like in the Knörr–Robinson formulation of Alperin's weight conjecture. The author is currently working on extending this viewpoint to encapsulate block decompositions, block fusion systems, and other ideas that should produce at least a numerical framework for talking about local invariants, if not a structural one.

Whatever the final form of an extension of Broué's conjecture to arbitrary defect groups is, one that will apply to all finite groups and not just those with abelian Sylow p-subgroups, it will almost certainly involve conceptually difficult categorical objects. The representation theory of finite groups has changed a lot in the last century.

Exercises

Exercise 6.1 Prove Brauer's height zero conjecture for tame blocks. If you know a priori that there are exactly four height-zero characters in a tame block, can you determine which they are from the decomposition matrix?

Exercise 6.2 Let B be a block with dihedral defect group. Show that the Morita equivalence class of B is determined by its ordinary character degrees. If B has quaternion defect group and $l(B) \neq 2$, does the same result hold?

Exercise 6.3 Let M and N be indecomposable modules for a dihedral 2-group D_{2^n}. If M and N are odd-dimensional, prove that $M \otimes N$ has a single indecomposable summand of odd dimension, and all other indecomposable summands are band modules. Deduce that the set of odd-dimensional indecomposable kD_{2^n}-modules takes on the structure of an abelian group under tensor product.

If M is a string module of even dimension, show that $M \otimes M$ has exactly two string modules as indecomposable summands, and that all other indecomposable summands are band modules.

The next two exercises develop a little of the theory of Brauer pairs.

Exercise 6.4 Let G be a finite group and let $Q \trianglelefteq R$ be p-subgroups of G. Let e be a block idempotent of $kC_G(R)$. Prove that there exists a unique R-stable block idempotent f of $kC_G(Q)$ such that $\mathrm{Br}_R(f)e = e$, and that if f' is any other R-stable block idempotent of $kC_G(Q)$, then $\mathrm{Br}_R(f)e = 0$.

We therefore impose a partial order on the set of Brauer pairs. If (Q, f) and (R, e) are Brauer pairs, write $(Q, f) \trianglelefteq (R, e)$ if $Q \trianglelefteq R$, f is R-stable, and $\mathrm{Br}_R(f)e = e$. Define $\leq$ to be the transitive closure of $\trianglelefteq$.

Exercise 6.5 Let G be a finite group and let $Q \leq R$ be p-subgroups of G. Prove that if (R, e) is a Brauer pair, there exists a unique block idempotent f of $kC_G(Q)$ such that $(Q, f) \leq (R, e)$.

Chapter 7
Clifford Theory

Clifford's theorem [118] gives us a description of the restriction of an irreducible representation to a normal subgroup N of finite index in a group G. Clifford theory is the general theory that tries to relate representations of N to representations of G, and to try to construct representations of G from representations of N.

If the characteristic of the base field k is 0 and G is finite then it was known to Frobenius back in 1898 that the restriction of an irreducible (ordinary) character to a normal subgroup is a sum of characters that are 'conjugate' to a given one. Clifford in one short paper [118] reproved this, but over all fields and for infinite groups as well, and then went about showing how to construct all irreducible representations of G that restrict to have a given (irreducible) representation of N as a composition factor.

Clifford theory is more like a philosophy than a theory. Given a problem in representation theory, you want to use Clifford theory to reduce the problem to a tightly defined class of finite groups (often quasisimple groups). Standard theorems usually get you some, but not all of the way, and the rest of it is bespoke, following the ideas that have gone before, but always having to produce a slightly different version of them.

We will first prove Clifford's theorem, then introduce group-graded algebras (which includes twisted group algebras), the main object used to describe representations of G in terms of those of N and those of G/N. We then discuss when a representation of N extends to a representation of G.

Finally, we give a few standard results from the Clifford theory of blocks, such as Fong's first and second reductions, and a slightly more detailed description of blocks that cover nilpotent blocks than was given in Sect. 6.3.

Clifford theory appears in several textbooks, but often not all of the theory we need. The book [339] proves much of the module theory in this chapter, in great generality. Both module and block Clifford theory appears in [33], and there is also Clifford theory in [394], with some characteristic 0 theory and some block theory in [447].

D. A. Craven, *Representation Theory of Finite Groups: a Guidebook*, Universitext,
https://doi.org/10.1007/978-3-030-21792-1_7

7.1 Representations and Normal Subgroups

We will prove Clifford's theorem. First, because it is quite easy to prove, and second because the proof is important to the understanding of the result. The weak form of Clifford's theorem is as follows.

Theorem 7.1.1 (Clifford's Theorem, Weak Version) *Let k be any field (including ones that are not necessarily algebraically closed), let G be a group and let N be a normal subgroup of finite index in G. If V is a simple kG-module then $V\downarrow_N$ is semisimple.*

Let V be a kG-module, and let N be a subgroup of G. The restriction of V to N can be a complete mess, even if V is irreducible. If $x \in G$ normalizes N and $W \leq V$ is a subspace that is a kN-submodule (but not necessarily a kG-submodule), then we can take the subspace

$$W^x = \{w \cdot x \mid w \in W\};$$

this is actually a kN-submodule when x normalizes N, because then every element of N is of the form h^x for some $h \in N$, and

$$w \cdot x \cdot h^x = (w \cdot h) \cdot x \in W^x.$$

Now we let V be irreducible, and let W be an irreducible kN-submodule of V, for N a *normal* subgroup of G. If X denotes a (right) transversal to N in G, then for all $x \in X$, W^x is a kN-submodule of V, and is obviously irreducible as well. The subspace

$$\sum_{x\in X} W^x$$

is actually a kG-submodule of V, hence is equal to V (as it is obviously non-zero).

We just need to show that this sum is actually a direct sum if we throw away some of the terms. To do this we simply order the elements of X as $x_1, \dots, x_n$, where $|G : N| = n$, and let

$$W_i = \sum_{j=1}^{i} W^{x_j}.$$

By induction W_i is semisimple, and if $W_{i+1} = W_i$ then W_{i+1} is semisimple. If $W_{i+1} > W_i$ then, as $W_i \cap W^{x_{i+1}}$ is a proper submodule of $W^{x_{i+1}}$ (as otherwise $W_{i+1} = W_i$) it must be zero, whence

$$W_{i+1} = W_i \oplus W^{x_{i+1}},$$

and we again have semisimplicity.

This is a complete proof of the weak form of Clifford's theorem that we gave above. To strengthen it we want to understand the kN-submodules W^x, when they are isomorphic, how many there are, and so on.

Let W be a simple kN-module. The *inertia subgroup*, $T(W)$, is the set of all $x \in G$ such that $W^x \cong W$. (This is clearly a subgroup, and contains N.) We can now state the full version of Clifford's theorem.

Theorem 7.1.2 (Clifford's Theorem, Strong Version [118]) *Let k be any field (including ones that are not necessarily algebraically closed), let G be a group and let N be a normal subgroup of finite index in G. If V is a simple kG-module then*

$$V{\downarrow_N} \cong \bigoplus_{x \in X} (W^x)^{\oplus e},$$

where W is any simple summand of $V{\downarrow_N}$, X is a transversal to $T(W)$ in G, and e is a positive integer, the ramification index *of V in N.*

This is the start of Clifford theory: we have related representations of N to representations of G, but now the question is to understand representations of G in terms of representations of N and representations of G/N (whose inflations to G are obviously going to appear in this picture as they are the representations whose restriction to N is trivial).

We say that a simple kG-module V *covers* W if W is a summand of $V{\downarrow_N}$ (or equivalently, in light of Clifford's theorem, a composition factor), and that an irreducible (ordinary or Brauer) character χ of G covers λ of N if the simple module corresponding to χ covers the simple module corresponding to λ. The model for Clifford theory is the representation theory of a direct product, so $G = N \times H$. In this case, the simple kN-modules always extend to simple kG-modules, if W is a kN-module then $T(W) = G$, and every simple kG-module that covers W is of the form $W \otimes U$, where we abuse notation slightly and write W for both the kN-module and its extension to G, and U is a kH-module inflated to G (via $H \cong G/N$).

At the other extreme, if G is the symmetric group S_3 or the dihedral group D_8 and χ is the irreducible ordinary character of degree 2, then its restriction to a normal subgroup N of index 2 is the sum of two different characters of degree 1, and if ϕ denotes one of them then $T(\phi) = N$ and $\chi = \phi{\uparrow^G}$.

These two examples illustrate the two extremes of $T(W) = N$ and $T(W) = G$. The first of these is by far the easiest to deal with: if W is irreducible and $T(W) = N$ then $W{\uparrow^G}$ is irreducible and is the only irreducible module covering W. (This is an easy consequence of Frobenius reciprocity in characteristic 0, and the Nakayama relation in characteristic p.)

Thus normally people concentrate completely on the case where $T(W) = G$, because once that problem is 'solved', you simply induce from $T(W)$ to G. In fact, this is true for indecomposable modules in general. (This is often called *Fong's first reduction*.)

Lemma 7.1.3 (Fong's First Reduction. See [33, Proposition 3.13.2]) *Let N be a normal subgroup of G, and let W be an indecomposable kN-module, with inertia subgroup H. If*

$$W{\uparrow}^H = V_1 \oplus V_2 \oplus \cdots \oplus V_r$$

is a direct sum decomposition into indecomposable summands, then each $V_i{\uparrow}^G$ is indecomposable, and $V_i{\uparrow}^G \cong V_j{\uparrow}^G$ if and only if $V_i \cong V_j$.

Apart from the final condition this all appears in [549]. The integer r is equal to 1 if and only if H/N is a p-group: one direction of this is Green's indecomposability theorem (Theorem 3.3.1). If H/N is cyclic of prime order different from p then it is an easy exercise to prove that $W{\uparrow}^H$ is decomposable. Thus by transitivity of induction, if $r = 1$ then H/N cannot have a subgroup of prime order other than of order p, i.e., H/N is a p-group.

The case where $T(W) = G$ is often called *stable Clifford theory*. If the characteristic of k does not divide $|G|$ (for example, $k = \mathbb{C}$) then things are a bit complicated, but there is an underlying theory that for example has helped us reduce the local-global conjectures from Chap. 4 to simple groups. In characteristic p there are even more problems.

If we write $\mathrm{Irr}(G)$ for the irreducible ordinary characters of G, and $\mathrm{IBr}(G)$ for the irreducible Brauer characters, then it is common to write $\mathrm{Irr}(G \mid \lambda)$ and $\mathrm{IBr}(G \mid \lambda)$ for the irreducible characters of G covering $\lambda \in \mathrm{Irr}(N)$ and $\lambda \in \mathrm{IBr}(N)$ respectively. In this case, we have that $\chi{\downarrow}_N = e(\chi)\cdot\lambda$, where $e(\chi)$ is the ramification index of χ. We are interested in the integers $e(\chi)$ for $\chi \in \mathrm{Irr}(G \mid \lambda)$, for which most is known, and occasionally $\mathrm{IBr}(G \mid \lambda)$: write

$$\lambda{\uparrow}^G = \sum_{\chi\in\mathrm{Irr}(G|\lambda)} e(\chi)\cdot\chi.$$

Thus by Frobenius reciprocity $\chi{\downarrow}_N = e(\chi)\cdot\lambda$, and putting these two together, and considering $\lambda{\uparrow}^G{\downarrow}_N$ we obtain

$$|G : N| = \sum_{\chi\in\mathrm{Irr}(G|\lambda)} e(\chi)^2.$$

Of course, this is clear in the case of the direct product $G = N \times H$, because the $e(\chi)$ were simply the degrees of the characters in $\mathrm{Irr}(H)$, and

$$\sum_{\psi\in\mathrm{Irr}(H)} \psi(1)^2 = |H| = |G : N|.$$

(In characteristic p this doesn't work, because if W is a simple kN-module then the induction $W{\uparrow}^G$ isn't semisimple in general, but it does work if $|G : N|$ is not divisible by p, as the induction is semisimple in this case (see [561, Satz 3.4] for example).)

Back in characteristic 0, notice that λ extends to an irreducible character χ if and only if $e(\chi) = 1$. In this case, the characters in $\mathrm{Irr}(G \mid \lambda)$ are simply the tensor products $\chi \otimes \phi$ for ϕ (the inflation of) a character for G/N. (This is originally due to Gallagher [239].) We also see from this that if $|G : N|$ has prime order r then either $T(\lambda) = N$, in which case $\mathrm{Irr}(G \mid \lambda)$ consists of $\lambda{\uparrow}^G$, or $T(\lambda) = G$, in which case there are r different extensions of λ to an irreducible character of G, related via tensoring by the 1-dimensional characters of G/N. This makes the character theory of soluble groups 'easy' to understand. If we work in characteristic p rather than characteristic 0, then the same result holds if $r \neq p$. If $r = p$ however, then we can have $T(G) = N$ and $\mathrm{IBr}(G \mid \lambda)$ consists of a single character $\lambda{\uparrow}^G$, or $T(G) = G$, but this time there is a unique extension of λ to an irreducible Brauer character of G.

Applying Clifford theory can quickly give us quite strong results, especially if there is a severe restriction on either the normal subgroup N or the quotient group G/N. For example, a result of Itô [316] concerns the case where N is abelian.

Theorem 7.1.4 (Itô) *Let G be a finite group and let N be a normal abelian subgroup of G. If χ is an irreducible ordinary character of G, then $\chi(1)$ divides $|G : N|$.*

There is a converse to this in a sense: a consequence of this theorem is that if G has a normal abelian Sylow p-subgroup then all ordinary characters have p'-degree. The converse is as follows.

Theorem 7.1.5 (Itô, Michler) *Let G be a finite group. A Sylow p-subgroup of G is normal and abelian if and only if all irreducible ordinary characters of G are of p'-degree.*

This was proved by Michler in [429]. For G a p-soluble group, this is again a theorem of Itô, but for G a general finite group it needs the classification of the finite simple groups, in particular the statement that for p odd, every finite simple group of Lie type has a block of defect zero [430].

For Brauer characters, we have to drop the word 'abelian': a finite group G has a normal Sylow p-subgroup if and only if, in characteristic p, it has no irreducible Brauer character of degree divisible by p. (This is also proved in [429] using [430].)

There is also a characteristic p version of Itô's theorem, due essentially to Swan [534, Theorem 6], but explicitly stated by Dade [140].

Theorem 7.1.6 (Fong–Swan, Dade) *Let G be a finite p-soluble group and let N be a normal abelian subgroup of G. If ψ is an irreducible Brauer character of G, then $\psi(1) \mid |G : N|$. More generally, if G is a p-soluble group then any irreducible Brauer character is the reduction modulo p of an irreducible ordinary character.*

Swan proved the second half of this result in [534] using results of Fong from [224], hence it is usually known as the *Fong–Swan theorem*. Dade proved this theorem from first principles in [140], then afterwards he noted that it follows from the Fong–Swan theorem (and Theorem 7.1.4).

Clifford in his 1937 paper [118] also gave a way to construct all elements of $\mathrm{Irr}(G \mid \lambda)$, in fact $\mathrm{IBr}(G \mid \lambda)$ for all primes p, including those dividing $|G/H|$. It involves the concept of a projective representation. (Note that this is *not* the same as a projective module, and a projective character is the character of a projective module, so the character of a projective representation is not a projective character. Even worse, every representation of G is a projective representation of G. For more examples of overused mathematical words, see 'normal', 'regular' and 'good'.)

Whereas a representation is a homomorphism $G \to \mathrm{GL}_n(k)$ for some G, n and k, a *projective representation* is a homomorphism $G \to \mathrm{PGL}_n(k)$, i.e., to the *projective* general linear group. Clifford proved in [118] that if $\chi \in \mathrm{Irr}(G \mid \lambda)$ and $G = T(\lambda)$ then there are projective representations A and B of G such that $\chi \cong A \otimes B$, that B is a projective representation of G/H, and that the dimensions of λ and A are the same. In order to proceed further, we will need the idea of a group-graded algebra, which is in the next section.

7.2 Group-Graded Algebras

A *G-grading* on a ring R is a set of additive subgroups R_g of R, one for each element $g \in G$, and such that if $r_g \in R_g$ and $r_h \in R_h$, then $r_g r_h \in R_{gh}$. We also require that R is the direct sum of the R_g as an additive group. A *homogeneous element* of R is an element lying in some R_g, and in this case g is its *degree*.

Of course, every ring is graded by every group, setting $R_1 = R$ and $R_g = 0$ for all $g \neq 1$. This is the *trivial grading*. Given a particular algebra, we may ask if it has a non-trivial G-grading for a particular group G. Popular candidates for gradings are the cyclic group C_2 and the integers $\mathbb{Z}$. The Laurent polynomial ring is (non-trivially) $\mathbb{Z}$-graded, for example, and algebras with a C_2-grading are often called *superalgebras*.

In our case algebras will be graded by a finite group G. An obvious example of a G-graded algebra is kG, with $(kG)_g = \{\lambda g \mid \lambda \in k\}$. More generally kG possesses a G/N-grading for any normal subgroup N, with the subalgebra kN being the 1-component and its cosets the other components.

There are others, importantly *twisted group algebras* A. Here A has a basis a_g for $g \in G$, each A_g is 1-dimensional and generated by a_g, but unlike kG, $a_g a_h \neq a_{gh}$ in general, but

$$a_g a_h = \zeta(g, h) a_{gh},$$

where $\zeta : G \times G \to k^\times$ is some map. For a fixed ζ, we write $k_\zeta G$ for this algebra. Clearly not every map ζ will yield an associative algebra. Exercise 7.3 asks you to verify that $k_\zeta G$ is associative if and only if ζ satisfies the identity

$$\zeta(g, h)\zeta(gh, j) = \zeta(g, hj)\zeta(h, j)$$

for all $g, h, j \in G$. This is called the 2-*cocycle condition*, and arises in group cohomology when done algebraically (rather than done homologically as we did in Sect. 3.4).

Thus $k_\zeta G$ is an associative algebra for 2-cocycles ζ. It also possesses a unit, namely $\zeta(1,1)^{-1}a_1$, which is checked in Exercise 7.3. We also introduce 2-*coboundaries*, which are cocycles obtained from maps $\alpha : G \to k^\times$. Note that

$$\zeta : G \times G \to k, \qquad \zeta(g,h) = \alpha(g)\alpha(h)\alpha(gh)^{-1},$$

is a 2-cocycle, as one can easily compute. Write $B^2(G, k^\times)$ for the 2-coboundaries and $Z^2(G, k^\times)$ for all 2-cocycles, so that $B^2(G, k^\times) \subseteq Z^2(G, k^\times)$. We want to turn the 2-cocycles into an abelian group, and then the 2-coboundaries will become a (normal) subgroup. The quotient

$$H^2(G, k^\times) = Z^2(G, k^\times)/B^2(G, k^\times)$$

is called the 2-*cohomology group* of G with coefficients in k. The group multiplication is simply multiplication of maps: $(\zeta\eta)(g,h) = (\zeta(g,h))(\eta(g,h))$. One need not have $k^\times$ as the codomain for the 2-cocycles, but any abelian group A. Thus we obtain $Z^2(G, A)$, $B^2(G, A)$ and $H^2(G, A)$ completely analogously.

The relationship between twisted group algebras, cohomology, and central extensions is as follows. Let G be a finite group and let $\hat{G}$ be a central extension of G; that is, a group $\hat{G}$ with a central subgroup A such that $\hat{G}/A \cong G$. Let $g, h \in G$ and choose preimages $\hat{g}, \hat{h} \in \hat{G}$ of g and h. Notice that $\hat{g}\hat{h}$ and $\widehat{gh}$ differ by an element of A, i.e.,

$$\hat{g}\hat{h}(\widehat{gh})^{-1} \in A.$$

Write $\zeta(g,h) = \hat{g}\hat{h}(\widehat{gh})^{-1}$. It is easy to check that ζ is a 2-cocycle, but obviously depends on the choice of preimages of the elements of G. However, while a different choice of preimages yields a different 2-cocycle, it yields the same 2-cohomology class. Thus a central extension $\hat{G}$ yields a class in $H^2(G, A)$, and this class corresponds to the direct product $G \times A$ (which is of course a central extension of G) if and only if ζ is a 2-coboundary.

Conversely, supposing that we have chosen $\zeta \in Z^2(G, A)$, we aim to construct a central extension of G. Let $\hat{G}$ denote the set $G \times A$, and define the multiplication on the set by

$$(g,a)(h,b) = (gh, \zeta(g,h)ab).$$

We claim that this is a group, with A a central subgroup and $\hat{G}/A$ isomorphic to G. We will do this one, having left all the others as exercises for the reader: on the one hand,

$$((g,a)(h,b))(j,c) = (gh, \zeta(g,h)ab)(j,c) = (ghj, \zeta(g,h)\zeta(gh,j)abc),$$

and on the other,

$$(g, a)((h, b)(j, c)) = (g, a)(hj, \zeta(h, j)bc) = (ghj, \zeta(g, hj)\zeta(h, j)abc).$$

Thus $\hat{G}$ is associative, by the cocycle condition. The identity is $(1, 1)$ of course, and the inverse of (g, a) is $(g^{-1}, \zeta(g, g^{-1})^{-1}a^{-1})$.

Thus given a central extension one obtains a cohomology class, and vice versa, with the two maps being mutually inverse. The relationship with twisted group algebras is the following (see, for example, [394, Proposition 1.2.17]).

Proposition 7.2.1 *Let G be a finite group and A an abelian group. Let $\zeta \in Z^2(G, A)$ be a 2-cocycle, and $\hat{G}$ the associated central extension of G. If $\phi : A \to k^\times$ is a homomorphism of groups, then $\zeta\phi$ is an element of $Z^2(G, k^\times)$, and there is a surjective algebra homomorphism*

$$k\hat{G} \to k_{\zeta\phi}G.$$

This proposition looks like it says that we can find all twisted group algebras as quotients of central extensions of G. Of course, since G is a finite group, the image of ζ in $k^\times$ is a finite subset, so contained in a finite subgroup of $k^\times$. Of course, $H^2(G, k^\times)$ could still be infinite: however, if A is a finite group, then the order of every element in $H^2(G, A)$ divides both $|G|$ and $|A|$, so therefore in particular $H^2(G, k^\times)$ is finite. (See, for example, [394, Proposition 1.2.9].) Thus if we have any 2-cocycle in $H^2(G, k^\times)$, we may exhibit it in a central extension with a finite centre, rather than all of $k^\times$ as a centre.

7.3 Extensions of Representations

There is a sizeable general theory of extensions from normal subgroups, spurred on by Dade in a series of papers [141–143, 146], with other contributions from, for example, Gallagher [239], Isaacs [311], Schmid [509] and Thévenaz [538]. (These are just the ones mentioned here.)

Now we have some theory of group-graded algebras, we are able to give a general criterion for when, given an irreducible character $\lambda \in \mathrm{Irr}(N)$, there exists some $\chi \in \mathrm{Irr}(G \mid \lambda)$ such that $e(\chi) = 1$, i.e., λ extends to a representation of G. By Fong's first reduction (Lemma 7.1.3), we may assume that $T(\lambda) = G$. We already know that λ extends to a projective representation of G, i.e., a representation of a central extension of G, by Clifford's results from Sect. 7.1, and the previous section gave us a theory related to central extensions.

Thus we must construct a central extension corresponding to a G-stable representation W of N. Since W is G-stable, W^g is isomorphic to W for all $g \in G$. Following [538], itself based on work of Dade [146], we write $\phi_g : W^g \to W$ for a fixed isomorphism, and then from this produce a corresponding isomorphism

$\psi_g : W \to W$: if $w\phi_g = w'$, then simply set $w\psi_g = w'$. Notice that, for all $x \in N$, if ρ is a representation whose module is W, we have for $g \in G$ and $x \in N$,

$$\psi_g^{-1}(x\rho)\psi_g = (x^g)\rho.$$

We may always choose the ϕ_g so that ψ_1 is the identity map, and $\psi_{gx} = \psi_g(x\rho)$ for all $g \in G$, $x \in N$, hence $\psi_x = x\rho$ if $x \in N$.

For $g, h \in G$, set

$$\zeta(g, h) = \psi_g \psi_h \psi_{gh}^{-1}.$$

The notation is suggestive, namely that ζ is a 2-cocycle, so an element of $Z^2(G, k^\times)$. However, what exactly is the image of ζ? The maps ψ_g, as we have seen above (they satisfy $\psi_g^{-1}(x\rho)\psi_g = x^g\rho$) are elements of $\mathrm{Aut}_{kG}(W)$, and this group is $k^\times$ if W is a simple kN-module by Schur's lemma. In other situations, $\mathrm{Aut}_{kG}(W)$ might be an abelian group, for example if W is a direct sum of non-isomorphic simple modules, or if W is an extension of a simple module by a non-isomorphic simple module, but in general it is of course not.

If W is simple, then we obtain an element of $Z^2(G, k^\times)$, and different choices of the ψ_g yield different elements of $Z^2(G, k^\times)$, but the same class of $H^2(G, k^\times)$. We write ζ_W for the element of $H^2(G, k^\times)$, and $\hat{G}_W$ for the corresponding central extension of G/N.

Theorem 7.3.1 *Assume that W is simple. There is an extension of W to G if and only if ζ_W is trivial, i.e., $\hat{G}_W$ is a split extension.*

There is a general version of this theorem where the assumption that W is simple is dropped. Now the normal subgroup of the extension is no longer $k^\times$ but $\mathrm{Aut}_{kN}(W)$, and there is an extension of W to G if and only if the extension $\hat{G}_W$, with quotient G and normal subgroup $\mathrm{Aut}_{kN}(W)$, is split. (See [538, Theorem 1.7].)

This theorem gives us a method to determine whether one may extend W to all of G, and can be used to deduce some previous results in extension theory. For example, it can be used to prove the following.

Theorem 7.3.2 *Suppose that G/N is a p-group, and that a Sylow p-subgroup of G is abelian. Any G-invariant character $\lambda \in \mathrm{Irr}(N)$ extends to a character of G.*

Isaacs proved the case where N is soluble in [311], when Dade picked up the thread and removed the solubility condition in [142], using the theory from [141]. In [142], Dade also proves the following.

Theorem 7.3.3 *Let G be a finite group and let N be a normal subgroup such that $|N|$ and $|G/N|$ are coprime. Any G-invariant irreducible ordinary or modular character of N may be extended to a character of G.*

The characteristic 0 version of this theorem is due to Gallagher [239, Theorem 6], and its extension to modular characters is what Dade proved.

7.4 Clifford Theory of Blocks

Exercise 3.12 asked you to prove that the blocks of kG are simply the indecomposable summands of kG as a (kG, kG)-bimodule, or equivalently a $k(G \times G)$-module. Because of this, one might expect there to be a Clifford theory of blocks to go alongside the Clifford theory of modules.

There is, but it is a lot less well developed. In Sect. 6.3 we defined the concept of one block covering another, and made a few steps towards a Clifford theory for nilpotent blocks.

Definition 7.4.1 Let N be a normal subgroup of a finite group G, and let e be a p-block idempotent of N. A p-block idempotent b of G *covers* e if $be = e$.

As we know, if $e_1, \dots, e_r$ are the block idempotents of kN, then

$$1 = e_1 + \cdots + e_r.$$

Since G acts by conjugation on N, G permutes the e_i, i.e., for $g \in G$, $e_i^g = e_j$ for some j. Write $f_1, \dots, f_s$ for the sums of the G-conjugacy classes of the e_i. Note that $f_i^g = f_i$ for all $g \in G$, so f_i is central in kG. Also, f_i is a sum of primitive central idempotents of N, so $f_i^2 = f_i$, and $f_i f_j = 0$ if $i \neq j$. Thus the f_i are sums of block idempotents of kG. Moreover, it is easy to see that if e is a block idempotent of kN that is a constituent of f_i, and b is a block idempotent of kG that is a constituent of f_i, then $be = e$.

Thus, up to G-conjugacy, any block of kG covers a unique block of kN. We can also talk about an inertia subgroup $T(e)$, which is the set of $g \in G$ such that $e^g = e$. As for simple modules, where if $T(W) = N$ then $W{\uparrow}^G$ is simple, we obtain a similar statement for blocks. In fact, we can get more or less perfect information in this situation, the analogue of Fong's first reduction—Lemma 7.1.3—for blocks.

Theorem 7.4.2 *Let N be a normal subgroup of the finite group G. Let e be a block idempotent of kN and let $H = T(e)$ be the inertia subgroup of e. Let b be a block idempotent of kG such that $be = e$.*

(i) *There is a one-to-one correspondence between blocks of kG covering kNe and blocks of kH covering kNe. If kHc is a block of kH covering kNe, then the corresponding block idempotent of kG is*

$$b = \sum_g c^g,$$

where the sum runs over a (right) transversal to H in G. In addition, $bc = c$ and $ce = c$.

(ii) *Every defect group of kHc is a defect group of kGb.*

(iii) *Induction of a kHc-module from H to G yields a kGb-module, and this map induces a Morita equivalence between the two blocks. Moreover, if $ikHi$ is a source algebra for kHc then $ikGi$ is a source algebra for kGb, and $ikHi$ and $ikGi$ are isomorphic.*

If k has characteristic 0 then the first part is simply Lemma 7.1.3. In fact, as we mentioned at the start of this section, one may use the interpretation of blocks of kG as indecomposable summands of kG as a (kG, kG)-bimodule (Exercise 3.12) to obtain both the first and second parts of this result (see Exercise 7.1).

This theorem started life as a result of Fong [224], in the case where N is a p'-subgroup. Reynolds in [474] proved (i) and (ii), and proved that the decomposition matrices of the two blocks were the same, with induction relating the two blocks. Because the language of Morita equivalences and source algebras was not available to him, he could not couch the result in these terms, which was later done by Puig in his work on p-soluble groups. However, I do not think he gave a proof in the literature: a full proof may be found in [394, Theorem 6.8.3].

Often results in Clifford theory break into two collections, as the users of Clifford theory break into two collections: 'structuralists' and 'quantitatives'. The structuralists, interested in Donovan's and Broué's conjectures (see Chap. 4), would like to understand the block structure of kG in terms of the structures of kN and $k(G/N)$, for example the decomposition and Cartan matrices, or the source algebras. The quantitatives on the other hand, often work in characteristic 0 and want information on the character degrees, fields of values, and so on, motivated by the McKay and $k(B)$-conjectures (again, see Chap. 4). This theorem satisfies the requirements of both schools, since it gives precise information relating both the blocks and characters. The end result of this is that we can always assume that a block of the normal subgroup is G-stable, both in structural and quantitative results.

The defect group of a block is obviously of fundamental importance in representation theory, so the next result is necessary.

Proposition 7.4.3 (Knörr [364]) *Let N be a normal subgroup of G, and let B be a p-block of G covering a block B_0 of N.*

If Q is a defect group of B_0 then there is a defect group D of B such that $Q = D \cap N$. Conversely, if D is any defect group of B then there exists some $g \in G$ such that $D \cap N$ is a defect group of $(B_0)^g$.

If B_0 is G-stable and D is any defect group of B, then $N \cap D$ is always a defect group of B_0, and for some choice of B among the covering blocks of B_0, DN/N is a Sylow p-subgroup of G/N.

If B_0 is not G-stable then there is no reason why a defect group of B_0 should be conjugate in N to a defect group of $(B_0)^g$; this is the reason why the first results are so specific, and why when G-stability is added one obtains a much cleaner result.

Now that we have an understanding of how defect groups interact with normal subgroups, the case $D \cap N = 1$ is obviously of special interest. This is where Fong's second reduction comes in. It is a result about blocks, but it can equally be thought of as a result about modules, because if B covers B_0 and $D \cap N = 1$, then B_0 is a block of defect zero, with a single ordinary character λ associated to it. Thus the characters of B lie in $\mathrm{Irr}(G \mid \lambda)$, and we can use standard Clifford theory. This is *Fong's second reduction*, and allows us to assume that the only normal subgroup of G that intersects the defect group trivially is the centre. The original version by Fong, with N a p'-subgroup, appears in [224]; Reynolds analysed the case where the defect group D of the block is normal in G in [474]. This more general form appears in, for example, [394, Theorem 6.8.13].

Theorem 7.4.4 (Fong's Second Reduction) *Let G be a finite group, B a p-block of G with defect group D, and N be a normal subgroup of G such that $N \cap D = 1$. There exists a central extension $\hat{G}$ of G/N, and a block $\hat{B}$ of $\hat{G}$, such that B and $\hat{B}$ are Morita equivalent.*

More specifically, if b_0 is the idempotent of the p-block of N covered by B, then kGb_0 (the sum of all p-blocks of G covering kNb_0) is isomorphic as a k-algebra to a tensor product

$$(kNb_0) \otimes_k k_\zeta(G/N),$$

where ζ is a specific 2-cocycle associated to G/N.

We will describe how to construct this 2-cocycle now. Let b_0 be the idempotent of a block of defect zero of a normal subgroup N. The first thing to note is that by Exercise 7.4, all k-algebra automorphisms of a matrix algebra over a field are inner, and of course, as a block of defect zero, $B_0 = kNb_0$ is a matrix algebra by Theorem 2.3.2. Thus, for any $g \in G$, there exists some $b_g \in B_0$ such that $g^{-1}\sigma g = b_g^{-1}\sigma b_g$ for all σ in the block B_0. It turns out that one may choose the b_g so that $b_g b_h = b_{gh}$ if one of g, h lies in N. This yields a 2-cocycle

$$\zeta(g,h) = b_{gh} b_g^{-1} b_h^{-1}.$$

(This looks like it is the wrong way round from the way that we usually write 2-cocycles. As one sees in the proof of the result in [394], one needs to change sides for the action at one point, hence everything gets inverted.)

The first point is that the image lies in $k^\times$. To see this, note that $b_g b_h$ and b_{gh} have the same action on B_0, so $\zeta(g,h)$ acts trivially on B_0; in particular, this means that as an element of B_0 it is a scalar matrix, hence (canonically) an element of $k^\times$. This also helps to prove the 2-cocycle condition, since we may move scalar matrices around inside matrix equations.

A block of defect zero has two properties: all automorphisms are inner, and it is nilpotent (Sect. 6.3). Thus Fong's second reduction theorem can be seen as a special case of extensions of nilpotent blocks. The theorem of Külshammer and Puig mentioned at the end of Sect. 6.3 looks a lot like Theorem 7.4.4. The source algebra of an extension B (with defect group D) of a G-stable nilpotent block can be factored as a tensor product

$$S \otimes k_\zeta E,$$

where E is a group with normal subgroup $D \cap N$ and quotient G/N, and $S = \mathrm{End}_k(V)$ for some endopermutation kD-module V with vertex D. For Morita equivalences only, rather than source algebras, one may ignore the S, so B is simply Morita equivalent to a block of a central extension of E. The complete details of this are difficult to write down, but see [394, Theorem 8.12.5] (and of course the original reference [377]) for more details.

Clifford theory relates the representations of normal subgroups of G to G itself, and the Brauer and Green correspondences relate representations of G to representations of normalizers of p-subgroups of G. Harris and Knörr proved that the concepts 'covering blocks' and 'Brauer correspondence of blocks' interact as well as possible. Our normal subgroup is H for this theorem, to avoid the undesirable notation '$N_N(Q)$'.

Theorem 7.4.5 (Harris–Knörr [275]) *Let H be a normal subgroup of the finite group G, and let D be a p-subgroup of H. Let B_0 be a p-block of H with defect group D, and let B_0' denote its Brauer correspondent in $N_H(D)$. If B' is a p-block of $N_G(D)$ covering B_0', then its Brauer correspondent B in G covers B_0.*

This is often described as there being a one-to-one correspondence between the p-blocks of G of defect group Q covering B_0 and those of $N_G(Q)$ covering B_0', given by the Brauer correspondence. Hidden in this statement is the fact that if D is a defect group of B_0 chosen so that $Q \cap H = D$, then $N_G(Q)$ is in fact contained in $N_G(D)$, so the Brauer correspondence of blocks of defect group Q between G and $N_G(D)$ makes sense in the first place.

Since blocks are modules for a direct product (Exercise 3.12), we can obtain the Harris–Knörr theorem as a consequence of a more general result of Alperin [5] on how Green correspondence and Clifford theory interact.

Theorem 7.4.6 (Alperin [5]) *Let H be a normal subgroup of the finite group G, and let Q be a p-subgroup of H. Let V be an indecomposable kH-module with vertex Q, with W its Green correspondent in $N_H(Q)$.*

There is a one-to-one correspondence between indecomposable kG-modules covering V and indecomposable $kN_G(Q)$-modules covering W.

Of course, we need to define what it means to cover an indecomposable module, rather than just a simple module. Let V be an indecomposable kG-module and let W be an indecomposable summand of $V\downarrow_H$ for a normal subgroup H. The Mackey formula (Theorem 3.2.14) shows that some vertex of W is contained in some vertex of V; if one may choose a vertex of W to be the exact intersection of H and a vertex of V, then V *covers* W. If we use the module-theoretic description of blocks from Exercise 3.12, then this definition of covering extends the block-theoretic version.

The technical details of the Clifford theory of blocks are too much to go into here, and would push the book to be too long. The results above give a flavour of what is known, and are the most commonly used ideas, but each question tends to need its own slightly different approach.

Exercises

Exercise 7.1 Prove the first two parts of Theorem 7.4.2 using the bimodule formulation mentioned after the theorem.

Exercise 7.2 Let N be a normal subgroup of the finite group G. Prove that if G/N is a p-group and B_0 is a p-block of N, then there exists a unique p-block B of G covering B_0.

Exercise 7.3 Let $\zeta : G \times G \to k^\times$ be a map, and let $k_\zeta G$ denote the twisted group algebra.

(i) Show that $k_\zeta G$ is associative if and only if ζ satisfies the condition

$$\zeta(g,h)\zeta(gh,j) = \zeta(g,hj)\zeta(h,j)$$

for all $g, h, j \in G$.

(ii) From now on suppose that ζ satisfies the 2-cocycle condition above. Prove that $\zeta(1,1) = \zeta(1,g) = \zeta(g,1)$ for all $g \in G$, and verify that $\zeta(1,1)^{-1}a_1$ is the unit of $k_\zeta G$, where a_1 is the basis element of $k_\zeta G$ corresponding to $1 \in G$ (that was used to define ζ in Sect. 7.2).

(iii) Prove that if two 2-cocycles ζ and η satisfy the relationship

$$\zeta(g,h)\eta(g,h)^{-1} = \alpha(g)\alpha(h)\alpha(gh)^{-1}$$

for some map $\alpha : G \to k^\times$, then $k_\zeta G$ and $k_\eta G$ are isomorphic under the map $g \mapsto \alpha(g)g$.

(iv) Deduce that every 2-cohomology class contains a 2-cocycle ζ such that a_1 is the unit of $k_\zeta G$.

Exercise 7.4 Prove that any k-algebra automorphism of a matrix algebra (over k) is inner. (This is a special case of the Skolem–Noether theorem.)

Exercise 7.5 Let G be a finite group and let N be a normal subgroup of G. Let B be a p-block of G and B_0 be a p-block of N, with block idempotents b and b_0 respectively. If χ is an irreducible character of B, show that every irreducible constituent of $\chi\downarrow_N$ belongs to a G-conjugate of B_0.

Exercise 7.6 Let N be a normal subgroup of the finite group G, and suppose that M is a relatively N-projective indecomposable module. Prove that M covers *every* summand of $M\downarrow_H$.

Exercise 7.7 Let G be a finite group and let N be a normal subgroup of G. If $\lambda \in \mathrm{Irr}(N)$, show that the number of characters in $\mathrm{Irr}(G \mid \lambda)$ is at most $k(G/N)$. (You might want Exercise 2.9.)

Chapter 8
Representations of Symmetric Groups

Representations of symmetric groups are intimately linked to the combinatorics of partitions. Even the number of irreducible complex characters of the symmetric group S_n is equal to the number of partitions of n. From there we can give an explicit bijection between the irreducible characters of S_n and the partitions of n: this bijection allows us to describe the character degrees using a combinatorial formula based on partitions, and there is even a recursive algorithm to compute the entire ordinary character table. We can also determine, for example, the constituents of the restrictions of irreducible characters of S_n to S_{n-1}, using these partitions.

It might be thought that the representation theory of symmetric groups is very advanced, and it is in characteristic 0. In characteristic p though, while there is a labelling of the Brauer characters, and the distribution of the ordinary and Brauer characters into blocks is known, the decomposition numbers, and even the Brauer character degrees, are not.

Despite this lack of information, the local-global theory of symmetric groups is pretty well understood: by a deep result of Chuang and Rouquier, all blocks of symmetric groups with the same (not necessarily abelian) defect group are derived equivalent, and a result of Chuang and Kessar proves that, for each abelian defect group, there is a particular block of a particular symmetric group that is Morita equivalent to the principal block of a wreath product of symmetric groups $S_p \wr S_w$, where the defect group has order p^w. This in turn is known to be derived equivalent to the principal block of the normalizer of the defect group, and hence Broué's conjecture is true for all blocks of symmetric groups (with abelian defect groups).

In 1991, Scopes proved that Donovan's conjecture is also true for symmetric groups, which was partially extended to blocks of classical groups using the same ideas. The Alperin–McKay and Alperin weight conjectures are also known for symmetric groups, painting a picture of a family of groups where a great deal is understood.

There are still some open problems in this area, not least the dimensions of the irreducible Brauer characters, but the theory we will summarize in this chapter

D. A. Craven, *Representation Theory of Finite Groups: a Guidebook*, Universitext,
https://doi.org/10.1007/978-3-030-21792-1_8

shows that these groups have a rich combinatorial structure that enables us to study them to a greater depth than almost any other class of groups.

We will do some characteristic 0 combinatorics for characters first, then look at the structure of the Specht modules in characteristic p in Sect. 8.2, including information on irreducibility of Specht modules and the modular branching rules. The section after that considers decomposition numbers, and the final section is a brief overview of the double covers of the symmetric groups.

In choosing topics for this chapter, I have stuck to the title of the book: I (almost always!) only discuss finite groups here. The theory of symmetric group representations is heavily connected both with the theory of reductive groups (infinite, so I won't talk about them in this setting, but see Sect. 9.1) and certain algebras such as the Schur algebra, (Iwahori–)Hecke algebra, Ariki–Koike algebra and KLR algebra (not even groups, so definitely beyond the scope). They are subjects deserving of a treatment in their own right, but the right place is not here.

8.1 The Combinatorics of the Character Table

It is well known that two elements of the symmetric group S_n are conjugate if and only if they have the same cycle type. Since the cycle type is a partition of n, we see that the conjugacy classes of S_n are naturally labelled by the partitions of n.

Normally, the conjugacy classes of a finite group and the irreducible ordinary representations are in bijection, because they have the same size, but there is no *natural* bijection. However, for symmetric groups there is a natural bijection, so we can label the irreducible ordinary characters of S_n by partitions of n.

In the next section we will construct modules, defined over $\mathbb{Z}$, one for each partition of n; when considered over $\mathbb{C}$ they form a complete set of irreducible $\mathbb{C}S_n$-modules, and this even works to some extent over $\mathbb{F}_p$ for primes p, which we will talk about in that section.

In this section we will treat the construction of these modules, called *Specht modules*, as a black box, so to every partition λ of n, we associate an irreducible $\mathbb{C}S_n$-module S^λ and an irreducible ordinary character χ^λ. One may construct the ordinary characters of S_n without going via Specht modules, as Frobenius did in 1900, but we will not go down this route.

For the main results of this section, we will give references to the books of James [324], James and Kerber [327] and Sagan [498], but they will appear in any most books on symmetric group representation theory.

Example 8.1.1 We have that $S^{(n)}$ is the trivial module for S_n, $S^{(n-1,1)}$ is the non-trivial summand of the permutation module for S_n, and $S^{(1^n)}$ is the sign representation of S_n, i.e., the non-trivial representation of degree 1 obtained by sending even permutations to (1) and odd permutations to (-1).

We need a little bit of combinatorics before we can describe our first result, which is to give the character degree $\chi^\lambda(1)$. This is given by the 'hook length formula', so in order to define it, we need to define a hook length.

We start by associating to every partition a *Young diagram*. This is a collection of boxes called *nodes*, with the number of nodes on the ith row being equal to the ith part of λ. This is best seen by example. If $\lambda = (4, 4, 3, 1)$ is a partition of 12, then the Young diagram of λ is

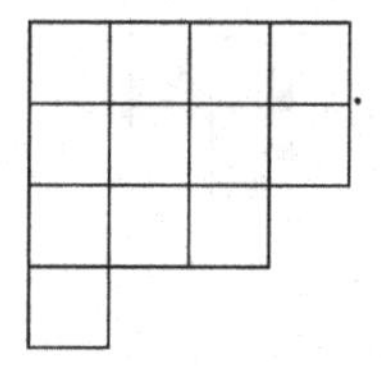

The node in row i and column j is simply labelled (i, j), so that $(2, 4)$ is in λ but $(3, 4)$ is not.

Example 8.1.2 The ordinary character table of S_5 was given in Example 2.2.8. There, the ordinary characters were labelled $\chi_1, \dots, \chi_7$, and the corresponding labelling by partitions is as follows:

χ_1	χ_2	χ_3	χ_4	χ_5	χ_6	χ_7
$\chi^{(5)}$	$\chi^{(1^5)}$	$\chi^{(4,1)}$	$\chi^{(2,1^3)}$	$\chi^{(3,2)}$	$\chi^{(2,1^2)}$	$\chi^{(3,1^2)}$

Notice that tensoring by the sign character $\chi_2 = \chi^{(1^n)}$ acts as a permutation on the set of irreducible characters, hence on the set of partitions of 5 (and on the partitions of n in general). It swaps (5) with (1^5), $(4, 1)$ with $(2, 1^3)$, and $(3, 2)$ with $(2, 1^2)$, while fixing $(3, 1^2)$. The easiest way to understand this map is by drawing the Young diagrams of the partitions:

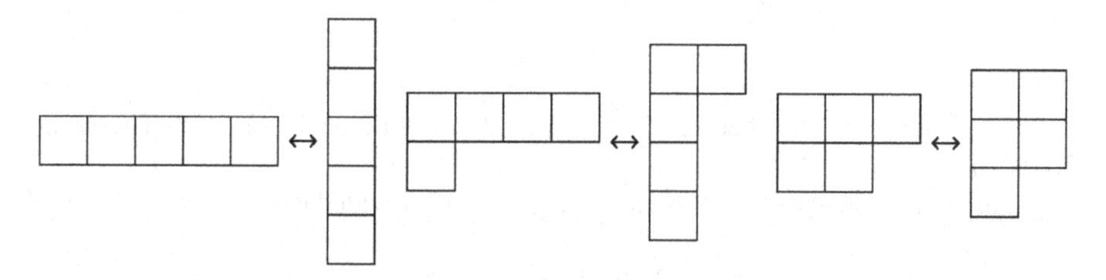

It is simply reflection in the diagonal line from top-left to bottom-right. Of course this action is an involution on the set of partitions, and the image of a partition under the map is called the *conjugate* partition or *transpose* partition. (Occasionally people call it the dual partition, but note that the simple kG-modules for symmetric groups are always self-dual.)

The conjugate partition will turn up regularly in the theory of symmetric groups.

The *hook* at a node $A = (i, j)$ consists of all nodes to the right of A, all nodes below A, and A itself. The *hook length* of the node A is the number of nodes in the hook at A. Later we will also need the arm and leg of a hook: the *arm* is all nodes to the right of A in the hook, and the *leg* is all nodes below A in the hook. Thus the hook length is the arm length, plus the leg length, plus 1. For example, the hook at node (2, 2) in the diagram above is as follows:

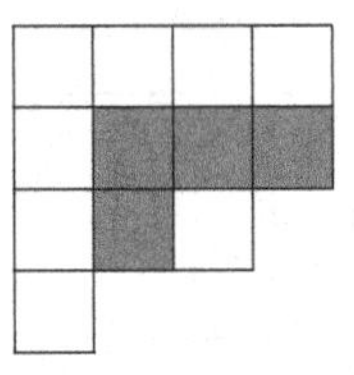

Of course, the hook length is 4, the arm length is 2 and the leg length is 1. Writing into each node the length of the hook at that node yields the following.

7	5	4	2
6	4	3	1
4	2	1	
1			

Write $h_{i,j}$ for the hook length at the node (i, j), and we simply write $(i, j) \in \lambda$ if the node (i, j) is in the Young diagram of λ, i.e., if $\lambda = (\lambda_1, \ldots, \lambda_r)$ then $i \geq r$ and $\lambda_i \geq j$. With this, we can now state the formula for $\chi^\lambda(1)$.

Theorem 8.1.3 (Hook Length Formula, Frame–Robinson–Thrall [235]) *If λ is a partition of n then*

$$\chi^\lambda(1) = \frac{n!}{\prod_{(i,j)\in\lambda} h_{i,j}}.$$

(See for example [324, Theorem 20.1], [327, Theorem 2.3.21], or [498, Theorem 3.10.2].)

Thus in the case above, the character degree is 12! divided by

$$7 \cdot 5 \cdot 4 \cdot 2 \cdot 6 \cdot 4 \cdot 3 \cdot 1 \cdot 4 \cdot 2 \cdot 1 \cdot 1,$$

i.e., 2970.

Example 8.1.4 Notice that $\chi^{(n)}(1) = 1$, $\chi^{(n-1,1)}(1) = n - 1$ and $\chi^{(1^n)}(1) = 1$. This correlates with our descriptions of $S^{(n)}$, $S^{(n-1,1)}$ and $S^{(1^n)}$ in Example 8.1.1.

We can also use hooks to describe the branching rule for S_n. In general, if H is a subgroup of G, then the *branching rule* from G to H is informally the description of how the irreducible characters of G decompose as sums of irreducible characters of H. More formally, assuming that the irreducible characters of H are ordered $\psi_1, \ldots, \psi_r$, one can describe a branching rule as a map from $\mathrm{Irr}(G)$ to $\mathbb{Z}^r$, which assigns to each $\chi \in \mathrm{Irr}(G)$ a tuple $(a_1, \ldots, a_r)$, such that

$$\chi\downarrow_H = \sum_{i=1}^{r} a_i \psi_i.$$

One is often interested in, for example, triples (G, H, χ) where $H \leq G$ and $\chi \in \mathrm{Irr}(G)$ such that $\chi\downarrow_H$ is irreducible, or the sum of at most two constituents, or all χ for H a maximal subgroup of G.

In our case, other than A_n, the most natural subgroup to consider is $S_{n-1} \leq S_n$. Here there is a complete answer.

Definition 8.1.5 Let λ be a partition of n. A node A of λ is *removable* if the diagram obtained from λ by removing A is still a partition. Equivalently, a removable node is one whose hook length is 1.

If one removes a node A from a partition λ of n, write $\lambda \setminus \{A\}$ for the resulting partition of $n-1$. The branching rule from S_n to S_{n-1} is as follows.

Theorem 8.1.6 (Branching Rule) *Let λ be a partition of n. The restriction of χ^λ to S_{n-1} is given by*

$$\chi^\lambda\downarrow_{S_{n-1}} = \sum_A \chi^{\lambda\setminus\{A\}},$$

where the sum runs over all removable nodes A of λ.

(See for example [324, Theorem 9.2], [327, Theorem 2.4.3], or [498, Theorem 2.8.3].)

Of course, this also allows us to determine the constituents of the induction of a character from S_{n-1} to S_n. This means that we should define an *addable node* to be a node A outside of the Young diagram of λ such that $\lambda \cup \{A\}$ is a partition, i.e., A is a removable node of a partition μ and $\lambda = \mu \setminus \{A\}$.

The branching rule can be seen as the start of understanding the (ordinary) character table of S_n. It allows you to determine the character value of any character at any element that has a fixed point—assuming you have the character table of S_{n-1}—simply by adding up the contributions from all $\lambda\setminus\{A\}$ for removable nodes A.

Thus we need only worry about fixed-point-free permutations x. The obvious case to concern ourselves with is the n-cycle $x = (1, 2, \ldots, n)$: if λ is a partition of n, then $\chi^\lambda(x) = 0$ if λ is not a *hook partition*, i.e., a partition of the form $(m, 1^{n-m})$ for some $1 \leq m \leq n$, and it is $(-1)^{n-m}$ in that case. In other words $\chi^\lambda(x)$ is equal to -1 to the power of the leg length of the hook partition.

We can now obtain the complete character table recursively by restricting χ^λ to products of symmetric groups. If n is an integer, then a *composition* of n is a sequence $(\lambda_1, \lambda_2, \ldots, \lambda_r)$ where $\sum \lambda_i = n$ and $\lambda_i \geq 1$ for all i. (If $\lambda_i \geq \lambda_{i+1}$ for all i then λ is a partition.)

Definition 8.1.7 Let $X_1, X_2, \ldots, X_r$ be a set partition of the integers $\{1, \ldots, n\}$. The *Young subgroup* corresponding to this set partition is the direct product

$$\mathrm{Sym}(X_1) \times \mathrm{Sym}(X_2) \times \cdots \times \mathrm{Sym}(X_r).$$

If λ is a composition of n, the *standard Young subgroup* S_λ of S_n is the direct product of symmetric groups $\mathrm{Sym}(X_i)$, where X_i is the set of all integers x such that

$$\sum_{j=1}^{i-1} \lambda_j < x \leq \sum_{j=1}^{i} \lambda_j.$$

Thus S_λ is isomorphic to the product

$$S_{\lambda_1} \times S_{\lambda_2} \times \cdots \times S_{\lambda_r},$$

with a particular embedding specified.

Given an element $x \in S_n$, x lies in the Young subgroup corresponding to the cycle decomposition of x, and we may conjugate x until it is a product of disjoint cycles of the form $(i, i+1, \ldots, i+j)$; then x lies in the standard Young subgroup S_λ corresponding to the cycle type λ of x.

If λ is a partition of n and A is a node of λ, we have already defined the hook ξ corresponding to A. We can remove this hook from λ, leaving two partitions, then push the smaller partition up and to the left, filling the hole left by the hook and creating a new, smaller partition.

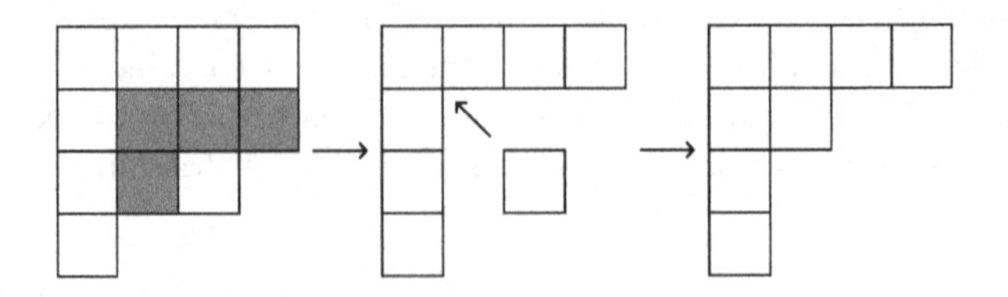

Alternatively one may reflect a hook along the boundary of the partition:

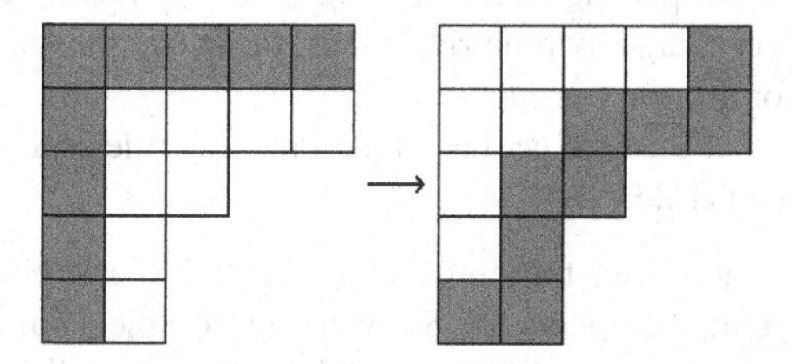

Doing this yields a *rim hook*. Removing a hook is the same thing as removing a rim hook, but without the annoying sliding up and left.

We can now state the Murnaghan–Nakayama rule. If λ is a partition and ξ is a set of boxes of λ, write $\lambda \setminus \xi$ for the result of removing the boxes in ξ from λ. This might or might not be a partition, depending on ξ. If ξ is a hook of λ, then write $\lambda \backslash \xi$ for the partition obtained by removing the rim hook corresponding to ξ from λ.

Theorem 8.1.8 (Murnaghan–Nakayama Rule) *Let λ be a partition of n and let $\mu = (\mu_1, \dots, \mu_r)$ be a composition of n, writing $\mu' = (\mu_1, \dots, \mu_{r-1})$, a composition of $n - \mu_r$. If $x \in S_n$ has cycle type μ, then*

$$\chi^\lambda(x) = \sum_\xi (-1)^{ll(\xi)} \chi^{\lambda\backslash\xi}(y),$$

where the sum runs over all hooks ξ of λ of length μ_r, $ll(\xi)$ denotes the leg length of ξ, and y is an element of $S_{n-\mu_r}$ with cycle type μ'.

(See for example [324, Theorem 21.1], [327, Theorem 2.4.7], or [498, Theorem 4.10.2].)

One way of proving this goes via the *Littlewood–Richardson rule*, which determines the integers $c^\nu_{\lambda,\mu}$ in the expression

$$\chi^\nu \downarrow_{S_n \times S_m} = \sum_{\lambda,\mu} c^\nu_{\lambda,\mu} \chi^\lambda \cdot \chi^\mu,$$

where λ is a partition of n, μ is a partition of m, and ν is a partition of $m + n$. The Littlewood–Richardson rule is not particularly complicated to state, but is very easy to get wrong. For example, the Littlewood–Richardson rule is first stated in [396], although the proof there is wrong, and wasn't corrected until around 40 years later, by Thomas [540, 541] and Schützenberger [515], although see [321] for another approach. Not only was the proof in [396] wrong, even the example of the rule in action given in [396] was wrong! To be on the safe side, I will not even state the rule here, and refer to [324, Section 16], [327, Section 2.8] and [498, Section 4.9].

The recursive Murnaghan–Nakayama rule is very quick to compute, meaning that a computer algebra package such as Magma or GAP can easily produce the character tables of very large symmetric groups, far above the size where it is easy to compute tables for arbitrary groups.

Notice that if $\mu_r = 1$ then we get back the branching rule as a special case of the Murnaghan–Nakayama rule.

One might say that, given the entire character table can be recursively constructed, the representation theory of S_n over the complex numbers is 'solved'. There are, however, still some difficult problems in the ordinary representation theory of symmetric groups, but the combinatorial theory that we have just seen equips us with the tools we need to attack many questions about the characters of S_n.

Just to give you an example of something that we cannot do, let m and n be positive integers, and consider the permutation modules for S_{mn} on the cosets of $S_n \wr S_m$ and $S_m \wr S_n$, both of which are natural subgroups of S_{nm}.

Conjecture 8.1.9 (Foulkes's Conjecture) Let m and n be positive integers with $m \leq n$. If λ is a partition of nm, then

$$\langle \chi^\lambda, 1_{S_n \wr S_m} \uparrow^{S_{nm}} \rangle \leq \langle \chi^\lambda, 1_{S_m \wr S_n} \uparrow^{S_{nm}} \rangle,$$

i.e., the difference $1_{S_m \wr S_n} \uparrow^{S_{nm}} - 1_{S_n \wr S_m} \uparrow^{S_{nm}}$ is a character of S_{nm}.

In 1942, Thrall proved Foulkes's conjecture for $m = 2$ [544], via an explicit calculation of the constituents of $1_{S_n \wr S_2} \uparrow^{S_{2n}}$. This was actually done before Foulkes's conjecture was made [234] in 1950. We had to wait until 2000 until the case $m = 3$ was resolved, by Dent and Siemons [159], although Brion proved Foulkes's conjecture for any fixed m and all but finitely many n in [65, p. 352]. McKay [428] proved the case $m = 4$, by constructing homomorphisms directly. For $m = 5$, Cheung, Ikenmeyer and Mkrtchyan produced a proof in 2017 [111], and if $m + n \leq 20$ then Foulkes's conjecture holds (for $m + n \leq 19$ this appears in the work of Paget and Wildon [462], using techniques of [204], and $m + n = 20$ appears on Wildon's website) but in general it remains open.

So for characteristic 0 there are still difficult open problems, but there is a fairly good theory. When the field k has characteristic p, that's a different story. We start looking at this in the next section.

8.2 Specht Modules

This section sees us construct the Specht modules S^λ, $\mathbb{Z}S_n$-modules that can therefore be defined over any ring, including fields of characteristic p. We will place a bilinear form on them, such that the radical $(S^\lambda)^\perp$ of the form is zero if the field has characteristic 0 or $p > n$, and for $p \leq n$, we have that $S^\lambda/(S^\lambda)^\perp$ is either 0 or is an irreducible representation, denoted D^λ. The D^λ form a complete set of (isomorphism classes of) simple modules for S_n, giving us a construction of all simple modules in all characteristics. (Note that the radical $(S^\lambda)^\perp$ of the form is not the radical in the sense of the Jacobson radical $\mathrm{rad}(S^\lambda)$. To see this, the Jacobson radical can never be the whole module, for example.) The only trouble is that this radical is difficult to construct in general, so it is not a great tool to actually compute decomposition numbers or the simple $\mathbb{F}_p S_n$-modules.

The construction of Specht modules in this way, and James's submodule theorem, come from [317] (the original construction may be found in [528]).

We have to define a few objects before we can construct the Specht modules. A *Young tableau* (often simply 'tableau') is a Young diagram with each node filled in with a positive integer. In general these integers can repeat, but for what we want them for right now, if t is a tableau with n nodes, the entries in the tableau will be the integers $1, \ldots, n$. For example,

$$\begin{array}{|c|c|c|}\hline 1&4&5\\\hline 2&3&6\\\hline 7&8\\\cline{1-2}\end{array}\,,\quad \begin{array}{|c|c|c|}\hline 1&3&4\\\hline 2&5&8\\\hline 6&7\\\cline{1-2}\end{array}\quad\text{and}\quad \begin{array}{|c|c|c|}\hline 3&1&2\\\hline 7&4&6\\\hline 8&5\\\cline{1-2}\end{array}$$

are tableaux of shape $(3, 3, 2)$. A tableau is *row standard* if the entries increase along each row, *column standard* if the entries increase down each column, and *standard* if it is both row and column standard. Thus the first tableau above is row standard, the third is column standard, and the second is standard.

Proposition 8.2.1 *If λ is a partition of n, then $\chi^\lambda(1)$ is equal to the number of standard tableaux of shape λ.*

(See, for example, [324, Corollary 8.5], [327, Corollary 7.2.8] and [498, Theorem 2.6.5].)

This suggests that the Specht module S^λ might have a basis (as a vector space) labelled by standard tableaux, and it does. But to construct this, we need another idea.

A *tabloid* is like a tableau, but where we ignore the ordering along the rows, so that

$$\begin{array}{|c|c|c|}\hline 1&2&3\\\hline 4&6\\\cline{1-2} 5\\\cline{1-1}\end{array}\quad\text{and}\quad \begin{array}{|c|c|c|}\hline 3&1&2\\\hline 6&4\\\cline{1-2} 5\\\cline{1-1}\end{array}$$

are ‘the same’. To demonstrate this equality, we remove the vertical bars separating the nodes:

$$\begin{array}{lll}\hline 1 & 2 & 3 \\ \hline 4 & 6 & \\ \cline{1-2} 5 & & \\ \cline{1-1}\end{array}\ .$$

If t is a tableau, denote by $\{t\}$ the associated tabloid.

We let S_n act on the right on tableaux and tabloids by permuting the entries of the tableau and tabloid respectively. Hence

$$\begin{array}{|c|c|c|}\hline 1 & 2 & 3 \\ \hline 4 & 6 \\ \cline{1-2} 5 \\ \cline{1-1}\end{array} \cdot (1,5,2)(3,6) = \begin{array}{|c|c|c|}\hline 5 & 1 & 6 \\ \hline 4 & 3 \\ \cline{1-2} 2 \\ \cline{1-1}\end{array}\ .$$

Of course, S_n acts regularly on the set of tableaux, but will not act regularly on the set of tabloids; the stabilizer of a given tabloid is a Young subgroup of S_n. If t is a tableau and $x \in S_n$, we write $t \cdot x$ for the tableau obtained under this action, and by extension $\{t \cdot x\}$ for the corresponding tabloid.

For t a tableau of shape λ, the *row stabilizer* of λ is the stabilizer in S_n of the tabloid $\{t\}$. The *column stabilizer* of t is the obvious analogue of the row stabilizer. Denote the row and column stabilizers of a tableau t by R_t and C_t respectively; hence $x \in R_t$ if and only if $\{t \cdot x\} = \{t\}$, and if $x \in C_t$ then $\{t \cdot x\} = \{t\}$ if and only if $x = 1$.

Given a tableau t, the *polytabloid* of t is the formal linear combination

$$\sum_{x \in C_t} \operatorname{sgn}(x)\{t \cdot x\},$$

where $\operatorname{sgn}(x)$ is the sign of the permutation x and as we said before $\{\cdot\}$ denotes the operation of taking a tabloid of a given tableau. For example, the polytabloid of the tableau

$$\begin{array}{|c|c|c|c|}\hline 1 & 2 & 3 & 4 \\ \hline 5 & 6 \\ \cline{1-2}\end{array}$$

is

$$\begin{array}{llll}\hline 1 & 2 & 3 & 4 \\ \hline 5 & 6 & & \\ \cline{1-2}\end{array} - \begin{array}{llll}\hline 1 & 3 & 4 & 6 \\ \hline 2 & 5 & & \\ \cline{1-2}\end{array} + \begin{array}{llll}\hline 3 & 4 & 5 & 6 \\ \hline 1 & 2 & & \\ \cline{1-2}\end{array} - \begin{array}{llll}\hline 2 & 3 & 4 & 5 \\ \hline 1 & 6 & & \\ \cline{1-2}\end{array}\ .$$

If t is a tableau, we denote by e_t the associated polytabloid. The element combination $\sum_{x\in C_t} \mathrm{sgn}(x)x \in \mathbb{Z}S_n$ will be denoted by κ_t, so that $e_t = \{t\} \cdot \kappa_t$.

We need somewhere concrete for these polytabloids to live in, as they are currently just formal linear combinations. If λ is a partition of n and k is any field (in fact any ring will do, but we need to be more careful about the definition of module then), let M^λ denote the permutation module for kS_n with basis all tabloids of shape λ; thus this is simply (up to isomorphism) the permutation module on the standard Young subgroup S_λ. Polytabloids are elements of M^λ, of course, so let S^λ be the kS_n-submodule generated by the polytabloids e_t for standard tableaux t of shape λ. The module S^λ is called the *Specht module* corresponding to λ.

The next theorem summarizes the most salient properties of Specht modules.

Theorem 8.2.2 *Let λ be a partition of n.*

(i) *The set of polytabloids for standard tableaux of shape λ is a basis for S^λ as a vector space.*
(ii) *If k has characteristic 0 and $S^\lambda \cong S^\mu$ then $\lambda = \mu$.*
(iii) *If k has characteristic 0 then S^λ has character χ^λ given by the Murnaghan–Nakayama rule.*
(iv) *If $\mathrm{char}\, k = 0$ or $\mathrm{char}\, k > n$ then S^λ is irreducible.*

(References for this include [324, Theorems 4.12, 8.4 and 21.1], [327, Theorems 2.4.7, 7.1.9 and 7.2.7] and [498, Theorems 2.4.6, 2.5.2 and 4.10.2].)

Of course, this theorem leaves open the case where $p = \mathrm{char}\, k \le n$, i.e., where $p \mid |S_n|$. Here the last part of the theorem is definitely false, but there is still something that can be done.

We place a bilinear form $(\cdot, \cdot)$ on M^λ by defining

$$(\{t\}, \{s\}) = \begin{cases} 1 & \{t\} = \{s\}, \\ 0 & \{t\} \neq \{s\}, \end{cases}$$

on the basis elements of M^λ and extending by bilinearity. This bilinear form allows us to get some pretty good information about S^λ.

Theorem 8.2.3 (James's Submodule Theorem [317]) *Let λ be a partition of n. If V is a submodule of M^λ then either V contains S^λ or V is contained in $(S^\lambda)^\perp$.*

(See for example [324, Theorem 4.8], [327, Theorem 7.1.7], or [498, Theorem 2.4.4].)

Applying this to a proper submodule of S^λ, we get the following corollary.

Corollary 8.2.4 *If λ is a partition of n then $S^\lambda/(S^\lambda \cap (S^\lambda)^\perp)$ is either 0 or irreducible.*

Write D^λ for this quotient. It turns out that D^λ is non-zero if and only if λ is p-regular: a partition $\lambda = (\lambda_1, \lambda_2, \ldots, \lambda_r)$ of n is *p-regular* if there does not exist

$i \leq r - p + 1$ such that $\lambda_i = \lambda_{i+1} = \cdots = \lambda_{i+p-1}$. In other words, λ is p-regular if no part appears at least p times.

The number of conjugacy classes of p-regular elements in S_n is equal to the number of p-regular partitions of n, a result from combinatorics known as Glaisher's theorem [249]. This means that, if they are all pairwise non-isomorphic, the modules D^λ form a complete set of irreducible modules for S_n as λ ranges over the p-regular partitions of n.

The simple modules D^λ are all self-dual, because all conjugacy classes of S_n are real. What about the Specht modules S^λ? In characteristic 0, they are simple and self-dual, but in characteristic p this is not true, and there is a more subtle relationship. This can be found in [324, Theorem 8.15], for example.

Proposition 8.2.5 *Let λ be a partition of n and let λ' denote the conjugate partition of λ. Writing* sgn *for the sign representation $S^{(1^n)}$, we have*

$$(S^\lambda)^* \cong S^{\lambda'} \otimes \text{sgn}.$$

The decomposition numbers of S_n are simply given by the composition factors of S^λ. Having found one constituent, D^λ, we now need to understand the composition factors of $S^\lambda \cap (S^\lambda)^\perp$, the radical of the form. To do this, we need another definition.

If $\lambda = (\lambda_1, \ldots, \lambda_r)$ and $\mu = (\mu_1, \ldots, \mu_s)$ are partitions of n, then λ *dominates* μ, written $\lambda \trianglerighteq \mu$, if $r \leq s$ and for all $1 \leq i \leq r$ we have

$$\sum_{j=1}^{i} \lambda_j \geq \sum_{j=1}^{i} \mu_j.$$

The dominance ordering is a partial ordering, with (n) as the maximum and (1^n) as the minimum. It has been introduced here because it is crucial to understanding the factors of $S^\lambda \cap (S^\lambda)^\perp$.

Proposition 8.2.6 *Let λ and μ be partitions of n. If D^λ is a composition factor of the module M^μ then λ dominates μ. Furthermore, if μ is p-regular then D^μ appears in M^μ exactly once.*

(See for example [324, Theorem 12.1] or [327, Theorem 7.1.14].)

So the decomposition number for D^λ in S^λ is 1, and if D^λ appears in S^μ then λ dominates μ. In particular, if we order the rows of the decomposition matrix by dominance ordering (with any ordering between incomparable partitions) then we may arrange the columns so that the matrix is lower triangular, and one such arrangement is by ordering the columns (which are labelled by p-regular partitions) using the same ordering as on the rows. This statement, originally proved by Farahat, Müller and Peel in [205], yields strong information about decomposition numbers, but of course there is still a great deal more work to be done.

One should also look at this proposition in characteristic 0. It says that the permutation module M^μ is the sum of Specht modules S^λ for $\lambda \trianglerighteq \mu$, with S^μ

appearing exactly once. The multiplicities of S^λ in M^μ, written $K_{\lambda,\mu}$, are called the *Kostka numbers*, and are, unsurprisingly, given by a combinatorial object associated to partitions.

However, we need to change the set from which the entries in a tableau are drawn, and we need to create a minor variant on standardness. A tableau t of shape λ has *type* μ for some composition $\mu = (\mu_1, \mu_2, \ldots, \mu_s)$ if there are exactly μ_i entries in t that are i. Thus

2	5	2
4	2	2
4		

is an example of a tableau of shape $(3, 3, 1)$ and type $(0, 4, 0, 2, 1)$. A tableau is *semistandard* if the entries weakly increase along the rows and strictly increase along the columns.

Theorem 8.2.7 (Young's Rule) *If λ and μ are partitions of n, then $K_{\lambda,\mu}$ is equal to the number of semistandard tableaux of shape λ and type μ.*

(See for example [324, Theorem 14.1], [327, Theorem 2.8.5], or [498, Theorem 2.11.2].)

If the characteristic of the field divides the order of S_n then we have a different situation. The permutation module M^μ is, by the Krull–Schmidt theorem, the direct sum of a finite number of specific indecomposable modules. Thus by James's submodule theorem, as not all summands can be in $(S^\mu)^\perp$, at least one, and therefore exactly one, must contain S^μ. Call this summand a *Young module*, and denote it by Y^μ. (Alternatively, we can use results from Chap. 3: since M^μ is a permutation module, its summands are trivial-source modules (Example 3.2.7), and by Proposition 3.2.8 trivial-source modules are liftable to characteristic 0. As the ordinary character of M^μ contains χ^μ, a lift of one indecomposable summand must also contain χ^μ, and this is Y^μ.)

It turns out that

$$M^\mu = \bigoplus_\lambda Y^\lambda,$$

where the sum runs over some collection of partitions λ that dominate μ (and each one may occur multiple times, except for μ, which occurs exactly once). The multiplicity of Y^λ in a direct sum decomposition of M^μ is called the *p-Kostka number*. There has been work on these numbers over the years, partly because of the connection between Young modules and the Schur algebra, which we won't be able to get into in this book.

One thing that we might be lulled into thinking when doing this is that in order for S^λ to always lie inside Y^λ, which is indecomposable, this means that S^λ is itself indecomposable. In odd characteristic this is true, and was proved by Peel in [465].

A more conceptual proof was later given using results of Carter and Lusztig [108] (but see James's write-up in [324, Section 13]), which prove that the dimension of $\mathrm{Hom}(S^\lambda, M^\mu)$, which is the number of semistandard tableaux of shape λ and type μ in characteristic 0 (Young's rule) is also the dimension in characteristic $p > 2$. James in [324] even pushes the method so that equality holds when $p = 2$ and λ is 2-regular. In particular, this means that $\mathrm{Hom}(S^\lambda, M^\lambda)$ is 1-dimensional, so S^λ cannot be decomposable. (This can be proved more easily using Exercise 8.2.)

However, this does leave the possibility that S^λ can be decomposable when $p = 2$ and λ is not 2-regular. James gave the first example of a decomposable Specht module in characteristic 2 in [322], where he showed that $S^{(5,1,1)}$ is decomposable. Murphy in [444] analysed the Specht modules labelled by hook partitions $\lambda = (n - r, 1^r)$, and proved that S^λ is always indecomposable when n is even, and if n is odd then S^λ is indecomposable if and only if $n - 2r - 1$ is divisible by a power of 2 greater than r (i.e., $n - 2r - 1 \equiv 0 \bmod 2^a$ for some a such that $r < 2^a$).

No more information was known until 2012, when Dodge and Fayers independently discovered new decomposable Specht modules [166] of the form $(n, 3, 1^m)$ for n, m even. They found lots of new decomposable Specht modules of this form, for example whenever $n + m \equiv 0, 2 \bmod 8$, and $n \geq 6$, $m \geq 4$. (There are even more cases in [166].)

There appear to be many decomposable Specht modules in characteristic 2. As I was writing this book, Donkin and Geranios released a paper [169] which gave a two-parameter family of decomposable Specht modules and showed that there is no bound on the number of summands that may appear in a direct sum decomposition.

If Specht modules are always indecomposable for odd primes, but the situation looks really hard for $p = 2$, a much more restrictive question is therefore to ask when Specht modules are in fact irreducible.

If λ is p-regular then we know that D^λ is a composition factor of S^λ, so in this case we are asking if $D^\lambda = S^\lambda$, or if its radical is 0. The first question for λ not p-regular is: can we identify any composition factor of S^λ at all? This can be answered using the process of regularization, introduced by James in [319]. This is an algorithm that takes a partition λ and returns a p-regular partition λ^R.

To run this algorithm, we place a label in the (i, j) box: the $(i, 1)$ box has label i, and the (i, j) and $(i - (p - 1), j + 1)$ box have the same label, so the label of box (i, j) is $i + (p - 1)(j - 1)$. The first few for $p = 4$ (there is no reason for p to be a prime) look as follows.

1	4	7	10
2	5	8	11
3	6	9	12
4	7	10	13
5	8	11	14

The collection of boxes labelled by a given integer i form a single line of slope $p - 1$, thought of as a runner. The *regularization* of a partition (which depends on p) is obtained by moving all boxes as far up their given runners as possible. For example, the 3-regularization of (4, 4, 3, 2, 2, 2, 2) is given in Fig. 8.1, and involves pushing the boxes labelled '8' and '9' as far up their runners as possible. The result of this process is always a p-regular partition. (It is easier to see that the output is p-regular than it is to see that it is a partition.)

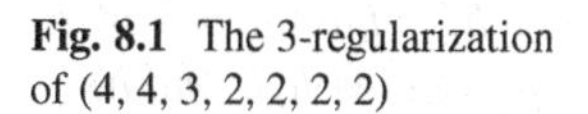
Fig. 8.1 The 3-regularization of (4, 4, 3, 2, 2, 2, 2)

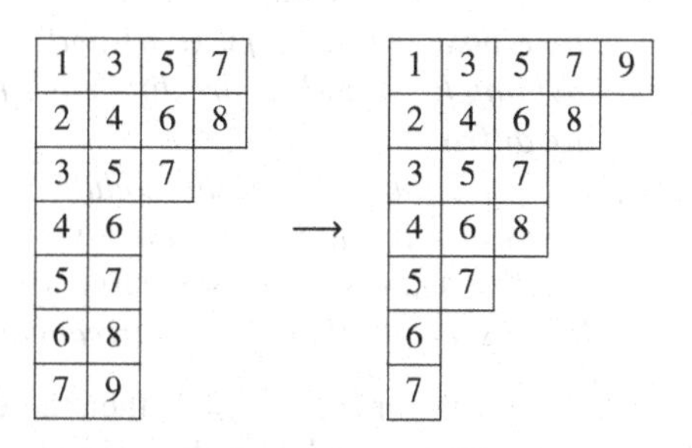

We have already seen in Proposition 8.2.6 that if D^μ is a composition factor of S^λ then μ dominates λ. In [319], James proved a tightening of this result to something stronger when λ is not p-regular.

Theorem 8.2.8 *If λ is a partition of n, then the composition factors of S^λ are D^μ for $\mu \trianglerighteq \lambda^R$, and D^{λ^R} appears with multiplicity* 1.

Thus if S^λ is irreducible then $S^\lambda = D^{\lambda^R}$. We deal with the case where $S^\lambda = D^\lambda$ first: Carter proposed a necessary and sufficient criterion for $S^\lambda = D^\lambda$ (for λ a p-regular partition of course), which was that each pair of hook lengths in each column have the same power of p dividing them. James proved this for $p = 2$ in [323], and for p odd he proved one half of the conjecture, that if some column of λ contains hook lengths with two different powers of p in it then S^λ is reducible, with the remaining part of the conjecture being proved by James and Murphy a year later in [330].

For λ an arbitrary partition though, it took another 20 years for progress to be made. Of course, S^λ is irreducible if and only if $S^{\lambda'}$ is (where λ' is the conjugate of λ) by Proposition 8.2.5, so we may assume that neither λ nor its transpose is p-regular in any description of the answer.

A complete solution was achieved by James and Mathas for $p = 2$ in [329]: if neither λ nor λ' is 2-regular, and S^λ is irreducible, then $\lambda = (2, 2)$. They gave a general conjecture, which is produced in [424, Conjecture 5.47], and is reproduced a bit more simply in [208, 407]. It states that S^λ is reducible if and only if there exists a box (i, j) in λ for which p divides the hook length $h_{i,j}$, and there are also two more boxes, one in the same row as (i, j) and one in the same column as (i, j), and their hook lengths do not have the same power of p dividing them as $h_{i,j}$.

If λ is p-regular then the box on the end of any row does not have a hook length divisible by p, so the condition on having a box in the row with a different power of p dividing its hook length is always fulfilled; thus for p-regular partitions this is simply Carter's conjecture above. Fayers [208], building on work of Lyle [407], proved that the condition on hook lengths implies reducibility. The converse was proved a short while later by Fayers in [209], so we have the following theorem.

Theorem 8.2.9 (James, James–Murphy, James–Mathas, Lyle, Fayers) *Let λ be a partition of n.*

(i) *If p is odd then the Specht module S^λ is reducible if and only if there are three boxes in λ, (i, j), (i, a) and (b, j), such that $p \mid h_{i,j}$, the powers of p dividing $h_{i,j}$ and $h_{i,a}$ are different, and the powers of p dividing $h_{i,j}$ and $h_{b,j}$ are different.*
(ii) *If $p = 2$ then the Specht module S^λ is reducible if and only if there are two boxes in λ in the same column with different powers of 2 dividing their hook lengths, and two in the same row with different powers of 2 dividing their hook lengths, except that $S^{(2,2)}$ is irreducible.*

If $S^\lambda \cong D^\mu$ for some $\mu \neq \lambda$, then there is a (non-zero) homomorphism from S^μ to S^λ. In general, understanding $\mathrm{Hom}(S^\lambda, S^\mu)$ (in positive characteristic!) should offer considerable insights into the structure of Specht modules.

As we saw above, Carter and Lusztig proved that the dimension of $\mathrm{Hom}(S^\lambda, M^\mu)$ is equal to the number of semistandard tableaux of shape λ and type μ for odd primes p. For $p = 2$ and λ a 2-regular partition, James proved the same result, as we mentioned above. For λ not 2-regular, this number only gives an upper bound, as the set of 'semistandard homomorphisms' spans, but is not always a basis for, the Hom-space [213, Lemma 2]. As S^μ is a summand of M^μ, this gives a (large) upper bound for the dimension of $\mathrm{Hom}(S^\lambda, S^\mu)$.

Carter and Payne in [109] gave a criterion for $\mathrm{Hom}(S^\lambda, S^\mu)$ to be non-zero. Let μ be obtained from λ by moving s boxes from the ath row of λ to the bth row. In other words, $\mu_i = \lambda_i$ for $i \neq a, b$, $\mu_a = \lambda_a - s$ and $\mu_b = \lambda_b + s$. If, for some $d > 0$ we have that $s < p^d$ and

$$\lambda_b - \lambda_a + a - b \equiv s \bmod p^d,$$

then $\mathrm{Hom}(S^\lambda, S^\mu) \neq 0$. In terms of general results like this, the most useful is a row-removal result: the original row-removal result is Theorem 8.3.9 in the next section, but in this section we have the following result of Fayers and Lyle [212].

Theorem 8.2.10 *Suppose that p is an odd prime, and let $\lambda = (\lambda_1, \lambda_2, \dots, \lambda_r)$ and $\mu = (\mu_1, \mu_2, \dots, \mu_s)$ be two partitions of n. Suppose that, for some $0 < i < \min(r, s)$, the sum of the first i parts of λ and μ are the same, and write $\bar{\lambda}$ and $\bar{\mu}$ for the partitions with the first i parts removed. We have that*

$$\mathrm{Hom}(S^\lambda, S^\mu) \cong \mathrm{Hom}(S^{\bar{\lambda}}, S^{\bar{\mu}}).$$

The same result holds upon removing the first i columns from λ and μ, with the analogous condition on the sizes of the columns.

Apart from some special cases, such as when μ is a hook partition, which for the partition (n) is considered in [324, Theorem 24.4], and in general is computed by Loubert in [398], and a general albeit fairly unwieldy algorithm to compute the Hom-space for fixed λ and μ in [324, p. 102], little else is known.

The previous topic, computing homomorphisms between Specht modules, is boring in characteristic 0 and hard in characteristic p, and the same is true of the next topic, although at least this one has been solved. In characteristic 0, we saw back in Example 8.1.2 that taking the tensor product of χ^λ by the 1-dimensional sign character $\chi^{(1^n)}$ yields the character $\chi^{\lambda'}$, where λ' is the conjugate of λ.

In Proposition 8.2.5, we said that in characteristic p the effect on S^λ of tensoring by the sign representation is $S^\lambda \otimes \mathrm{sgn} = (S^{\lambda'})^*$. The remaining question is: what is $D^\lambda \otimes \mathrm{sgn}$? It must be D^μ for some μ, since it is irreducible, but what is μ? As a quick example, even sgn is a simple module, so $\mathrm{sgn} = D^\mu$ for some μ. In this case, the answer is given in Theorem 8.2.8, as $S^{(1^n)} = D^\mu$, and we know that $S^{(1^n)}$ has the regularization $D^{(1^n)^R}$ as a constituent, so $\mu = (1^n)^R$.

For $p = 3$, Mullineux conjectured an algorithm that computes the bijection induced by tensoring by the sign character in [443]. He came up with the general case, for all p, in [442]: although this appeared before [443], it was submitted a month after, so the timing does make sense. The bijection itself is a bit of a mess to write down; one can read Mullineux's original paper for a description of it, but since then a few different approaches for computing the bijection have appeared, unfortunately none particularly easy.

In 1996, Kleshchev produced a (verified, not conjectural) algorithm for computing the partition labelling the simple module $D^\lambda \otimes \mathrm{sgn}$, but the algorithm was not Mullineux's. A year later though, Ford and Kleshchev proved [233] that the two algorithms gave the same answer, proving Mullineux's conjecture at last, although quite indirectly.

There have been some improvements in the method of proof since then, including a third algorithm given by Xu in [571], which was used by Brundan and Kujawa to give a different proof of the Mullineux conjecture in [81].

Another topic that is easy in characteristic 0, with a nice, simple answer, but a real headache in characteristic p, is the branching rule. Of course, branching for the Specht modules remains as before, Theorem 8.1.6 (at least in terms of Brauer characters, although the induction of a Specht module does have a filtration by Specht modules that is independent of p [324, Section 17]), so we mean branching for the simple modules.

Branching can either be thought of on the level of Brauer characters (or composition factors of the restriction) or can be thought of as the module structure of the restriction, but both questions are in a very real sense too hard. If we could understand the composition factors of $D^\lambda \downarrow_{S_{n-1}}$ then this yields by induction the Brauer character values of D^λ on all (p-regular) conjugacy classes of S_n

that have a fixed point. This doesn't quite look like enough to determine the decomposition matrix, but actually it is a theorem of Kleshchev [361] that knowing the decomposition numbers for symmetric groups and knowing the composition factors of the restrictions of simple modules to S_{n-1} are equivalent problems. (The extra piece of information that allows you to bridge the gap is the fact that the distribution of simple modules into blocks is very rigid.)

What is known about branching is largely a result of the work of Kleshchev. The first suggestions that something interesting could be said, at least in characteristic 2, were made by Benson [32]. For example, he conjectured that $D^\lambda \downarrow_{S_{n-1}}$ is irreducible if and only if all of the parts of λ have the same parity: as an example, he gives the fact that $D^{(8,6,2)}$ restricts to $D^{(7,6,2)}$. He went further, and actually made the following conjecture: if $\lambda = (\lambda_1, \ldots, \lambda_r)$ is 2-regular, then $D^\lambda \downarrow_{S_{n-1}}$ is semisimple if and only if λ_{2i} and λ_{2i+1} are congruent modulo 2 for all i, and he gives a precise decomposition into irreducibles, namely the sum of $D^{\lambda'}$, where λ' ranges over all partitions of $n-1$ obtained from λ by removing a box from the ith row, whenever $\lambda_{i-1} \not\equiv \lambda_i$ modulo 2. (In other words, there is one extra composition factor every time the congruence of the λ_i modulo 2 changes.) This isn't quite right, though: as we will see later there are only two summands to the restriction, although the criterion for semisimplicity is correct. This sort of conjecture suggests that some removable nodes in Young diagrams are more special than others, something that Kleshchev made formal in his papers on the subject.

How bad can the branching rule get in general? James produced a branching result back in 1976 [319] that showed that the multiplicities of composition factors in the restriction of simple modules from S_n to S_{n-1} can (and do) take each non-negative integer value. James showed that if λ is p-regular and A is the highest removable node such that $\lambda_A = \lambda \setminus \{A\}$ is p-regular, then the constituents of $D^\lambda \downarrow_{S_{n-1}}$ are D^μ for μ dominating λ_A, and D^{λ_A} appears with multiplicity exactly equal to the number of removable nodes of λ that are not below A (i.e., above A and including A). In [334], Jantzen and Seitz reprove that the multiplicities are unbounded, and also give a suggestion for when the restriction $D^\lambda \downarrow_{S_{n-1}}$ is irreducible for odd primes. They prove the following.

Theorem 8.2.11 *Let $\lambda = (\lambda_1^{a_1}, \ldots, \lambda_r^{a_r})$ be a p-regular partition of n with $\lambda_i > \lambda_{i-1}$ for all i. If, for all $i \leq r-1$, $\lambda_i + a_i + a_{i+1} \equiv \lambda_{i+1} \bmod p$, then $D^\lambda \downarrow_{S_{n-1}}$ is irreducible, and is equal to D^{λ_A}, where A is the highest removable node of λ (i.e., in row a_1).*

They then conjectured that these are all modules with irreducible restriction, which encapsulates Benson's conjecture on irreducible restrictions above. This was proved by Ford and Kleshchev in a series of three papers [232, 355, 356].

Benson's conjectures, and the result above, focus our attention on specific types of removable node. We fix a prime p as we are only considering positive characteristic now. If $A = (i, j)$ is a node, then its *residue* is $(j - i) \bmod p$. Often people write *c-node* for a node of residue c, signifying how important the residue of a node is. A removable c-node A in row i of a partition λ is *normal* if, starting from row $i - 1$ and moving upwards in the Young diagram, at no point do we count more addable c-nodes than removable c-nodes. Another way to say this is that, in the rows $1, \ldots, i-1$, to every addable c-node we can associate a removable c-node that is below it, and distinct addable c-nodes have distinct removable c-nodes associated to them. A removable node is *good* if there are no normal nodes below it in the Young diagram with the same residue. (Exercise 8.4 is to show that every p-regular partition has at least one good node, and that if λ is p-regular and A is a good node, then $\lambda \setminus \{A\}$ is always p-regular.)

In [357], Kleshchev proved that $\mathrm{soc}(D^\lambda \downarrow_{S_{n-1}})$ is the sum of all $D^{\lambda\setminus\{A\}}$ as A ranges over all good nodes. In addition, if λ is a p-regular partition of n and μ a p-regular partition of $n-1$, then $\mathrm{Hom}_{kS_{n-1}}(S^\mu, D^\lambda)$ is 0 unless $\mu = \lambda \setminus \{A\}$ for A a normal node, and is 1-dimensional in this case.

In [358] he pushes the results further, by proving the following.

Theorem 8.2.12 *Let λ be a p-regular partition of n, and let $A_1, \ldots, A_r$ be all good removable nodes of λ. The restriction of D^λ to S_{n-1} has exactly r indecomposable summands*

$$D^\lambda \downarrow_{S_{n-1}} = M_1 \oplus M_2 \oplus \cdots \oplus M_r.$$

Each M_i is self-dual, with socle (and top) $D^{\lambda\setminus\{A_i\}}$, and M_i and M_j belong to different p-blocks of S_{n-1} if $i \neq j$.

This theorem focuses attention on the M_i, the indecomposable summands of the restriction, or equivalently the block decomposition of the restriction. We can label them by good removable nodes, since the socle (and top) of each one is $D^{\lambda\setminus\{A\}}$ for some good removable c-node A, so write M_A for this summand, D_A for the simple module $D^{\lambda\setminus\{A\}}$, and D for $D^\lambda \downarrow_{S_{n-1}}$.

In [359], Kleshchev shows that the multiplicity of D_A in D (or equivalently in M_A) is equal to the number of normal removable c-nodes in λ. Thus we see that M_A is simple if and only if there are no normal c-nodes other than A itself. We may therefore recover a precise condition on when D is semisimple (all normal nodes are good) and then when D is irreducible (exactly one normal node). We also see why Benson's original conjecture was slightly off: there are at most p good nodes (as each has a residue and this is taken modulo p) and so there are at most p summands of D in characteristic p.

We can even say something about the endomorphisms of D: of course, the dimension of $\mathrm{End}_{kS_{n-1}}(D)$ is the sum of the dimensions of $\mathrm{End}_{kS_{n-1}}(M_A)$, since there are no non-zero homomorphisms between modules belonging to different blocks. We can see one non-trivial homomorphism (if M_A is not simple), which sends the top D_A to the D_A in the socle of M_A. Moreover, and what is less clear, for every copy of D_A inside M_A, there is an endomorphism $\phi : M_A \to M_A$ where the image of the top D_A is that copy of D_A. In fact, there is an endomorphism ϕ such that $\phi^m = 0$, where m is the number of normal c-nodes in λ, and $\phi, \phi^2, \ldots, \phi^{m-1}$, together with the identity map, form a basis (as a vector space) for $\mathrm{End}_{kS_{n-1}}(M_A)$ [360]. In particular, we see that our obvious map sending the top of M_A to the socle of M_A is ϕ^{m-1}.

The papers of Kleshchev also give results about the induction of D^λ to S_n where λ is a p-regular partition of $n-1$, and we refer to the summary [361] for what is known about them.

As an example of the power of these rules, they were enough for Sheth, together with the information on the composition factors of the Specht modules $S^{(n-r,r)}$ from Theorem 8.3.7 in the next section, to determine the complete structure of $D^\lambda\downarrow_{S_{n-1}}$ whenever λ is a two-part partition.

We will end this section by talking about some local properties of Specht and simple modules, namely vertices and sources. Since Specht modules are indecomposable, they have vertices, sources and Green correspondents. This is a large area, almost completely unexplored, and progress appears difficult. There seem to be few general results: Wildon [560] proved a lower bound for the order of a vertex of a Specht module by finding a non-trivial p-subgroup (dependent on the particular partition) that always lies in the vertex; this result was improved by Giannelli in [248].

In 2003, Wildon proved [558] that the only Specht modules with cyclic vertex lie in blocks with cyclic defect group, and in the same paper managed to find the vertices of Specht modules for hook partitions if $p\nmid n$. If $\lambda=(n-m,1^m)$ is a hook, $p\nmid n$, and S^λ is indecomposable (so $p\geq 3$ for example) then a vertex of S^λ is a Sylow p-subgroup of $S_{n-m-1}\times S_m$. For the other cases, Müller and Zimmermann in [441] did a few more, then Danz in [149], until the last cases were solved by Danz and Giannelli in [151]. If all that work was needed just for hook partitions, it is of no surprise that little is known in this area.

The simple modules also have vertices of course; if $\dim(D^\lambda)\leq 1000$ then Danz computed the vertex of D^λ in [150], which is usually the defect group (especially if the defect group is abelian by Knörr's theorem, Theorem 3.2.11). For symmetric groups up to degree 14, Danz, Külshammer and Zimmermann determined the vertices of all simple modules [153]. There is still a lot of work to be done in this direction before any sensible theory can emerge though.

8.3 Blocks and Decomposition Numbers of Symmetric Groups

The Murnaghan–Nakayama rule requires us to remove rim hooks from a partition. This is also how we determine the blocks of symmetric groups. If λ is a partition of n, then λ is a *d-core* if λ possesses no rim d-hooks, i.e., rim hooks of length d. The partition obtained from λ by repeatedly removing rim d-hooks from λ until no more can be removed is well defined, and called the *d-core* of λ. (Exercise 8.3 asks you to prove that this is well defined.) The number of rim d-hooks removed from λ to obtain the d-core is called the *d-weight* of λ.

Theorem 8.3.1 (Nakayama Conjecture) *If λ and μ are partitions of n, then χ^λ and χ^μ belong to the same block of kS_n if and only if λ and μ have the same p-core.*

The Nakayama conjecture was made in 1941 by Nakayama in [446], and it was proved by Brauer [52] and Robinson [490] in 1947. This only deals with ordinary characters. However, the same statement holds for Brauer characters: if λ is p-regular, then the simple module D^λ belongs to the same block as χ^λ does. (This follows since D^λ is a composition factor of S^λ, and S^λ has Brauer character the restriction $(\chi^\lambda)^0$ of χ^λ to p-regular elements.)

Since all of the irreducible characters in a p-block B come from partitions with the same p-weight, we say that B itself has that particular weight. The defect group of a block of weight w is isomorphic to a Sylow p-subgroup of S_{pw} (see for example [327, Theorem 6.2.45]), so a direct product of iterated wreath products of the cyclic group C_p. In particular, if $w < p$ then the defect group is elementary abelian.

What about the alternating group? If χ^λ is an ordinary character of S_n, then by Clifford's theorem $\chi^\lambda \downarrow_{A_n}$ is the sum of two irreducible characters if λ is self-conjugate, and remains irreducible otherwise. The situation for blocks is easiest to describe using the block idempotents, and is as follows (the proof is Exercise 8.6).

Theorem 8.3.2 *Let b be a block idempotent of kS_n, and let μ denote the p-core associated to b. If μ is not self-conjugate, let b' denote the block idempotent of kS_n associated to the conjugate partition μ'.*

(i) *If b has defect zero, then b is a p-block idempotent of A_n if μ is not self-conjugate, and if μ is self-conjugate then b is the sum of two block idempotents of kA_n, each of defect zero.*
(ii) *If b has positive defect and μ is not self-conjugate then $b + b'$ is a block idempotent of A_n. In particular, if p is odd then kS_nb and $kA_n(b + b')$ are Morita equivalent.*
(iii) *If b has positive defect and μ is self-conjugate then b is a block idempotent of kA_n.*

In all cases, the defect groups of the blocks of A_n are the intersections of those of S_n with A_n.

At this point there are a few different avenues of inquiry:

- Can we say anything about blocks of small weight?
- Can we relate blocks of the same weight?
- What is known about the decomposition numbers for a given block?

We will explore all these questions. In order to understand how blocks relate to one another, we need the notion of an abacus and a Scopes move. We start with the abacus. If $\lambda = (\lambda_1, \ldots, \lambda_r)$ is a partition of n, then a *β-set* for λ is the set of first-column hook lengths

$$X = \{\lambda_1 + r - 1, \lambda_2 + r - 2, \ldots, \lambda_r\}.$$

We can allow partitions to end with 0s and be 'the same' partition, so that the β-sets $\{4, 2, 1\}$ and $\{5, 3, 2, 0\}$ label the same partition, namely $(2, 1, 1)$. Another way to approach it is to say that a β-set is any finite set of non-negative integers, and place an equivalence relation on β-sets generated by

$$X \sim \{0\} \cup \{i + 1 \mid i \in X\}.$$

In each equivalence class of β-sets there is a unique one that does not contain 0, and these are the first-column hook lengths for the partition corresponding to these β-sets. (Note that $\emptyset$ is the β-set for the empty partition.)

The *abacus* is best defined by example. It consists of p runners, labelled 0 to $p - 1$, and each runner has places on it, labelled by all non-negative integers. If λ is a partition of n, with β-set X, then we place a bead on the ith place of the jth runner if and only if $pi + j$ lies in X. The example is if λ is the partition $(9, 8, 5, 5, 2, 2, 1, 1)$. The β-set with no 0 in it is $\{16, 14, 10, 9, 5, 4, 2, 1\}$, and an equivalent β-set is $\{18, 16, 12, 11, 7, 6, 4, 3, 1, 0\}$. Drawn on a 5-abacus, these two β-sets are given in Fig. 8.2.

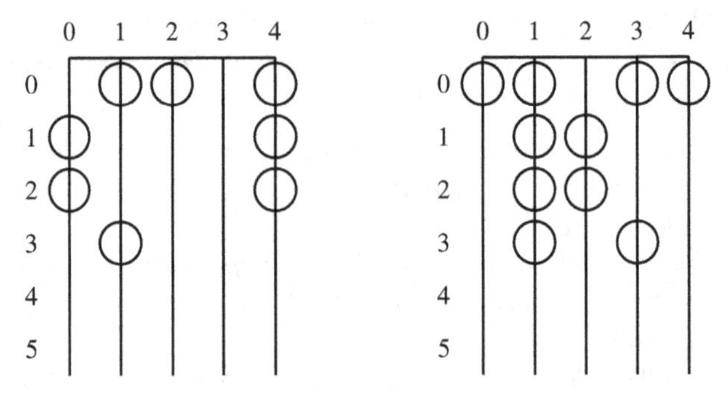

Fig. 8.2 Two equivalent β-sets drawn on the 5-abacus

Some of our results, on the branching rule and the well-definedness of the p-core of a partition, have equivalent statements on the p-abacus.

Removing a removable box from λ is equivalent to subtracting 1 from one of the first-column hook lengths. On the abacus, this is simply moving the corresponding bead one runner to the left (or if it lies on the 0th runner, move it onto the $(p - 1)$th runner and then up a place). More interesting is the removal of a p-hook. On the p-abacus, this is simply moving a bead up a place on its runner. (Exercise 8.3 asks you to prove this.)

A *Scopes move*[1] is swapping two adjacent runners of the abacus. This is a bijection on the set of all β-sets, but it doesn't preserve the equivalence relation

[1] I appear to be the only one to actually call this a Scopes move. Other people seem to call it 'swapping two adjacent runners of the abacus'. I think my name is catchier.

above, so it initially doesn't seem that useful. However, if we replace a β-set X by another equivalent one X', swapping two adjacent runners of X will correspond to swapping two, probably different, adjacent runners of X'. This is also the 'right' way to swap the 0th and $(p-1)$th runners of X, which are of course adjacent, but it is a little confusing to swap them directly, so replace X by an equivalent β-set so that the 0th runner moves to a different one. Thus the set of all p different Scopes moves *is* an invariant of the equivalence class of a β-set, so we might obtain something interesting after all.

Example 8.3.3 Let λ be the partition from the previous example, with the first associated β-set. The result of swapping the 0th and 1st runners of the abacus is given in Fig. 8.3.

Let X be a β-set corresponding to a partition λ of n, and consider swapping the ith and $(i+1)$th runners of the abacus for X to produce a β-set X' of a partition λ'. Suppose that there are l more beads on the ith runner than the $(i+1)$th runner. The first thing to note is that λ' is a partition of $n+l$, and it can be obtained from λ by removing some removable boxes and adding some addable boxes, depending on which beads are on the ith and $(i+1)$th runners.

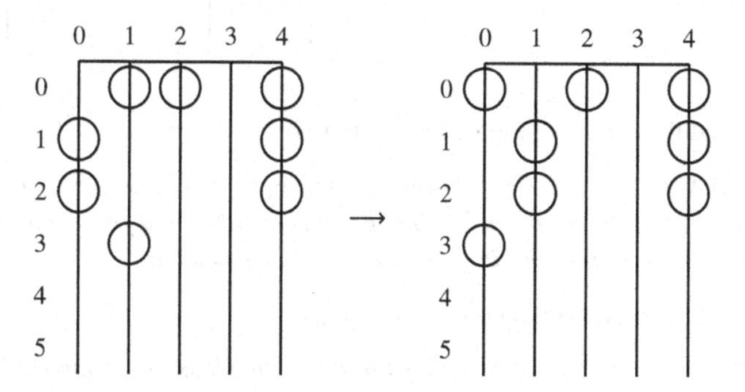

Fig. 8.3 The result of swapping the 0th and 1st runners of the abacus

If λ is a p-core then so is λ', so Scopes moves allow us to move between blocks of S_n and blocks of S_{n+l} for various n and l. In addition, if λ is a p-core, then all of the irreducible ordinary characters in a block of S_{n+pw} of weight w and p-core λ have a β-set obtained from X by moving various beads a single space w times, and one may apply the Scopes move to these β-sets $X_1, \ldots, X_r$, obtaining β-sets $X'_1, \ldots, X'_r$. The partitions with β-sets X'_i are exactly those whose irreducible ordinary character of S_{n+pw+l} belongs to the block labelled by λ', so Scopes moves not only send blocks of weight w to blocks of weight w, they also induce bijections between the irreducible ordinary characters of those blocks.

Write ϕ for this bijection. Using the branching rule, one may check (Exercise 8.7) that if $l \geq w$ then induction from S_{n+pw} to S_{n+pw+l}, when considered just for the blocks in question, yields

$$(S^\mu)\uparrow^{S_{n+pw+l}} = (S^{\mu\phi})^{\oplus l!}$$

for χ^μ in the block of S_{n+pw}, i.e., only $\chi^{\mu\phi}$ from the target block appears in the induction, and it appears exactly $l!$ times. This doesn't work if $l < w$: for example, let $p = 3$, $w = 2$ and $l = 1$, and swap the 1st and 2nd runners with β-set $\{1, 2, 4\}$. If we consider the effect of swapping the runners on the character with β-set $\{2, 4, 7\}$, it becomes the character with β-set $\{1, 5, 8\}$, but induction does not send the one character to the other. (See Fig. 8.4.)

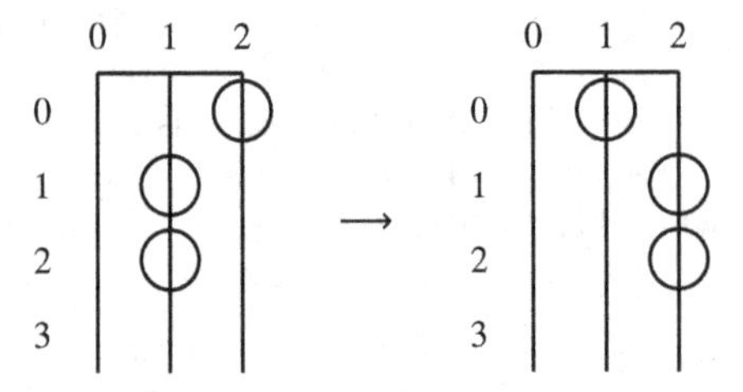

Fig. 8.4 Swapping runners when $l < w$ does not always yield a Morita equivalence

Scopes used this to prove the following in [516].

Theorem 8.3.4 *Let B_1 and B_2 be p-blocks of weight w in S_n and S_{n+l} respectively, and suppose that the partition labelling B_2 may be obtained from that of B_1 by a Scopes move. If $l \geq w$ then B_1 and B_2 are Morita equivalent.*

From this, it is easy to deduce the following.

Corollary 8.3.5 *Donovan's conjecture holds for the class of symmetric groups.*

If $w = 1$ then all Scopes moves are Morita equivalences, and since every p-core can be connected to any other by a series of Scopes moves (to see this, notice that there is always a Scopes move that reduces the size of the partition, so there is a series of Scopes moves that relate any partition to the empty partition) we see that all blocks with cyclic defect groups are Morita equivalent. We know this from the theory of Chap. 5, since all characters of S_n are real so all blocks are lines, but it is nice to see the same results proved using a more general theory.

This forms the basis for the Chuang–Rouquier proof of Broué's conjecture for symmetric groups in 2008 [115]. Unfortunately it is a bit outside the scope of this book as the details can get a bit technical, but we can say something about it. Induction from S_{n-1} to S_n can be thought of as taking a partition λ and mapping it to a linear combination of all possibilities for adding an addable node to λ. Since we have the idea of an addable i-node, we can define i-induction, often denoted $E_{i,n}$, from $\mathrm{Irr}(S_{n-1})$ to $\mathrm{Irr}(S_n)$, which takes S^λ to $\bigoplus_A S^{\lambda\cup\{A\}}$, where A runs over all addable i-nodes of λ. We can similarly break up restriction into its component

i-restrictions $F_{i,n}$. Exercise 8.5 asks you to prove that if $\lambda \cup \{A_1\}$ and $\lambda \cup \{A_2\}$ have the same p-core then A_1 and A_2 have the same residue, so i-induction interacts well with the block decompositions of $\mathrm{Irr}(S_n)$. Writing $E_i = \bigoplus_{n\geq 1} E_{i,n}$ and $F_i = \bigoplus_{n\geq 1} F_{i,n}$, the operators E_i and F_i for $i \in \mathbb{F}_p$ act on the set of all partitions, and actually if we take all formal finite k-linear combinations of partitions, we get an action of the affine Lie algebra $\widehat{\mathfrak{sl}_p}$. Kashiwara showed in [340] that the affine Weyl group acts on the 'crystal graph', and from [386] we see that this yields a transitive action of the Weyl group of $\widehat{\mathfrak{sl}_p}$ on all p-blocks of all symmetric groups of a fixed weight.

From here, the theory of $\mathfrak{sl}_2$-categorification, developed by Chuang and Rouquier in [115], proves that the action of a simple reflection of this Weyl group on a p-block induces a derived equivalence between it and its image. This shows that all algebras in the same orbit of the Weyl group are derived equivalent, but that means all p-blocks of a fixed weight in all symmetric groups are derived equivalent. (This includes those weights w that are at least p, where the p-blocks have non-abelian defect group.)

To get from this to Broué's conjecture, one needs to know that all of the Brauer correspondents of all of these blocks are derived equivalent, but this is fairly easy (they are Morita equivalent), and one needs to prove Broué's conjecture for one block of weight w for each $w < p$. This was done for a so-called RoCK block (named after Rouquier, Chuang and Kessar). Conjectured by Rouquier and proved by Chuang and Kessar in [113], the statement is: the RoCK block is Morita equivalent to the principal block of the group $S_p \wr S_w$; this is derived equivalent to the group algebra of $(C_p \rtimes C_{p-1}) \wr S_w$ (the normalizer of a Sylow p-subgroup of $S_p \wr S_w$); this is Morita equivalent to the Brauer correspondent of the original block of S_n.

So what is a RoCK block? It is a block B with p-core λ, such that λ has an abacus representation with $w - 1$ more beads on runner $i + 1$ than runner i for all $0 \leq i \leq p - 1$. (For $p = 3$ and $w = 2$, this is the partition $(3, 1, 1)$, so the RoCK block is in S_{11}.) Looking at Scopes's theorem above, we see that the RoCK block is the block of the largest symmetric group S_n such that there is no Scopes move to a block of a smaller symmetric group S_{n-l} with $l \geq w$ (so it would be a Morita equivalence). (There is a more flexible definition of RoCK block, where it is any block with *at least* $w - 1$ more beads on successive runners, which can be more useful in practice, but the definition here means that it is *the* RoCK block, not *a* RoCK block.)

When $p > w$, the RoCK block is Morita equivalent to the principal block of $S_p \wr S_w$, but when the defect group is non-abelian, i.e., $p \leq w$, this is no longer the case. A technical-looking conjecture as to its structure was made by Turner in [547]; in 2017, Evseev proved [202] part of this conjecture, that the RoCK block B, when 'cut' by an idempotent f, in other words, the algebra fBf, becomes Morita equivalent to the principal block of $S_p \wr S_w$, even for $w \geq p$. Along with Kleshchev, he also gave a proof in [203] of the full-fat version of Turner's conjecture, i.e., an explicit description of an algebra that is Morita equivalent to B itself, not just fBf.

The description is too complicated to go into here, but gives some hint as to how difficult the non-abelian theory might be.

We can bounce around the blocks of a given weight using only derived equivalences, but derived equivalences definitely do not preserve decomposition matrices, so if we want information about decomposition numbers then we need other techniques. It is not clear at the moment whether $\mathfrak{sl}_2$-categorification can be used to provide an algorithm to enable us to determine decomposition matrices: the idea would be that, if you knew the one decomposition matrix, and you are given a derived equivalence with another, to use information about that derived equivalence to compute the decomposition matrix of the target block. Nobody has done this yet, but in future this might give a recursive algorithm for computing decomposition numbers.

The first attempt that I know of to write a general algorithm for computing decomposition numbers for symmetric groups is by Robinson in the last couple of chapters of [491], but as Kerber and Peel in [342], and later James in [318, 320] pointed out, there are errors in the method that mean that it cannot work in general.

The most promising avenue for computing decomposition matrices in a general way is James's conjecture. I have tried to avoid the mention of other algebras in this chapter, but now I'm going to have to bring one out. The *Hecke algebra* $\mathcal{H}_{k,q}(S_n)$ is an algebra dependent on $n \in \mathbb{N}$, k a field, and $q \in k$ a parameter. The smallest positive integer e such that $1 + q + \cdots + q^{e-1} = 0$ is called the *quantum characteristic*, setting $e = \infty$ if no such e exists. We refer to [424] for the definition of a Hecke algebra, and also for the theory we will mention now. To each partition λ of n we may associate a Specht module S^λ of $\mathcal{H}_{k,q}(S_n)$, and this yields decomposition matrices for any field k and any q. (An analogue of Nakayama's conjecture holds, and the simple modules are parametrized by e-regular partitions, so we can label our decomposition matrices in the same way as for S_n and the blocks are labelled by the same rows and columns.)

If $q = 1$ then $\mathcal{H}_{k,q}(S_n)$ is the symmetric group algebra kS_n (and $e = \operatorname{char} k$ in this case), so understanding the decomposition numbers of the Hecke algebras is a more general question. If $k = \mathbb{C}$ and the quantum characteristic is e then an algorithm exists, called the LLT algorithm (after Lascoux et al. [386]), which was conjectured to produce the decomposition numbers, and proved to do so by Ariki in [20]. (Grojnowski also announced a proof of this in the mid-1990s, but it has not yet appeared.) Of course, what we want is the decomposition matrix of kS_n for $q = 1$.

It turns out by a result of Geck that if D denotes the decomposition matrix of a block of $\mathcal{H}_{k,q}(S_n)$ (quantum characteristic e) and D_0 denotes the decomposition matrix of the 'same' (as in the same e-core labelling it) block of $\mathcal{H}_{\mathbb{C},\zeta}(S_n)$ (for ζ a primitive eth root of 1) then there is a matrix A with non-negative integer entries such that $D = D_0 A$. The matrix A, as it performs the adjustment from the case $k = \mathbb{C}$ to the case $k = \mathbb{F}_p$ (but importantly with the same quantum characteristic), is called the *adjustment matrix*.

Since D_0 is computable by LLT, we have that D and A are 'equally' difficult to compute. James, working from the decomposition matrices of $\mathrm{GL}_n(q)$ and S_n for $n \leq 10$, was led to conjecture the following in [326, Section 4].

Conjecture 8.3.6 (James's Conjecture) If B is a block of $\mathcal{H}_{k,q}(S_n)$ with weight less than p, then the adjustment matrix for B is the identity matrix.

There is something that should be said about this conjecture: it is false. However, it would be hard to directly find a counterexample, because the known ones are so vast computers will not handle them. If the block has weight 1 then the Brauer tree is a line (we will explicitly determine it later in this section), and this is the same as for the Hecke algebra, so James's conjecture is true for $w = 1$. If $w = 2$ (and so $p \geq 3$) then everything is known about the decomposition matrix of a block of kS_n, and indeed the Hecke algebra, by the work of Scopes [517] and Richards [475], and so James's conjecture can be verified directly. For blocks of weight 3 (and $p \geq 5$), Fayers proved James's conjecture holds [210], and then proved that the decomposition numbers for such blocks of Hecke algebras (and therefore the symmetric groups) are at most 1. For $p = 2, 3$ (outside the range of James's conjecture) the decomposition numbers can be equal to 2, as the principal blocks of S_6 and S_{11} have entries equal to 2 for $p = 2$ and $p = 3$ respectively. (In particular, the entries for $D^{(6)}$ in $S^{(4,1^2)}$, and $D^{(11)}$ in $S^{(8,2,1)}$, are equal to 2.)

Fayers managed to prove that James's conjecture also holds for weight 4 blocks [211]. For weight 5, the prime must be at least 7, so James's conjecture holds for $n \leq 34$. The order of S_{35} is roughly 10^{40}, and this is only just the first case not covered by these general results, not necessarily a counterexample.

The counterexamples appear because the Lusztig conjecture is also false. This will be talked about more in the next chapter, but Williamson [563] proved that the dimensions of representations of algebraic groups (like the dimensions of simple modules for symmetric groups) are not as 'nice' as previously expected. Counterexamples to Lusztig's conjecture for GL_n translate through to counterexamples to James's conjecture for symmetric groups, although the counterexample that is generated using Williamson's method in [563] is the principal 2237-block of weight 780 of $S_{1744860}$. The smallest known counterexample to James's conjecture is now a weight 561 block of S_{467874} in characteristic 839. I don't think this is the smallest counterexample, but we are far above the practical computational range to be able to actually check this.

So what is going wrong? The analogy is that the 'awfulness' is growing exponentially, but from a very small base, and the degree of the symmetric group is growing linearly, so it takes a while but eventually the awfulness catches up with the group and overwhelms it. And then the adjustment matrix becomes more complicated.

Is there any hope? Chuang, Miyachi and Tan in [114] offer a perhaps Panglossian view that the awfulness is confined to small parts of the decomposition matrix. Nobody is sure whether this is still too optimistic, or whether the awfulness can be kept at bay in the matrix A. I would be incredibly wary of guessing the large-scale behaviour of these things, given that we know that the most attractive ideas are wrong, both in terms of Lusztig's conjecture from the next chapter, and Guralnick's conjecture from Chap. 3. It is of course fun that things are much more complicated than we first thought, but at least right now the situation looks perhaps a little daunting.

If general (quick) algorithms look altogether too hopeful, what about computing particular decomposition numbers? The statement that if D^μ is a composition factor of S^λ then μ dominates λ (Proposition 8.2.6) is a powerful one for certain λ: for example, if λ has at most m parts then so does μ.

Since there is only one one-part partition, the first case that might be approached in this way is two-part partitions. In 1976, James determined the modular constituents of Specht modules $S^{(n-a,a)}$, which must be $D^{(n-b,b)}$ for some $b \leq a$. As in [318] and [324, Definition 24.12], if a and b are non-negative integers, then write

$$a = a_0 + a_1 p + a_2 p^2 + \cdots + a_r p^r, \quad b = b_0 + b_1 p + b_2 p^2 + \cdots + b_s p^s$$

for the p-adic expansions of a and b, i.e., $0 \leq a_i, b_i < p$ for all i. Say that a contains b to the base p if $s < r$ and each b_i is either 0 or equal to a_i. The next result was proved by James in [320], although only stated there for $p = 2$, and noted in [318] that the same proof works for all primes. (See also [324, Theorem 24.15].)

Theorem 8.3.7 *Let n, a, b be integers. The simple module $D^{(n-b,b)}$ is a composition factor of $S^{(n-a,a)}$ if and only if $n-2b+1$ contains $a-b$ to the base p, in which case it appears exactly once as a composition factor.*

This wasn't the first class of partitions whose Specht modules had been decomposed into irreducibles though. A few years earlier Peel [464] determined the decomposition numbers for S^λ when λ is a hook partition $(n-m, 1^m)$, at least over fields of odd characteristic.

Theorem 8.3.8 *Let $\lambda = (n-m, 1^m)$ be a partition of n, and let p be an odd prime.*

(i) *If $p \nmid n$ then the Specht module S^λ is irreducible.*
(ii) *If $p \mid n$ and $m = 0$ or $m = n-1$, then S^λ is irreducible (and 1-dimensional).*
(iii) *If $p \mid n$ and $1 \leq m < n-1$, then S^λ possesses exactly two composition factors: the regularizations D^{λ^R} and D^{μ^R} for $\mu = (n-m+1, 1^{m-1})$.*

For $p = 2$ things are different, see [324, p. 93]. Of course, $S^{(n-1,1)}$ is irreducible for n odd and has composition factors $D^{(n-1,1)}$ and $D^{(n)}$ if n is even, either an easy exercise or using Theorem 8.3.7. The module $S^{(n-a,1^a)}$ is always reducible for $a > 1$. In fact, $S^{(n-a,1^a)}$ has the same composition factors as the sum

$$S^{(n-a,a)} \oplus S^{(n-a+2,a-2)} \oplus S^{(n-a+4,a-4)} \oplus \cdots.$$

There have been a few other attempts at understanding special classes of partitions. After two-part partitions come three-part partitions. James determined the composition factors of $S^{(n-m,m-1,1)}$ for $p = 2$ in [320], and for p odd they were determined by To Law [545]. In 2006, Williams [562] produced the decomposition numbers for all three-part partition Specht modules whose third part is less than p.

A couple of general tools are available for computing decomposition numbers for symmetric groups. The first was proved by James in [325], and was referred to in the previous section as a row-removal result. We saw Theorem 8.2.10 then, and mentioned that there was an original version for decomposition numbers. It was conjectured by James in a remark after [324, Corollary 24.21], and proved a couple of years later.

Donkin [168] generalized James's row and column removal theorems to allow the removal of multiple rows, and it is this version that we now give. (James's result is $i = 1$ in this theorem.)

Theorem 8.3.9 *Let $\lambda = (\lambda_1, \lambda_2, \dots, \lambda_r)$ and $\mu = (\mu_1, \mu_2, \dots, \mu_s)$ be two partitions of n. Suppose that, for some $0 < i < \min(r, s)$, the sum of the first i parts of λ and μ are the same, and write $\bar{\lambda}$ and $\bar{\mu}$ for the partitions with the first i parts removed. We have that*

$$[S^\lambda : D^\mu] = [S^{\bar{\lambda}} : D^{\bar{\mu}}],$$

where $[S^\lambda : D^\mu]$ denotes the multiplicity of D^μ as a composition factor of S^λ.

The same result holds upon removing the first i columns from λ and μ, with the analogous condition on the sizes of the columns.

Of course, the similarity between this and Theorem 8.2.10 should not go unnoticed. Using Kleshchev's branching rule, Theorem 8.2.12, James and Williams produce 'node removal' rules [331], but these are a little more complicated, so we refer to their paper.

At least once you have found a row of the decomposition matrix, you know that row can never appear again. Unlike, for example, in blocks with cyclic defect groups (Chap. 5) and tame blocks (Sect. 6.2), if p is odd then no two rows of the decomposition matrix are the same, by a result of Wildon [559]. If $p = 2$, then the only rows that coincide are those corresponding to a partition and its conjugate.

We must also talk about Schaper's formula, which appeared in an unpublished form in [508], and Schaper never officially published anywhere. The first published account that I know of is in [32], and we will talk about the result as it appears there.

In the previous section, we placed a bilinear form on a Specht module S^λ, and declared D^λ to be the quotient by the radical in characteristic p. This is the same as saying, over $\mathbb{Z}$, that D^λ is the quotient by the set of $u \in S^\lambda$ such that, for all $v \in S^\lambda$, (u, v) is divisible by p.

Working over $\mathbb{Z}$ now, we define the ith *Schaper submodule* to be the submodule

$$S^\lambda_{(i)} = \{u \in S^\lambda \mid p^i | (u, v) \text{ for all } v \in S^\lambda\}.$$

We then reduce each submodule $S^\lambda_{(i)}$ modulo p to obtain modules $\overline{S^\lambda_{(i)}}$ and consider their successive quotients, referred to as *Schaper layers*. As we just said, $\overline{S^\lambda_{(0)}}/\overline{S^\lambda_{(1)}}$ is either 0 or D^λ. Schaper's formula gives us the Brauer character of the sum of all of the $\overline{S^\lambda_{(i)}}$ as a $\mathbb{Z}$-linear combination of characters χ^μ for μ strictly dominating λ. Of course, that isn't quite what we would like, but then if it just gave us $\overline{S^\lambda_{(1)}}$, then we would have the decomposition numbers of symmetric groups, so that's asking too much.

Let $\lambda = (\lambda_1, \dots, \lambda_r)$ be a partition of n, with an associated β-set $X = \{x_1, \dots, x_r\}$ (with $x_i > x_{i+1}$). Write $\chi^\lambda_{(i)}$ for the Brauer character of $\overline{S^\lambda_{(i)}}$. Let $1 \le a < b \le r$ and let c lie between 1 and λ_b. Writing $h^\lambda_{i,j}$ for the hook length of the (i, j) box in λ (so that $x_i = h^\lambda_{i,1}$), consider the β-set $X(a, b, c)$, obtained from X by replacing x_a by $x_a + h^\lambda_{b,c}$, and x_b by $x_b - h^\lambda_{b,c}$. If this is a genuine β-set, i.e., no two elements are repeated, then we can write $\lambda(a, b, c)$ for the corresponding partition. If $X(a, b, c)$ is not a genuine β-set, then write $\lambda(a, b, c) = 0$.

The Schaper formula is the sum of $\chi^{\lambda(a,b,c)}$ as a, b, c range over the possibilities above, but each of these occurs with a multiplicity. First, there is the term $v_p(h^\lambda_{a,c}) - v_p(h^\lambda_{b,c})$, which is a difference in p-adic valuations. (If $p \nmid r$, then define $v_p(p^m r) = m$.) The second is a sign: let $\varepsilon(a, b, c)$ denote the number of elements of X lying in the set

$$\{x_a + 1, \dots, x_a + h^\lambda_{b,c} - 1\} \cup \{x_b - h^\lambda_{b,c} + 1, \dots, x_b - 1\}.$$

Theorem 8.3.10 (Schaper Formula) *Let $\lambda = (\lambda_1, \dots, \lambda_r)$ be a partition of n. Then*

$$\sum_{s>0} \chi^\lambda_{(s)} = \sum_{1\le a<b\le r} \sum_{c=1}^{\lambda_b} (-1)^{\varepsilon(a,b,c)} \cdot \left(v_p(h^\lambda_{a,c}) - v_p(h^\lambda_{b,c})\right) \cdot \chi^{\lambda(a,b,c)}.$$

It looks like a mess, but the point is that the right-hand side is labelled by partitions $\lambda(a, b, c)$, each of which strictly dominates λ, and so by induction we may assume that we understand how to break up $S^{\lambda(a,b,c)}$ into simple modules D^μ. The only thing missing from the left-hand side is $\overline{S^\lambda_{(0)}}/\overline{S^\lambda_{(1)}}$, which is simply D^λ. The problem is that we have overcounted each Schaper layer other than $\overline{S^\lambda_{(1)}}/\overline{S^\lambda_{(2)}}$, but note that this formula gives you upper bounds for $[S^\lambda : D^\mu]$, and $[S^\lambda : D^\mu] > 0$ *if and only if* D^μ appears in the Schaper formula.

Example 8.3.11 Let $\lambda = (n - 1, 1)$. The corresponding β-set is $\{n, 1\}$, so for the integers a, b, c in the Schaper formula, we only have $a = 1, b = 2, c = 1$. In this case, $h^\lambda_{2,1} = 1$, and $X(1, 2, 1) = \{n + 1, 0\}$, which is the partition (n). The sign $\varepsilon(1, 2, 1)$ is 0, and the valuation difference $v_p(h^\lambda_{1,1}) - v_p(h^\lambda_{2,1})$ is $v_p(n)$.

If $p \nmid n$ then the sum of the $\chi^\lambda_{(s)}$ is 0, so χ^λ is irreducible. If $v_p(n) = t$, then the sum of the $\chi^\lambda_{(s)}$ is $t \cdot \chi^{(n)}$.

We know that, in reality, $S^{(n-1,1)}$ has two composition factors: $D^{(n-1,1)}$ and $D^{(n)}$. This means that $\chi^{(n-1,1)}_{(0)}/\chi^{(n-1,1)}_{(1)}$ has the character of $D^{(n-1,1)}$, $\chi^{(n-1,1)}_{(t)}$ is the trivial character, and $\chi^{(n-1,1)}_{(s)} = \chi^{(n-1,1)}_{(s+1)}$ for $1 \leq s < t$.

In Exercise 8.8, you will have to use the Schaper formula and column removal to prove the decomposition numbers for Specht modules of two-part partitions given in Theorem 8.3.7. In [207], Fayers did the opposite, and used the Schaper formula, together with the known decomposition numbers, to describe the Schaper layers for two-part and hook partitions, and then understand when the first Schaper layer $\overline{S^\lambda_{(1)}}/\overline{S^\lambda_{(2)}}$ is non-zero (for partitions that are not p-regular).

Normally, however, the Schaper formula is used to get information on decomposition numbers, particularly in characteristic 2, for example in [32] and [438]. Characteristic 2 is nice because S^λ and $S^{\lambda'}$ have the same character, where λ' is the conjugate partition, as we saw in the statement of Wildon's result on the rows of decomposition matrices. In general, using the Schaper formula on S^λ where λ is *not* p-regular can get you a little bit of extra information, which is just what is needed for particular decomposition numbers.

We come to the last topic, that of blocks of small weight, which we will take to mean weight at most 4. If B is a block of defect zero in a symmetric group, then the unique irreducible complex character belonging to B must be χ^λ for λ a p-core. Thus the blocks of defect zero are easy to understand.

Notice that 2-cores are simply triangles: $(a, a-1, a-2, \dots, 2, 1)$. Thus it is easy to count blocks for $p = 2$. For $p = 3$ it is significantly more complicated, but we can state whether or not S_n has a 3-block of defect zero: it has no such 3-block if and only if $3n + 1$ can be written as $m^2 r$ where r is square-free and divisible by a prime congruent to 2 modulo 3 [261, Corollary 2].

For $d \geq 4$, it was a long-standing conjecture that for any n there exists a partition of n that is also a d-core, until a proof was given by Granville and Ono [261]. Thus for $p \geq 5$, for any symmetric group there is always a block of defect zero in S_n. In fact, more should be true, as conjectured by Stanton [531, Conjecture 10].

Conjecture 8.3.12 (Stanton's Monotonicity Conjecture) If $4 \leq d < n - 1$ then the number of partitions of n that are $(d + 1)$-cores is at least the number of partitions that are d-cores.

I proved this for the easy cases $n/2 \leq d < n - 1$ purely combinatorially [124] (Granville–Ono's proof of the existence of d-cores uses modular forms, although Kiming found a proof that avoids them [354]), and there are results for small d [353] and when n is much greater than d [17], but as far as I know nobody has made much headway with the general case.

Having considered blocks of defect zero, i.e., blocks of weight zero, we now consider blocks of weight 1, which have cyclic defect group and hence have a Brauer tree. This is very easy to describe: it is a straight line with p vertices. Of course, since we have labels for the irreducible ordinary and modular characters, we want to actually know the labels for each vertex and edge. That theorem is as follows.

Theorem 8.3.13 *Let B be a p-block of S_n, and suppose that B has weight* 1. *Let $\chi_1, \dots, \chi_p$ denote the irreducible ordinary characters of B. Let λ^i be the partition associated to χ_i.*

(i) *The χ_i can be ordered so that the unique p-hook of λ^i has leg length i. In the remaining parts of this result we will use this ordering.*
(ii) *The Brauer tree of B is a line, with χ_i attached to χ_{i-1} and χ_{i+1}, for all $1 < i < p$.*
(iii) *All λ^i are p-regular except for λ^p.*
(iv) *The label for the edge between χ_i and χ_{i+1} is D^{λ^i}.*

This theorem has been known for a very long time, and it is difficult to know to whom it should be attributed, although probably Robinson, as it already occurs in his book [491, Section 7.1]. Exercises 8.10 and 8.11 ask you to prove this result in two different ways. We give a couple of examples to see this in action now.

Example 8.3.14 Let B be the principal p-block of S_p, which has associated p-core the empty partition $\emptyset$. The ordinary characters in B are χ^λ for λ a hook partition, i.e., $(p-m, 1^m)$ for $m = 0, \dots, p-1$. The Brauer tree is as follows:

$$\chi^{(p)} — \chi^{(p-1,1)} — \chi^{(p-2,1^2)} — \chi^{(p-3,1^3)} — \circ \cdots \chi^{(2,1^{p-2})} — \chi^{(1^p)}$$

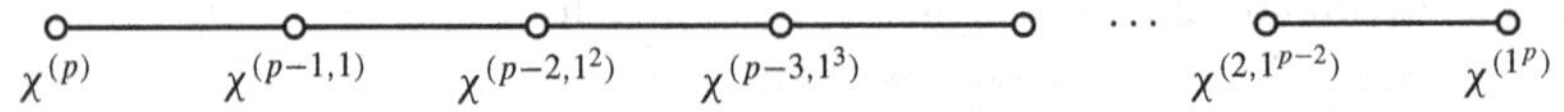

All other partitions of p are p-cores, and so yield blocks of defect zero. If B is the principal p-block of S_{p+1}, with associated p-core (1), then the complex characters belonging to B are labelled by the partitions

$$(p+1), \quad (i, 2, 1^{p-1-i}) \ (2 \le i \le p-1), \quad (1^{p+1}).$$

The Brauer tree is as follows:

$$\chi^{(p+1)} — \chi^{(p-1,2)} — \chi^{(p-2,2,1)} — \chi^{(p-3,2,1^2)} — \circ \cdots \chi^{(2^2,1^{p-3})} — \chi^{(1^{p+1})}$$

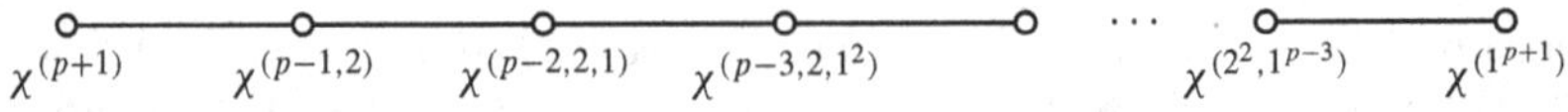

Having dealt with blocks of defect 0 and 1, we would like to move on to weight 2. Here there is good information, mostly produced by Scopes in [517] for odd primes. (For $p = 2$, the defect group is dihedral, so we should look at Sect. 6.2.) Using the Scopes moves given at the start of this section, she was able to give detailed information about a block of weight 2 for a symmetric group.

Theorem 8.3.15 *Let p be an odd prime and let B be a block of weight* 2 *in a symmetric group.*

(i) *The decomposition numbers and dimensions of* Ext^1 *between simple B-modules are either* 0 *or* 1.
(ii) *An irreducible ordinary character in B decomposes into a sum of at most five irreducible Brauer characters.*
(iii) *The diagonal entries in the Cartan matrix are at least* 3, *and the off-diagonal entries are at most* 2.
(iv) *There is a (quick) algorithm to determine the decomposition numbers of B, dimensions of* Ext^1 *between simple B-modules and the socle layers of the projective indecomposable B-modules.*

This result is proved by showing that a Scopes move does not affect the properties in this theorem, and indeed the last part is proved by showing that one may deduce the results for an arbitrary block of weight 2 once one knows them for the principal block of S_{2p}, using a sequence of Scopes moves. Thus we just need to understand the decomposition matrix for S_{2p}. This was done by Robinson in [491, Section 7.2], and is given explicitly in [421, Figure 8]. In addition, Martin proved Theorem 8.3.15 for this block in [420, 421], using among other tools the Schaper formula from earlier in this section. The Scopes moves yield even more information: for example, the author used them to show that the sources of simple modules in blocks of weight 2 are isomorphic to sources of simple modules for either the principal block of S_{2p} or to the RoCK block [129]. The combinatorics of weight 2 blocks was improved by Richards in [475], and a complete description of the decomposition numbers (i.e., not an algorithm that depended on Scopes moves) was given for blocks of weight 2.

On to weight 3. Here, Fayers gave the first complete proof that the decomposition numbers are at most 1 for $p \geq 5$, going via James's conjecture, as we mentioned earlier in this section. Since the decomposition numbers are known to be 0 or 1, for any given block one may use the Schaper formula to give a reasonably fast algorithm to determine the decomposition matrix. For $p = 2, 3$, there are a finite number of blocks to check by Scopes's proof of the Donovan conjecture for symmetric groups, and Fayers and Tan computed the adjustment matrices for these primes in [214], so the decomposition numbers are explicitly known for $p = 2, 3$, and known for $p \geq 5$ using the Schaper formula.

For weight 4, decomposition numbers can be greater than 1, even for $p \geq 5$. Because of this, the Schaper formula cannot always give us perfect information on the decomposition numbers. However, Fayers again proved that James's conjecture holds for such blocks, so we can run the LLT algorithm to find the decomposition numbers, at least.

For blocks of higher weight, little is known beyond what is given above. Of course, the decomposition matrices for S_n are known for some small n. The 2- and 3-decomposition matrices for S_n for all $n \leq 13$ were given in the appendix to James's book [324].

In 1987, Benson, using representations of the double cover $2 \cdot S_n$ of S_n and Schaper's sum formula, managed to produce some more 2-decomposition numbers in [32]: all of them for $n = 14$, and all but one unknown for $n = 15$, which he could show was either 0 or 1.

Müller [438] dealt with the unknown for $p = 2$ and $n = 15$, and went further, completing the 2-decomposition matrices for S_{16} and S_{17}. Nowadays, the decomposition matrices for all primes up to S_{17} are known and available on the *Modular Atlas* website [64], but pushing much further than this looks to be too ambitious.

8.4 The Double Cover of S_n

In Chaps. 4, 5 and 7 we saw that many questions about finite groups can be reduced to questions about quasisimple groups, or at the very least questions about representations of quasisimple groups cannot in general be reduced to questions about the simple quotient. The simple alternating groups, with the exception of A_6 and A_7, all have Schur multiplier of order 2. (The two other cases have Schur multiplier of order 6.) Thus we would like to understand the representation theory of the groups $2 \cdot A_n$, just as we want to understand the representation theory of A_n. One may start by understanding representations of $\tilde{S}_n = 2 \cdot S_n$, the double cover of S_n, instead, but the first problem is that $\tilde{S}_n$ is not well defined, and there are two non-isomorphic groups of the form $\tilde{S}_n$. In many respects these groups have the same behaviour though, so we may treat them as one group. (They are isoclinic, which we saw in Sect. 1.2; isoclinic groups have the same character degrees, as mentioned in [122, Section 6.7], but not the same character table in general.)

Faithful representations of $\tilde{S}_n$ are called *spin representations* of S_n, and their characters are called *spin characters*. (The reason is that the usual permutation representation of S_n embeds S_n into an orthogonal group: the double cover of an orthogonal group is called the spin group, and so we obtain a spin representation of the symmetric group, a faithful representation of $\tilde{S}_n$, from this.) Ideally, we would have a combinatorial theory of spin representations as detailed as that of standard representations, but at the moment it is much less well developed. Much of the early theory is due to Schur and later Morris, who continued the development in a series of papers. We will use [298] as a reference for some basic facts in this theory.

Schur in [513] computed the number of conjugacy classes of $\tilde{S}_n$ and $\tilde{A}_n$, but to express these we need a few definitions. A *strict partition* $\lambda = (\lambda_1, \ldots, \lambda_r)$ of n is a partition of n into distinct parts. The *parity* $\varepsilon(\lambda)$ of a partition λ is the quantity $(|\lambda| - r) \bmod 2$, so $\varepsilon(\lambda) \in \{0, 1\}$. Call λ *even* if $\varepsilon(\lambda) = 0$, and *odd* if $\varepsilon(\lambda) = 1$. It is an easy exercise to see that a strict partition is even if and only if it has an even number of parts of an even size, and this is the same condition as for an element of S_n with cycle type λ to be an even element.

If χ is a character of $\tilde{S}_n$ then so is the tensor product $\chi \otimes \mathrm{sgn}$ with the sign character, and we either have that $\chi \otimes \mathrm{sgn} = \chi$ or $\chi \otimes \mathrm{sgn} \neq \chi$.

Theorem 8.4.1 (Schur) *Given a strict partition λ of n, there exists a unique irreducible spin character χ_λ of S_n with the following properties:*

(i) *If λ is even then $\chi_\lambda = \chi_\lambda \otimes \mathrm{sgn}$.*
(ii) *If λ is odd then $\chi_\lambda \neq \chi_\lambda \otimes \mathrm{sgn}$.*
(iii) *The irreducible characters χ_λ for λ an even strict partition, and χ_λ and $\chi_\lambda \otimes \mathrm{sgn}$ for λ an odd strict partition, yield a complete set of irreducible spin characters of the symmetric group S_n.*
(iv) *The restriction of χ_λ to $\tilde{A}_n$ is irreducible if λ is odd, and splits as the sum of two distinct irreducible characters if λ is even. All irreducible spin characters of A_n appear in this way.*

(This may be found in [298, Theorem 8.6].) The character $\chi_\lambda \otimes \mathrm{sgn}$ is called the *associate character* of χ_λ. Schur also produced some information on the character values in [513]. In particular, for even λ, $\chi_\lambda(x) = 0$ whenever the image of x in S_n has an even cycle in its cycle type. If λ is odd then the same statement holds, except $\chi_\lambda(x)$ can also be non-zero if the image of x has cycle type λ itself. In this case,

$$\chi_\lambda(x) = \pm \mathrm{i}^{(n-r+1)/2} \cdot \frac{\sqrt{\prod \lambda_i}}{2}, \tag{8.1}$$

where $\lambda = (\lambda_1, \dots, \lambda_r)$ and $|\lambda| = n$ (see [298, Theorem 8.7]). An analogue of the Murnaghan–Nakayama rule for the spin characters can be used to compute the value of $\chi_\lambda(x)$ for x a product of odd cycles, which are the only cases whose values are left open from the above remarks. There is an explicit description of this in [298, Theorem 10.1], based on [433]. (Note that, because there are two groups $\tilde{S}_n$, there are two different character tables! They coincide for some elements, but not others, and the orders of the elements of the two groups are different. For example, one of the groups $\tilde{S}_5$ has a non-central element of order 2, and one of the groups does not.)

We can also give the degree of an irreducible spin representation χ_λ, in a formula that looks almost exactly like the hook formula, Theorem 8.1.3. To give this precisely, we need to produce the 'shifted' Young diagram. The usual Young diagram has all of the rows lined up on the left-hand side. The *shifted Young diagram* is obtained from the Young diagram by shifting the ith row by i boxes to the right. Thus the unshifted and shifted Young diagrams of $(7, 6, 4, 2)$ are

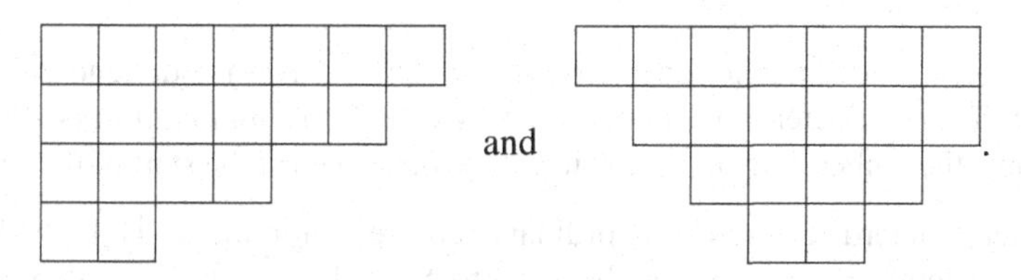

We place a copy of the shifted Young diagram, reflected diagonally like the conjugate partition, just to the left of this partition. Then fill in the hook lengths of the shifted Young diagram, as usual. (See Fig. 8.5.) If (i, j) is a box in the shifted Young diagram, let $h_{i,j}$ denote the hook length as above.

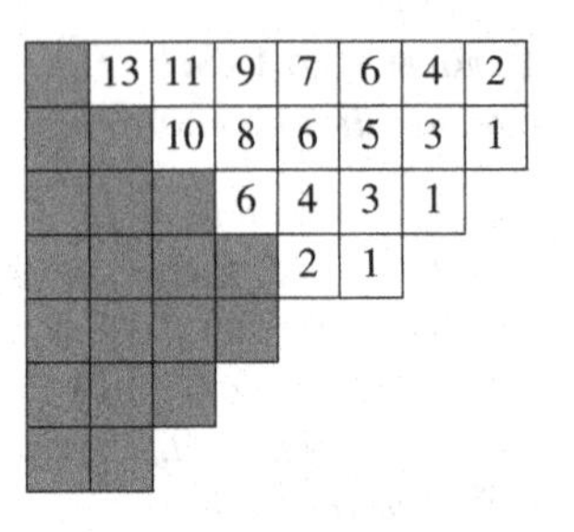

Fig. 8.5 Shifted Young diagram with hook lengths entered

Theorem 8.4.2 (Hook Length Formula for Spin Characters) *Let λ be a strict partition of n with r parts. We have*

$$\chi_\lambda(1) = 2^{(n-r-\varepsilon(\lambda))/2} \cdot \frac{n!}{\prod_{(i,j)} h_{i,j}},$$

where the product runs over all boxes in the shifted Young diagram of λ.

(See [298, Proposition 10.6 and Theorem 10.7].)

Example 8.4.3 The strict partitions of 5 are (5), (4, 1) and (3, 2), and $\varepsilon(\lambda) = 0, 1, 1$ respectively. Thus there is one spin character labelled by (5), and two each by (4, 1) and (3, 2). The power of 2 at the start of the degree is 4, 2 and 2 respectively. The hook lengths for the shifted Young diagrams are

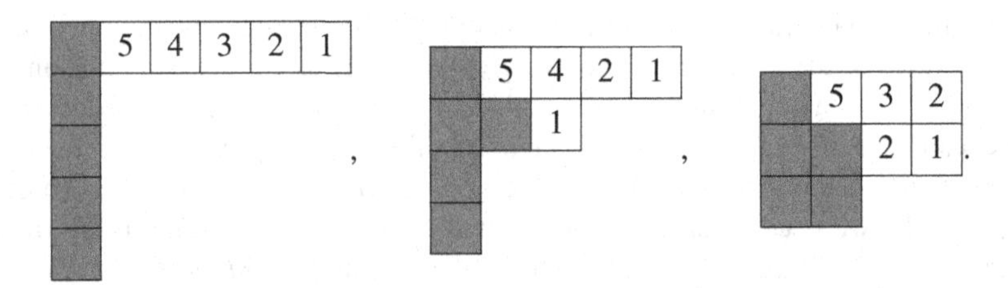

Thus the spin character degrees are 4, 4, 4, 6, 6. These restrict to $\tilde{A}_5$ with character degrees 2, 2, 4, 6.

Example 8.4.4 The character degree of $\chi_{(n)}$ is $2^{(n-1)/2}$ if n is odd, and $2^{(n-2)/2}$ if n is even. This character is normally known as the *basic spin character*. (It is the character of the embedding into the spin group mentioned at the start of the section.)

Another standard feature of the ordinary representation theory is the branching rule for restricting representations from S_n to S_{n-1}. For $\tilde{S}_n$, this is complicated a little by the presence of associated characters. Write $\bar{\chi}_\lambda$ for χ_λ if λ is even, and for $\chi_\lambda + \chi_\lambda \otimes \operatorname{sgn}$ if λ is odd. If λ is even then χ_λ is unambiguous, whereas if λ is odd then there are two characters with that partition.

Theorem 8.4.5 (Branching Rule for Spin Representations) *Let $\lambda = (\lambda_1, \ldots, \lambda_r)$ be a strict partition of n. If $\lambda_r > 1$, then*

$$\chi_\lambda \downarrow_{\tilde{S}_{n-1}} = \sum_A \bar{\chi}_{\lambda \setminus \{A\}},$$

where the sum runs over all removable boxes of λ such that $\lambda \setminus \{A\}$ is strict.

If $\lambda_r = 1$ then

$$\chi_\lambda \downarrow_{\tilde{S}_{n-1}} = \chi_\mu + \sum_A \bar{\chi}_{\lambda \setminus \{A\}},$$

where the sum runs over all removable boxes of λ such that $\lambda \setminus \{A\}$ is strict and with r parts. The partition μ is $(\lambda_1, \ldots, \lambda_{r-1})$, and we have chosen χ_λ and χ_μ so that $\chi_\lambda(x) = \chi_\mu(x)$ for $x \in \tilde{S}_{n-1}$ of cycle type μ.

The first question is: does this make sense? If $\lambda_r > 1$ then everything is fine, but what about $\lambda_r = 1$? In this case, notice that λ and μ have the same parity: if λ (and hence μ) is even, then again the sum is well defined, and if λ and μ are odd we need to make a choice of what we mean by χ_λ and χ_μ. This is done by using the value of χ_λ and χ_μ at an element $x \in \tilde{S}_{n-1}$ whose image in S_n has cycle type μ. Since λ is odd, it has an even part. If $\nu \neq \mu$ is a strict partition of $n-1$, then we mentioned just after Theorem 8.4.1 that $\chi_\nu(x) = 0$. Thus we are forced by the formula for the restriction of χ_λ in the branching rule to have $\chi_\lambda(x) = \chi_\mu(x)$. On the other hand, the character values $\chi_\lambda(x)$ and $\chi_\mu(x)$ are given in (8.1), and from that formula we see that $\chi_\lambda(x) = \pm\chi_\mu(x)$, so our choice of μ makes sense.

This result first appeared, together with branching for $\tilde{A}_n$, in [155, 156]. See also [298, Theorem 10.2].

We should also talk about the irreducible representations, rather than the irreducible characters. Again, as there are two groups $\tilde{S}_n$, we must consider two possible groups, or at least fix one of them, when constructing representations. The first construction of the irreducible spin representations, rather than just the characters, was by Nazarov in 1990 [454]. (One may also see the exposition in [298, Section 11].) The technology behind this construction starts to get far beyond the scope of this book, so we won't give it here. This construction, and work by Sergeev [520], pushed the spin theory in a more Lie-theoretic direction, and particularly uses the theory of superalgebras and supermodules.

Having considered ordinary representation theory, we will look at some of the modular theory. If we are trying to mimic our results for the standard representations of symmetric groups, we also need an analogue of the Nakayama conjecture.

Using the central character from Sect. 2.2, it is easy to see that if χ and ψ are two characters of a finite group that lie in the same p-block, and z is a central p'-element, then $\chi(z)/\chi(1) = \psi(z)/\psi(1)$. The upshot of this is that, if χ is a character of S_n and ψ is a faithful character of $\tilde{S}_n$, then χ and ψ cannot lie in the same p-block for

p odd (as $\chi(z) = \chi(1)$ and $\psi(z) = -\psi(1)$). (See also Exercise 3.14 for another method.) Thus there are 'spin blocks', consisting solely of spin characters, for p odd. For $p = 2$, the central element of order 2 in $\tilde{S}_n$ is no longer a p'-element, and so this argument doesn't work. Indeed, by Theorem 2.4.1, the simple kS_n-modules coincide with the simple $k\tilde{S}_n$-modules if k has characteristic 2. Since any 2-block must have some simple modules, the 2-blocks of S_n and $\tilde{S}_n$ are the same (see Exercise 3.8).

Consider p odd first, and suppose that we want to distribute the spin characters into spin blocks. This distribution was first proposed by Morris [434], and then proved by Humphreys in [309], with later proofs by Cabanes [84] and Olsson [460]. It involves the idea of p-bars, the spin analogue of rim p-hooks.

Let $\lambda = (\lambda_1, \dots, \lambda_r)$ be a strict partition of n and let p be an odd integer (of course, it will later be a prime). A *p-bar* of λ is a collection of p boxes of the Young diagram of λ, which has one of three possible forms.

(i) It can consist of all of row i of the Young diagram, if we have $\lambda_i = p$.
(ii) It can consist of all of rows i and j of the Young diagram, if we have $\lambda_i + \lambda_j = p$.
(iii) It can consist of the last p boxes of row i of the Young diagram, if removing them and then reordering the rows into a partition results in a strict partition, instead of just a partition.

Example 8.4.6 Let λ be the partition $(10, 7, 5, 3, 2, 1)$.

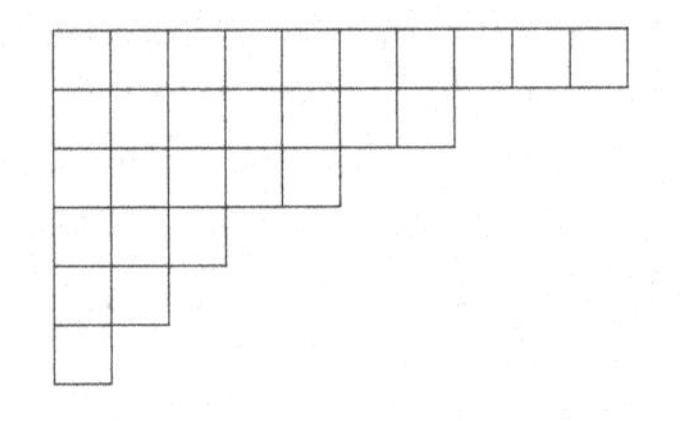

There are three possible 3-bars: removing three boxes from the second part; removing the fourth part; removing the fifth and sixth parts. There are two 7-bars: removing the second row; removing the third and fifth rows.

Just as removing all rim p-hooks from a Young diagram results in a well-defined partition, the p-core, removing all p-bars from a strict Young diagram results in a well-defined strict partition, the *p-bar core*. (This was proved in [435]. That paper also introduces a version of the abacus for this combinatorics, which we will see later in this section.) For example, the 3-bar core of the partition in the previous example is (1): we have a chain of 3-bar removals

$$(10, 7, 5, 3, 2, 1) \to (10, 7, 5, 3) \to (10, 7, 5) \to (10, 7, 2) \to (10, 4, 2) \to$$

$$(10, 2, 1) \to (10) \to (7) \to (4) \to (1).$$

The 7-bar core is $(10, 3, 1)$.

Theorem 8.4.7 (Humphreys [309]) *Let p be an odd prime. Let λ and μ denote two strict partitions of n.*

(i) *If λ is a p-bar core then χ_λ lies in a block of defect zero.*
(ii) *If λ is odd and not a p-bar core then both χ_λ and its associate character $\chi_\lambda \otimes \mathrm{sgn}$ lie in the same p-block of $\tilde{S}_n$.*
(iii) *If λ is not a p-bar core, then λ and μ have the same p-bar core if and only if χ_λ and χ_μ lie in the same p-block of $\tilde{S}_n$.*

The reason why this has to be so much more complicated than the Nakayama conjecture is that we have associate characters: the statement that χ_λ and χ_μ lie in the same p-block if and only if they have the same p-bar core is true, but tells us nothing about the associate characters, so we need (ii) for that. Of course, the assertion in (ii) is false if λ is a p-bar core: because both χ_λ and its associate character must lie in blocks of defect zero, they must lie in different blocks.

This is not the right answer for $p = 2$ though. In particular, the 2-bar core does not determine the 2-blocks of $\tilde{S}_n$. Knörr and Olsson produced a conjecture in 1986 [459], based on the computations by Benson in [32]. This was proved in 1993 by Bessenrodt and Olsson [39].

Theorem 8.4.8 *Let $\lambda = (\lambda_1, \ldots, \lambda_r)$ be a strict partition, and let* $\mathrm{dbl}(\lambda)$ *(the double of λ) denote the partition*

$$\left(\left\lfloor\frac{\lambda_1+1}{2}\right\rfloor, \left\lfloor\frac{\lambda_1}{2}\right\rfloor, \left\lfloor\frac{\lambda_2+1}{2}\right\rfloor, \left\lfloor\frac{\lambda_2}{2}\right\rfloor, \ldots, \left\lfloor\frac{\lambda_r+1}{2}\right\rfloor, \left\lfloor\frac{\lambda_r}{2}\right\rfloor\right).$$

The spin character χ_λ and the standard character $\chi^{\mathrm{dbl}(\lambda)}$ belong to the same 2-block of $\tilde{S}_n$, and in particular χ_λ and χ_μ belong to the same 2-block if and only if $\mathrm{dbl}(\lambda)$ *and* $\mathrm{dbl}(\mu)$ *have the same 2-core.*

Having distributed the ordinary characters into blocks, we would now like to label the irreducible modular spin representations. As we said earlier, by Theorem 2.4.1 there are no irreducible modular spin representations for $p = 2$, so this is a question for p odd.

The first obstruction is that spin blocks need not have lower triangular decomposition matrices (the associate characters again cause a problem).

Example 8.4.9 Let $n = 9$. There is a unique spin block for $\tilde{S}_9$ in characteristic 3. The decomposition matrix of this is given in Table 8.1. (Recall from Sect. 2.2 the convention of using . instead of 0 in a decomposition matrix.)

Here, χ^a_λ is the associate character to χ_λ. Notice that this matrix cannot be written in a lower triangular shape, since there is no row with a single non-zero entry. This is the smallest example of a decomposition matrix for $\tilde{S}_n$ that is not lower triangular.

Looking at the decomposition matrix in this example, it is not lower triangular, but that is only because of associate characters. The fact that associate characters exist means that we should 'pair up' characters and their associates, and expect the matrix to be block lower triangular, where the blocks have at most two rows and columns (but not necessarily the same number of rows and columns).

Character	Degree	ψ_1	ψ_2	ψ_3	ψ_4	ψ_5	ψ_6
$\chi_{(9)}$	16	1	1	.	.	.	.
$\chi_{(8,1)}$	56	1	.	1	.	.	.
$\chi^a_{(8,1)}$	56	.	1	.	1	.	.
$\chi_{(7,2)}$	160	1	.	1	.	1	.
$\chi^a_{(7,2)}$	160	.	1	.	1	.	1
$\chi_{(6,3)}$	224	1	1	.	.	1	1
$\chi^a_{(6,3)}$	224	1	1	.	.	1	1
$\chi_{(5,4)}$	112	1	.	.	.	1	.
$\chi^a_{(5,4)}$	112	.	1	.	.	.	1
$\chi_{(6,2,1)}$	240	2	2	.	.	1	1
$\chi_{(5,3,1)}$	336	2	2	1	1	1	1
$\chi_{(4,3,2)}$	96	.	.	1	1	.	.

Table 8.1 3-decomposition matrix of a faithful block of $\tilde{S}_9$

Another issue is that for representations of S_n, the proof that the number of p-regular conjugacy classes equalled the number of p-regular partitions is based on Glaisher's theorem. We need a 'spin analogue' of this.

How many p-regular conjugacy classes are there in $\tilde{S}_n$? We can use [298, Theorem 3.8]: the preimage of any conjugacy class of S_n in $\tilde{S}_n$ forms a single class or splits into two classes. This result states that those classes that split are exactly those of S_n whose cycle type (which is a partition of n) either has no even parts or has odd parity and is strict.

To count the number of Brauer spin characters, we must count the number of p-regular classes of $\tilde{S}_n$, then subtract the number of p-regular classes of S_n, and so we count exactly those p-regular classes of S_n that split. The number of p-regular classes whose cycle type is an odd strict partition is simply equal to the number of odd strict partitions with no part divisible by p. The number of partitions into odd parts is equal to the number of strict partitions by a famous partition identity of Euler—Exercise 8.13 asks you to prove this—and as we can see from this exercise this equality carries those partitions with no part divisible by p to those partitions with no part divisible by p.

The conclusion of all of this is that the number of p-modular spin characters is equal to the number of characters associated to strict partitions with no part divisible by p. Example 8.4.9 suggests this isn't a good set though, because there is no obvious way to assign the ψ_i to $\chi_{(8,1)}$, $\chi_{(7,2)}$ and $\chi_{(5,4)}$ and their associates. An old partition identity of Schur from 1926 [514] shows that the number of strict partitions with no part divisible by 3 is equal to the number of strict partitions $\lambda = (\lambda_1, \ldots, \lambda_r)$ such that $\lambda_i - \lambda_{i+1} \geq 3$ for all i, and $\lambda_i - \lambda_{i+1} > 3$ if $3 \mid \lambda_i$, called *Schur regular* in [38]. The Schur regular strict partitions of 9 are (9), (8, 1) and (7, 2). With respect to the lexicographic ordering on Schur regular partitions, the decomposition matrix now becomes block lower triangular, with the blocks being of size at most 2 and corresponding to associate characters if they have size 2.

Bessenrodt, Morris and Olsson gave this labelling in [38], and also produced a conjectural analogue of this for $p = 5$, which was proved by Andrews, Bessenrodt, and Olsson [18] shortly after. In this case, the number of strict partitions with no part divisible by 5 is equal to the number of partitions $\lambda = (\lambda_1, \ldots, \lambda_r)$ such that

(i) $\lambda_i - \lambda_{i+2} \geq 5$ for all i,
(ii) $\lambda_i - \lambda_{i+2} > 5$ if $5 \mid \lambda_i$ or $5 \mid (\lambda_i + \lambda_{i+1})$, and
(iii) there are no subsequences $(5j+3, 5j+2)$, $(5j+6, 5j+4, 5j)$, $(5j+5, 5j+1, 5j-1)$ or $(5j+6, 5j+5, 5j, 5j-1)$ in λ for some $j \geq 0$.

The increase in complexity from $p = 3$ to $p = 5$ suggests it would be difficult to proceed further. Indeed, there are also theoretical reasons that stymie proceeding past $p = 5$, because the recursive method that generated the sets for $p = 3$ and $p = 5$ in [38] can be shown to fail for all $p > 5$, in the sense that it does not produce enough strict partitions for it to be in bijection with strict partitions none of whose parts is divisible by p.

Brundan and Kleshchev used a different approach. The first thing to note is that it is not necessary for the labels for the simple modules in characteristic p to be a subset of those in characteristic 0. This is true for representations of S_n, but there is no reason why it should hold for spin representations. Indeed, although the partitions above for $p = 3$ are strict, for $p = 5$ they are not always strict. In [78], they gave a parametrization of the irreducible spin representations in characteristic p by restricted p-strict partitions. A partition is *p-strict* if all repeated parts have size divisible by p, and a p-strict partition is *restricted* if $\lambda_i - \lambda_{i+1} \leq p$, and $\lambda_i - \lambda_{i+1} \neq p$ if $p \mid \lambda_i$. We also need to worry about associate characters: defining $h_{p'}(\lambda)$ to be the number of parts of λ not divisible by p, if $n - h_{p'}(\lambda)$ is even then λ labels one character, and if $n - h_{p'}(\lambda)$ is odd then λ labels two.

It is not clear that the number of restricted p-strict partitions is equal to the number of strict partitions no part of which is divisible by p, but this must be true by Theorem 2.2.1. (This was done in [79, Lemma 2.1].) If $p > n$ (which includes the case $p = 0$) then we simply obtain strict partitions of n. Indeed, the Brundan–Kleshchev labelling for $p > n$ (and $p = 0$) matches the labelling by strict partitions given by Morris that we have been using so far (see [78, Remark 9.15]).

Recently [546], Tsuchioka and Watanabe overcame the issues that prevented the methods from [38] extending to larger primes, and have proved a generalization of the bijections from [18, 38] to all primes, removing the 'restricted' part of the Brundan–Kleshchev labelling by p-strict partitions. One of the conditions they replace it with is $\lambda_i - \lambda_{i+h} \geq p$, and $\lambda_i - \lambda_{i+h} \neq p$ if $p \mid \lambda_i$, where $p = 2h + 1$, exactly as in the definition of Schur regularity and the $p = 5$ definition above from [18]. The other conditions are that, for all $1 \leq d < h$ and all $j \geq 1, l \geq 0$, there is no i such that any of the following hold:

- $\lambda_i = pj + d, \lambda_{i+d} = pj - d$,
- $\lambda_i = pl + h + d, \lambda_{i+d} = pl + h + 1 - d$,
- $\lambda_i = pj + p + d, \lambda_{i+h+d} = pj - d$.

At this stage, I don't think it is known absolutely for sure which of these two labellings is best from a decomposition matrix point of view, as there are too few decomposition matrices for odd primes known. However, the evidence that does exist, particularly recent work of Fayers on weight 2 spin blocks that is as-yet unpublished, suggests that the Brundan–Kleshchev labelling is the more natural one. If there is a spin analogue of James's submodule theorem, or something like it, then this is likely to seal the fate of one of the two labellings, and it should become completely clear whether to use restricted p-strict partitions or this more complicated set.

The parametrization in [78] was obtained by using Clifford theory, particularly the twisted group algebra from Sect. 7.2. Using a Morita equivalence of Sergeev from [520], classifying the simple modules of the twisted group algebra is equivalent to classifying simple modules for an object called the 'Sergeev superalgebra'. This was achieved in [78], and then again, using a different method, in [79]. Thus there are two different parametrizations of the simple modules for $\tilde{S}_n$, with the same parametrizing set. The differing techniques of [78] and [79] meant that different theorems could be proved about the simple $k\tilde{S}_n$-modules, but they could not be combined because the two parametrizations were not known to be the same unless $p = 0$, where they both coincided with the standard one. In [80] for example, Brundan and Kleshchev produce a spin analogue of James's regularization result, Theorem 8.2.8, but only for the parametrization from [79]. Finally, in [362], it was proved that both parametrizations yield the same branching theorems, and therefore they are actually the same parametrization. This means that both techniques can be brought to bear on modular spin representations, and the combination of the two should mean deep and powerful results can be proved in future.

Just as for symmetric groups, there are only finitely many Morita equivalence classes of spin blocks of a given weight. This was proved by Kessar in [343], so Donovan's conjecture (and Puig's refinement to source algebras) holds for $\tilde{S}_n$ and $\tilde{A}_n$.

We can produce a version of the abacus for spin characters as well. This time, instead of the first-column hook lengths, we place the parts of the partition on the abacus. Removing a p-bar can be seen to be one of three types of moves: moving a bead up a runner; removing a bead from the top of the 0th runner; removing two beads, from the tops of the ith and $(p-i)$th runners, simultaneously. The result of performing all possible such moves must be the (abacus representation of the) p-bar core of λ. The definition of weight is the same as for symmetric groups, that is, how many moves it takes to go from a strict partition to its p-bar core, or equivalently the size of the partition, minus that of its p-bar core, divided by p.

Suppose that λ is a p-bar core and consider its abacus representation. From the description of bar removal on the abacus, it is clear that if the ith runner has a bead on it then the $(p-i)$th runner has no beads on it. We therefore see that the Scopes moves from Sect. 8.3 do not preserve being a p-bar core. Instead of swapping the ith and $(i+1)$th runners, we perform one of the following moves:

(i) move the $(p-1)$th runner to the first runner, and move the first runner to the $(p-1)$th runner, but push it up one place (so the bead at position $ap+1$ moves to $(a-1)p+(p-1)$);
(ii) swap the $(p-1)/2$th and $(p+1)/2$th runners;
(iii) swap the ith and $(i+1)$th runners, and at the same time the $(p-i)$th and $(p-i-1)$th runners, if $1 \leq i \leq (p-3)/2$.

These moves do send p-bar cores to p-bar cores. In the standard case, as long as there are at least w more beads on the one runner than the other, where w is the weight of the block, then the two corresponding blocks are Morita equivalent. This time, the same result holds, but more than w extra beads are required. However, it is still a function of w, and so there are finitely many Morita equivalence classes and Donovan's conjecture holds.

Along with Morita equivalences come derived equivalences. For symmetric groups, the Chuang–Rouquier theorem [115] proved that all blocks of symmetric groups of the same weight were derived equivalent. This is certainly not true for spin blocks, for the simple reason that not all such blocks have the same number of simple modules. The associate characters play a role here, and in fact for blocks of a given weight, there are two possibilities for the number of simple modules.

The same statement holds for blocks of $\tilde{A}_n$. In fact, the spin blocks of symmetric and alternating groups are a lot more closely related than the standard blocks of the two groups. Theorem 8.4.1 stated that a strict partition λ labels one spin character of S_n if λ is even and two if λ is odd, and the opposite for spin characters of A_n. In [352], Kessar and Schaps noted that in fact the Scopes moves on the abacus can induce maps between spin blocks of S_n and spin blocks of A_m, rather than sticking to either symmetric or alternating groups, and would do so if the move switched the parities of the partitions. This 'crossover' property turns out to mean that, conjecturally at least, there are exactly two derived equivalence classes of spin blocks of weight w, of both alternating and symmetric groups together, and both types appear in both symmetric and alternating groups.

Example 8.4.10 Let $n = 8$ and $p = 5$. There are six strict partitions of 8, and four strict partitions of 8, no part of which is divisible by 5. (There are nine ordinary spin characters and six modular spin characters.) There are two 5-bar cores, which are (3) and (2, 1), and the strict partitions are separated according to 5-bar core as

$$\{(8),\ (5,3),\ (4,3,1)\}, \quad \text{and} \quad \{(7,1),\ (6,2),\ (5,2,1)\}.$$

The former of these has two odd partitions and one even, so five ordinary spin characters, and the latter has one even and two odd, so four ordinary spin characters.

Brunat and Gramain proved [76] the character-theoretic shadow of the derived equivalences between the spin blocks, namely the perfect isometries from Sect. 4.4, do exist. Thus there are two perfect isometry classes of spin blocks of symmetric and alternating groups of a given weight. Livesey used this to show that the perfect isometry version of Broué's conjecture holds for these blocks [397]. However, at the moment, neither step in the proof of Broué's conjecture—proving that spin blocks of different symmetric groups are derived equivalent, or finding a 'good' (RoCK-like) block that is derived equivalent to a block of a wreath product, as in the Chuang–Rouquier proof—has been done at this stage.

As with the symmetric groups, we know the decomposition matrices of spin blocks for certain small n, and also for blocks of weight 1. The defect group of a block of weight w is a Sylow p-subgroup of $\tilde{S}_{pw}$, as was proved by Cabanes in [84], so blocks of weight 1 have cyclic defect groups, hence Brauer trees. As we saw in Sect. 5.3, these were determined by Müller in [439].

Yaseen, in his thesis [572], and in published work with Morris [436], computed decomposition matrices for spin blocks up to $n = 11$, and Maas [408] pushed things further, and computed the decomposition matrices of spin blocks of S_n for $14 \leq n \leq 18$ and p odd. (The cases $n = 12, 13$ appear not to have been published, and are available at the *Modular Atlas* website [64].) However, the decomposition numbers even for χ_λ with λ a strict two-part partition are not known in general.

There is still much that is even quite basic left to understand for these groups.

Exercises

Exercise 8.1 Prove that the hook length formula is equivalent to Frobenius's original formula for the character degrees:

$$\chi^\lambda(1) = \frac{n!}{\prod_{i=1}^{r} x_i!} \cdot \prod_{i<j}(x_i - x_j),$$

where $X = \{x_1, \ldots, x_r\}$ is a β-set for λ. One should also show that this formula is independent of the β-set chosen.

Exercise 8.2 Show that, if λ is p-regular, the top of S^λ is D^λ, and in particular S^λ is indecomposable.

Exercise 8.3 Show that the effect of removing a p-hook from a partition is the same as moving a bead up one place on its runner of the p-abacus, and hence that the p-core is well defined.

Exercise 8.4 Show that every p-regular partition has at least one good node. If λ is p-regular and A is a good node, show that $\lambda \setminus \{A\}$ is p-regular.

Exercise 8.5 Prove that if $\lambda \cup \{A_1\}$ and $\lambda \cup \{A_2\}$ have the same p-core then the nodes A_1 and A_2 have the same residue.

Exercise 8.6 Using the Clifford theory of blocks from Sect. 7.4, prove Theorem 8.3.2.

Exercise 8.7 One of the key points in the proof of Theorem 8.3.4 was the equation

$$(S^{\mu})\uparrow^{S_{n+pw+k}} = (S^{\mu\phi})^{\oplus k!},$$

mentioned just before the statement of that theorem. Prove this statement.

Exercise 8.8 Using the Schaper formula and column removal, prove Theorem 8.3.7.

Exercise 8.9 Let λ be a p-core, and let X be a β-set of λ. Let Y denote the subset of X consisting of the bottom bead on each runner. Write $Y = \{y_1, \dots, y_p\}$ with $y_i > y_{i+1}$.

Suppose that μ is another partition, and that removing a single p-hook of leg length i from μ results in λ. Show that a β-set of μ is

$$X \cup \{y_i + p\} \setminus \{y_i\}.$$

In other words, adding a p-hook of leg length i to λ is the same as adding p to y_i.

Exercise 8.10 This exercise will prove the description of the Brauer tree given in Theorem 8.3.13. This proof uses Scopes moves to reduce to the case where we consider the principal p-block of S_p, so we start by proving the tree in Example 8.3.14.

Using the branching rule, show that there is a projective module for S_p whose Brauer character is the reduction modulo p of $\chi^{(p-m,1^m)} + \chi^{(p-m+1,1^{m-1})}$, and hence these two characters are adjacent on the Brauer tree of the block. (Hint: induce from S_{p-1}.)

Now use induction on the degree of the symmetric group, a Scopes move, and Exercise 8.9 to show that every Brauer tree of every block (with cyclic defect groups, i.e., of weight 1) of every symmetric group has the form of Theorem 8.3.13. (Hint: the induction of the character of a projective module is the character of a projective module.)

Exercise 8.11 We now produce a second proof of the structure of the Brauer tree. This one requires the Littlewood–Richardson rule. Read the definition of the Littlewood–Richardson rule first, in for example [324, 327, 498] or somewhere else.

Let λ be a p-core of size n, and let χ_i denote the character $\chi^{(p-i,1^i)}$ of S_p for $0 \leq i \leq p-1$. By the first part of Exercise 8.10, $\chi_i + \chi_{i+1}$ is the character of a projective module of S_p. Thus $\chi^{\lambda} \otimes (\chi_i + \chi_{i+1})$ is the character of a projective module for $S_n \times S_p$.

Induce $\chi^{\lambda} \otimes (\chi_i + \chi_{i+1})$ from $S_n \times S_p$ to S_{n+p} (using the Littlewood–Richardson rule) and use this to prove that the Brauer tree of the block of S_{n+p} with p-core λ has the form stated in Theorem 8.3.13.

Exercise 8.12 Show that the basic spin character (and its associate if it is odd) is the faithful (ordinary) character of $\tilde{S}_n$ of smallest degree.

Exercise 8.13 Euler proved, using generating functions, that the number of strict partitions is equal to the number of partitions into odd parts. Glaisher gave a bijective proof of this.

Let λ be a partition into odd parts, with r_i parts of size i. (Of course, $r_{2i} = 0$ for all i.) The image $f(\lambda)$ is a partition with a part of size $2^a \cdot i$ if and only if there is a 1 in the ath position of the binary expansion of m_i.

Show that $\lambda \mapsto f(\lambda)$ is a bijection from the set of all partitions of n with only odd parts to the set of strict partitions of n, and f preserves the property of having no parts divisible by some odd prime p.

Chapter 9
Representations of Groups of Lie Type

The finite groups of Lie type form the bulk of the finite simple groups, and as such we are interested in their representation theory, both in its own right and because of the myriad questions about finite groups that can be reduced to finite simple groups.

The trouble is this theory is quite hard, even to talk about. For the symmetric groups, it was possible to give a broad overview of the theory as it stands without introducing too many new concepts, whereas here we first need an understanding of the groups themselves. That's a topic that would take far too long to cover here, so I will have to assume that the reader knows something about the theory of algebraic groups and groups of Lie type, say what you can find in [417].

The irreducible representations of groups of Lie type split into two cases, both very difficult: the first is when the characteristic of k is equal to the characteristic of the group, called the defining characteristic or equicharacteristic case; the second is where the characteristic of k is not the characteristic of the group, including the case where k has characteristic 0, called the non-defining or cross-characteristic case.

The defining characteristic case is in some sense easier, at least to talk about. The number of irreducible representations is easy to define, and there is a formula for the irreducible characters that works for very large primes (if one fixes the type of group, say GL_n).

In the cross-characteristic case things are more complicated to set up. There is a theory of unipotent characters, an analogue of unipotent classes, but the details, and even the definition of a unipotent character, are technical. We will give a rough overview of the state of the theory of unipotent characters, unipotent blocks, and then discuss the Bonnafé–Rouquier theory that relates arbitrary blocks of groups of Lie type to so-called 'isolated' and 'quasi-isolated' blocks. There are many technicalities in this theory, and the precise version of the group you work with ($\mathrm{GL}_n(q)$, $\mathrm{PGL}_n(q)$, $\mathrm{SL}_n(q)$), called the *isogeny type* of the group (see [107, Section 1.11])—although note that $\mathrm{GL}_n(q)$ technically is not isogenous to $\mathrm{SL}_n(q)$ or $\mathrm{PGL}_n(q)$—has a great impact on the theory. We necessarily are sometimes a bit vague about exactly which group we are talking about, for space considerations.

D. A. Craven, *Representation Theory of Finite Groups: a Guidebook*, Universitext,
https://doi.org/10.1007/978-3-030-21792-1_9

When it is really important we mention it, and we provide references where the reader who really needs to know the exact scope of the result can confirm whether the result is or is not proved for their group.

We unfortunately do not consider the Φ_d-theory here, as it is another field in its own right, with some serious technical difficulties to get to where we would want to be, and instead refer to [417] for the basics. From there it is possible to read some of the earlier papers in the area, most notably [71].

9.1 Defining-Characteristic Representations

This section attempts to give a (very) brief overview of roughly what happens in the defining-characteristic case. Entire books have been written about this, for example [333] is a canonical choice, and [307] is a survey assuming people have read [333], so since this section assumes people have read neither, we will necessarily have to skip a lot of material.

For this section, $\mathbf{G}$ is a semisimple linear algebraic group defined over our algebraically closed field k, so that the characteristic of the group $\mathbf{G}$ is p. We start by defining or reminding the reader about the notion of a weight. For this we need a maximal torus $\mathbf{T}$ of $\mathbf{G}$. Write $X(\mathbf{T})$ for the set of characters of $\mathbf{T}$, i.e., homomorphisms $\mathbf{T} \to k^\times$. This is a free abelian group of the same rank as $\mathbf{T}$. If V is a $\mathbf{G}$-module (i.e., there is a (rational) representation $\mathbf{G} \to \mathrm{GL}(V) = \mathrm{GL}_n(k)$ for some n) then we may write V as a direct sum

$$V = \bigoplus_{\alpha \in X(\mathbf{T})} V_\alpha,$$

where V_α consists of all $v \in V$ such that $v \cdot t = v(t\alpha)$ for all $t \in \mathbf{T}$, i.e., we decompose the restriction of V to $\mathbf{T}$ as a sum of irreducible $\mathbf{T}$-modules, and then group isomorphic modules together. Most (all but finitely many!) of the V_α will be 0; those α for which V_α is non-zero are called the *weights* of V, and the V_α are called the *weight spaces* of V. Since the Weyl group W of $\mathbf{G}$ acts on $\mathbf{T}$ it acts on $X(\mathbf{T})$ and on $E = \mathbb{R} \otimes_{\mathbb{Z}} X(\mathbf{T})$, and permutes the weights of V. The weights of the action of $\mathbf{G}$ on its Lie algebra $L(\mathbf{G})$ (which is called the *adjoint representation* of $\mathbf{G}$) are called *roots* of $\mathbf{G}$.

It turns out that only some elements of $X(\mathbf{T})$ can be the weights of $\mathbf{G}$-modules, so-called *abstract weights*. These are the weights λ such that $\langle \lambda, \alpha \rangle = 2(\lambda, \alpha)/(\alpha, \alpha)$ is an integer for all roots α. The abstract weights form a lattice in E. Given a base $\alpha_1, \dots, \alpha_r \in X(\mathbf{T})$ of simple roots of the root system Φ of $\mathbf{G}$ with respect to $\mathbf{T}$, there is a set $\lambda_1, \dots, \lambda_r \in X(\mathbf{T})$ of *fundamental dominant weights*, where $\langle \lambda_i, \alpha_j \rangle = \delta_{i,j}$. The *dominant weights* are the non-negative integral linear combinations of fundamental dominant weights; every abstract weight is conjugate under W to a unique dominant weight.

We may (partially) order weights by saying that $\lambda \leq \mu$ if $\mu - \lambda$ is a non-negative integral linear combination of positive roots. If $\mathbf{B}$ is a Borel subgroup of $\mathbf{G}$ containing $\mathbf{T}$ (corresponding to the base of Φ), then $\mathbf{B}$ must stabilize at least one 1-space on V; any non-zero vector v in such a 1-space is called a *maximal vector*. Clearly every element of $\mathbf{T}$ acts on v as a scalar, so v lies in V_λ for some weight λ, so we simply say that v has weight λ. If V is generated as a $\mathbf{G}$-module by a maximal vector of weight λ, then V is a *highest weight module* with highest weight λ. In this case, λ is dominant, the weight space V_λ is 1-dimensional, and $\lambda \geq \mu$ for all other weights μ of V.

We can then partially order the set of all highest weight modules with a given weight by whether one is a quotient of another. With this ordering, the set contains a unique minimal element $L(\lambda)$ and a unique maximal element $W(\lambda)$. We call $W(\lambda)$ the *Weyl module*, and since $L(\lambda)$ is irreducible we don't give it another special name. (If $\mathbf{G}$ is defined over k of characteristic 0, we will see below that $L(\lambda) \cong W(\lambda)$.)

The $L(\lambda)$ form a complete set of irreducible representations of $\mathbf{G}$. We summarize this in a theorem now.

Theorem 9.1.1 *Let* $\mathbf{G}$ *be a semisimple algebraic group.*

(i) *Given any dominant weight* λ, *there exists a unique (up to isomorphism) irreducible highest weight module* $L(\lambda)$ *with highest weight* λ.
(ii) *Every (finite-dimensional) irreducible* $\mathbf{G}$*-module is a highest weight module.*

(See, for example [417, Theorem 15.17] for the first part. The second part follows easily from the Lie–Kolchin theorem [417, Theorem 4.1].)

Thus for SL_2, for example, the dominant weights are labelled by non-negative integers, the simple module $L(0)$ is the trivial module, and the simple module $L(1)$ is the natural 2-dimensional module. If $p = 0$ then for all i the module $L(i)$ is simply the ith symmetric power of $L(1)$. However, if $p > 0$ then $L(i)$ is only the ith symmetric power of $L(1)$ for $0 \leq i < p$, and above that the symmetric power is reducible. What about $L(i)$ for $i \geq p$?

For this we need the idea of a Frobenius endomorphism. As is well known, if k has characteristic p then for $x, y \in k$, $(x + y)^p = x^p + y^p$ and $(xy)^p = x^p y^p$, so the map $x \mapsto x^p$ is a field endomorphism of k. Because matrix multiplication involves adding and multiplying the entries of the matrices, by raising each matrix entry to the power p we obtain a group homomorphism $F : \mathrm{GL}_n(k) \to \mathrm{GL}_n(k)$. Given a basis for a $\mathbf{G}$-module V, we therefore obtain a new module where instead of sending $g \in \mathbf{G}$ to a given matrix, it is sent to the matrix with those entries raised to the power p. The isomorphism type of this module does not depend on the basis chosen for V, so this is a reasonable construction. The module is usually written $V^{[1]}$, and often referred to as the *Frobenius twist* of V, abbreviated simply to *twist*. We iterate this, and write $V^{[i]}$ for $(V^{[i-1]})^{[1]}$.

If $V = L(\lambda)$ for some highest weight $\lambda = \sum_{i=1}^r a_i \lambda_i$ (with $a_i \in \mathbb{Z}_{\geq 0}$) then $V^{[1]} = L(p\lambda)$, where $p\lambda = \sum_{i=1}^r p a_i \lambda_i$. By choosing a basis for V so that the action of $\mathbf{T}$ is diagonal, this is not hard to see. What is less easy to see is that from this we can fill in the rest of the simple modules $L(\lambda)$. We often write just the coefficients a_i instead of $\sum_{i=1}^{r} a_i\lambda_i$, and write $\lambda = (a_1, \dots, a_r)$, so $L(1, 0, \dots, 0)$ and $L(\lambda_1)$ are the same module.

Theorem 9.1.2 (Steinberg's Tensor Product Theorem [532]) *Let p be a prime, let $\mathbf{G}$ be a connected, reductive algebraic group in characteristic p, and let $\lambda = (a_1, \dots, a_r)$ be a dominant weight for $\mathbf{G}$. Write*

$$a_i = \sum_{j\geq 0} a_{i,j}\, p^j,$$

where $0 \leq a_{i,j} \leq p - 1$. We have

$$L(\lambda) = \bigotimes_{j\geq 0} L(\mu^{(j)})^{[j]},$$

where $\mu^{(j)} = (a_{1,j}, a_{2,j}, \dots, a_{r,j})$.

(See, for example [417, Theorem 16.12].)

In other words, if we define the *p-restricted* weights to be those dominant weights λ such that each a_i lies between 0 and $p - 1$, then every simple module is the tensor product of various modules $V^{[j]}$ for p-restricted simple modules V.

For a fixed algebraic group, say SL_n, the dimension of the simple module $L(\lambda)$ can depend on p. For example, the natural module for SL_n, which is labelled by the dominant weight $(1, 0, \dots, 0)$, has dimension n in all characteristics. Its dual is also of dimension n, and is labelled by the dominant weight $(0, \dots, 0, 1)$; their tensor product has dimension n^2, and is $L(0) \oplus L(1, 0, \dots, 0, 1)$, i.e., the sum of the trivial module and the $(n^2 - 1)$-dimensional Lie algebra module, if $p \nmid n$, and is a uniserial module with socle and top $L(0)$ and heart (radical modulo socle) $L(1, 0, \dots, 0, 1)$ if $p \mid n$. Thus this simple module, which we can also write as $L(\lambda_1 + \lambda_{n-1})$, has dimension $n^2 - 1$ if $p \nmid n$ and $n^2 - 2$ if $p \mid n$.

This book is meant to be about representations of finite groups though, so why are we talking about algebraic groups? It turns out that the simple modules for $\mathrm{SL}_n(q)$ are the restrictions of simple modules for $\mathrm{SL}_n(k)$, where q is a power of p. The dominant weights involved are $\lambda = (a_1, \dots, a_{n-1})$ such that $0 \leq a_i \leq q - 1$.

This isn't just luck, but part of a more general phenomenon. Recall that any group of Lie type is of the form $\mathbf{G}^F$ for F a *Steinberg endomorphism*, i.e., a surjective homomorphism $\mathbf{G} \to \mathbf{G}$ that fixes only finitely many points (see [417, Chapter 22]). If F_p is the standard Frobenius map $x \mapsto x^p$, then there is some d, e such that $F^d = F_p^e$, then let $q = p^{d/e}$.

Theorem 9.1.3 (Steinberg's Restriction Theorem [532]) *Let $\mathbf{G}$ be a simple algebraic group over k, let F be a Steinberg endomorphism on $\mathbf{G}$, and let q be as above.*

(i) *Every simple $k\mathbf{G}^F$-module is the restriction of a simple $\mathbf{G}$-module.*

(ii) *If F is a Frobenius endomorphism then the restrictions of q-restricted simple $\mathbf{G}$-modules form a complete set of simple $k\mathbf{G}^F$-modules, with every module appearing once.*

The situation for Steinberg endomorphisms is a bit more complicated. For this you need to consider q^2-restricted highest weight modules, but not all of them. For ${}^2B_2$ and ${}^2G_2$, the simple modules for the algebraic group are labelled by highest weights (a_1, a_2), and we want the restrictions of modules $(a_1, 0)$ for all $0 \leq a_1 \leq q^2$. For ${}^2F_4$, the simple modules for the algebraic group are labelled by highest weights (a_1, a_2, a_3, a_4), and we want the restrictions of modules $(a_1, a_2, 0, 0)$ for all $0 \leq a_1, a_2 \leq q^2$.

Thus we have a parametrization of the simple kG-modules, where $G = \mathbf{G}^F$. In particular, we see that there are q^r of them, where r is the rank of a maximal torus of $\mathbf{G}$ (with suitable modifications for Steinberg endomorphisms).

Obvious questions include: can we compute the dimension of $L(\lambda)$? Can we determine the Brauer character value of $L(\lambda)$ on a particular semisimple element $g \in \mathbf{G}$, hence in G? What about $\mathrm{Ext}^1(L(\lambda), L(\mu))$ for the algebraic group and the finite group? How do they differ? What about the Cartan matrix? Structure of projectives? Things are pretty safe for $\mathbf{G} = \mathrm{SL}_2$, in the sense that we have a good idea what is going on. For other groups though things are very difficult.

The first question is: what are the dimensions of the simple kG-modules? Equivalently (via Steinberg's tensor product theorem, Theorem 9.1.2) we should be asking for the dimensions of the simple $\mathbf{G}$-modules for p-restricted dominant weights. In this case, there are lots of results. If $p = 0$ then we have a complete answer, in the shape of the *Weyl character formula* (originally proved by Weyl in [551–553] using analysis, then by Freudenthal algebraically [236]). The dimension of $L(\lambda)$ in characteristic 0 (equivalently the dimension of $W(\lambda)$, since $W(\lambda) = L(\lambda)$ in characteristic 0) is

$$\dim(L(\lambda)) = \prod_{\mu \in \Phi^+} \frac{\langle \lambda + \rho, \mu \rangle}{\langle \rho, \mu \rangle},$$

where ρ is half the sum of all positive roots (which is a weight) and Φ^+ is the set of positive roots in Φ. Kostant gave a multiplicity formula for the weights in $W(\lambda)$ [374], which implies the Weyl character formula.

Example 9.1.4 Let $\mathbf{G} = \mathrm{SL}_2(k)$, and assume k has characteristic 0. A maximal torus $\mathbf{T}$ can be chosen to be all diagonal matrices with entries a and a^{-1}, for all $a \in k^\times$. A character χ_i in $X(\mathbf{T})$ is labelled by an integer i, where

$$\chi_i : \begin{pmatrix} a & 0 \\ 0 & a^{-1} \end{pmatrix} \mapsto a^i.$$

The action of this matrix on the Lie algebra of $\mathbf{G}$ is

$$\begin{pmatrix} a^2 & 0 & 0 \\ 0 & 1 & 0 \\ 0 & 0 & a^{-2} \end{pmatrix}$$

(with respect to a suitable basis) so the characters $\chi_{\pm 2}$ are roots. We see that every character in $X(\mathbf{T})$ must satisfy

$$\langle \chi_i, \chi_2 \rangle = 2\frac{(\chi_i, \chi_2)}{(\chi_2, \chi_2)} \in \mathbb{Z}.$$

In particular, the natural representation, with weights $\chi_{\pm 1}$, must be the highest weight module $L(1)$. The Lie algebra is the highest weight module $L(2)$.

As we mentioned before, the ith symmetric power $S^i(L(1))$ of $L(1)$ is also simple, although this is a bit more difficult to see. It has dimension $i + 1$, and it is fairly easy to see that the weights of it are $\pm i$, $\pm(i - 2)$, $\pm(i - 4)$ and so on, so $S^i(L(1)) = L(i)$. Let us check the formula above for the dimension of $L(i)$: let α denote the fundamental dominant weight. Then the roots are $\pm 2\alpha$, and $\rho = (2\alpha)/2 = \alpha$. We therefore must compute, for $\lambda = i\alpha$,

$$\prod_{\mu \in \Phi^+} \frac{\langle \lambda + \rho, \mu \rangle}{\langle \rho, \mu \rangle} = \frac{\langle i\alpha + \alpha, 2\alpha \rangle}{\langle \alpha, 2\alpha \rangle} = i + 1.$$

The Weyl character formula, which yields this dimension, in fact gives the entire character of the representation in characteristic 0. To compute the (Brauer) character of $L(\lambda)$ in characteristic p is easy once you know the composition factors of $W(\lambda)$. Fixing $\mathbf{G}$ and $\mathbf{T}$, and fixing a dominant weight λ, $L(\lambda) = W(\lambda)$ for all but finitely many primes, but the set of exceptions is in general difficult to understand.

There are two dominant weights for which the dimension of $L(\lambda)$ is very easy: $\lambda = (0, \dots, 0)$ and $\lambda = (p - 1, p - 1, \dots, p - 1)$. The first of these is the trivial weight, so of course $L(0)$ has dimension 1. The second of these is the *Steinberg weight*, also written as $(p - 1)\rho$, where as above ρ is half the sum of the positive roots. The module $L(\lambda)$ for this λ is a simple module of dimension exactly $p^{|\Phi^+|}$, i.e., p to the power the number of positive roots. Using Steinberg's tensor product theorem, if $q = p^a$, we see that

$$\dim(L(q - 1, q - 1, \dots, q - 1)) = q^{|\Phi^+|}.$$

If $G = \mathbf{G}^F$ is a finite group, with F a standard Frobenius endomorphism F_q, then the module $L(q - 1, \dots, q - 1)$ is called the *Steinberg module*, and is a projective simple module for G. Since it is a projective simple module, it lies in a block of defect zero, and hence must be the reduction modulo p of an irreducible ordinary character, often denoted St.

For both the trivial and Steinberg weights λ, we have $L(\lambda) = W(\lambda)$ (the former is obvious from the Weyl character formula, the latter is Exercise 9.1). For other dominant weights, the question of the composition factors of $W(\lambda)$, or equivalently understanding $L(\lambda)$, is perhaps the most important question in all of the theory of representations of algebraic groups. Certainly, since λ is a highest weight of $W(\lambda)$, the weights of $W(\lambda)$, and of all composition factors of $W(\lambda)$, are μ for $\mu \leq \lambda$. In fact, by [305, Proposition 21.3] the weights of $W(\lambda)$ are exactly those dominant weights μ such that $\mu \leq \lambda$, together with all conjugates by the Weyl group. By a theorem of Premet [466] the same is true for $L(\lambda)$, provided $p \neq 2$ for groups of types B_n, C_n and F_4, and $p \neq 2, 3$ for type G_2.

If $W(\lambda) \neq L(\lambda)$, then there must be an indecomposable quotient module M such that M has two composition factors: $L(\lambda)$ as a quotient and $L(\mu)$ as a submodule. In particular this means that $\mathrm{Ext}^1_{\mathbf{G}}(L(\lambda), L(\mu)) \neq 0$ (see Theorem 9.1.8). A necessary condition on when this extension group can be non-zero is called the 'linkage principle', and in order to define it we need to extend the Weyl group action on abstract weights to an action of an 'affine Weyl group'.

As we stated earlier in this section, the Weyl group acts on abstract weights, and given any weight there is a unique dominant weight in the orbit under W. Given a prime p, we add in one more set of symmetries of the weight lattice: $\lambda \mapsto \lambda + p\alpha^\vee$ for α a simple root. Here $\alpha^\vee = (2/(\alpha, \alpha))\alpha$, called the *coroot* corresponding to α. The coroots form a root system as well, and it has the same type as Φ, except that B_n and C_n are swapped.

The group generated by these translations, together with the Weyl group W, is denoted W_p and called the *affine Weyl group*. (See Exercise 9.2 for a different way to define the affine Weyl group.)

Since W_p acts on the root lattice as a reflection group, it also acts on the space into which the root lattice embeds, $X(\mathbf{T}) \otimes_{\mathbb{Z}} \mathbb{R}$. For λ an element of this space and $w \in W_p$, write λ^w for the image of λ under this element. Notice that W_p is the semidirect product of the translations by the Weyl group. (The translations are adding $p\alpha^\vee$ for α a root, as we saw before.) Thus we may express λ^w as $\lambda^t + v$, where t is an element of the Weyl group and v is p times an element of the coroot lattice (i.e., all $\mathbb{Z}$-linear combinations of coroots).

If w is an element of W_p and λ is an element of $X(\mathbf{T}) \otimes_{\mathbb{Z}} \mathbb{R}$, then define $\lambda \cdot w$, called the *dot action* of w on λ, to be

$$(\lambda + \rho)^w - \rho,$$

so applying an element of the affine Weyl group, but with the 'origin' of the transformation shifted by ρ. Actually computing this isn't too bad: we may write this as $(\lambda + \rho)^t + v - \rho$, and then as $\lambda^t + v + (\rho^t - \rho)$ since W is a set of linear transformations of $X(\mathbf{T}) \otimes_{\mathbb{Z}} \mathbb{R}$.

Theorem 9.1.5 (Linkage Principle) *If $\mathrm{Ext}^1_{\mathbf{G}}(L(\lambda), L(\mu)) \neq 0$, then λ and μ lie in the same orbit of the affine Weyl group W_p under the dot action. Consequently, the same holds if $L(\lambda)$ and $L(\mu)$ are composition factors of some indecomposable module.*

Weight	$\dim(L(\lambda))$	$\dim(W(\lambda))$	Weight	$\dim(L(\lambda))$	$\dim(W(\lambda))$
10	3	3	32	39	42
11	8	8	33	63	64
20	6	6	40	15	15
21	15	15	41	35	35
22	19	27	42	60	60
30	10	10	43	90	90
31	18	24	44	125	125

Table 9.1 Dimensions of simple and Weyl modules for A_2 in characteristic 5

This result was originally conjectured by Verma; Humphreys proved the result for p greater than the Coxeter number h [304], before Kac and Weisfeiler gave a full proof [338].

Example 9.1.6 Let $\mathbf{G} = A_2$. The natural 3-dimensional module is labelled by $L(1, 0)$, and its dual $L(0, 1)$. In general, $L(i, j)$ is the dual of $L(j, i)$. For $p = 5$, the dimensions of $L(\lambda)$ and $W(\lambda)$ are given in Table 9.1. (Write ij for (i, j), so that 11 is the weight $(1, 1)$, for example.) From this table, we see that $L(\lambda) = W(\lambda)$ in most cases, but not for the highest weights 22, 31, 32 and 33. For weight 33, since $\dim(W(33)) - \dim(L(33)) = 1$, we must have an extension between $L(33)$ and $L(00)$ (as it is the only possible module of dimension 1). Thus there must exist an element $w_1 \in W_5$ such that $00 \cdot w_1 = 33$.

Let us prove this. First, let t be an element of the Weyl group W. If $\lambda = 00$ then $\lambda + \rho = \rho$, so $(\lambda + \rho)^t$ is a root of Φ (since ρ lies in Φ). In particular, ρ is both a dominant weight and a root, so we may choose t so that $(\lambda + \rho)^t = -\rho$. Now let $w_1 = t + v$, where $v = 5\rho$. Thus $\rho \cdot w_1 = \rho^{w_1} - \rho = 3\rho$, which is the weight 33, as claimed.

What about the other highest weight modules? We have that $\dim(W(32)) - \dim(L(32)) = 3$, so either there are three trivial composition factors in $W(32)$ or a single copy of $L(10)$ (or its dual), and the latter is the case, as 32 does not lie in the orbit of 00. (The orbit of 00 under the affine Weyl group has 5-restricted dominant weights 00, 04, 40, 33, 34 and 43: check this!) Thus there should be an element $w_2 \in W_5$ such that $10 \cdot w_2 = 32$.

Next, we have $\dim(W(31)) - \dim(L(31)) = 6$, and indeed the other composition factor is a single composition factor $L(20)$. Finally, $\dim(W(22)) - \dim(L(22)) = 8$, and we find a composition factor $L(11)$. Thus there are two more elements $w_3, w_4 \in W_5$ such that $31 = 20 \cdot w_3$ and $22 = 11 \cdot w_4$.

To fill in the details of this example, see Exercise 9.4.

In Sect. 8.3 we discussed Schaper's formula for Specht modules of symmetric groups. This is often called the Jantzen–Schaper formula, because it is very similar to (and can be derived from) the Jantzen sum formula for Weyl modules. It first appeared in [332] but see also, for example, [307, Section 3.9] or [333, II.8.19].

Although we have not discussed it, there is a bilinear form that may be placed on the Weyl module $W(\lambda)$, and this bilinear form induces a filtration of $W(\lambda)$

$$W(\lambda) = W(\lambda)^0 \geq W(\lambda)^1 \geq \cdots$$

such that $W(\lambda)^i = 0$ for some i, and $W(\lambda)^0/W(\lambda)^1 = L(\lambda)$. This filtration is important because, like the Schaper formula, there is an explicit formula, as a $\mathbb{Z}$-linear combination of characters of Weyl modules, for the character of

$$\bigoplus_{i\geq 0} W(\lambda)^i.$$

This means that the character $\mathrm{ch}(L(\lambda))$ of $L(\lambda)$ can be expressed as a $\mathbb{Z}$-linear combination of $\mathrm{ch}(W(\mu))$ for $\mu \leq \lambda$. Each of these characters may be computed by the Weyl character formula, so if the coefficients were understood then we would understand $L(\lambda)$.

Of course, knowing the coefficients cannot be easy. But Lusztig's conjecture gives us these coefficients for certain λ, namely those in the orbit of 0 under the dot action of W_p, that are 'not too far away' from 0 in a technical sense we will see.

The *Jantzen region* is the collection of all dominant weights $0 \cdot (-w)$ for those $w \in W_p$ with

$$\langle -w\rho, \alpha_0^\vee \rangle \leq p(p - h + 2),$$

where α_0 is the highest short root and $\alpha_0^\vee$ is the coroot of α_0, and h is the Coxeter number of $\mathbf{G}$.

Conjecture 9.1.7 (Lusztig Conjecture [403][1]) Assume that $p > h$ and $w \in W_p$. If $0 \cdot (-w)$ lies in the Jantzen region, then

$$\mathrm{ch}(L(0 \cdot (-w))) = \sum_{u\leq w} (-1)^{\ell(w)-\ell(u)} P_{u,w}(1) \cdot \mathrm{ch}(W(0 \cdot (-u)),$$

where only those u such that $0 \cdot (-u)$ is dominant are allowed in the sum, ℓ denotes the length function on the affine Weyl group W_p, and $\leq$ is the Bruhat ordering on W_p. In this formula, $P_{u,w}$ is the Kazhdan–Lusztig polynomial.

If, for a type of group, p is large, then Lusztig's conjecture is true, by work of Andersen, Jantzen and Soergel [15], but no explicit definition of 'large' is given there. Fiebig gave [222] a massive lower bound on p in terms of the Coxeter

[1] In [403, Problem IV] one will not find $p > h$ as the condition. First, the condition there is not explicitly given as a condition on the Coxeter number at all. In [333, II.8.22], we see that, via work of Kato [310], the original conjecture of Lusztig is equivalent to the one here with $p \geq 2h - 3$. The conjecture with $p > h$ became the accepted version; see also the footnote in [563].

number h of $\mathbf{G}$. In addition, the conjecture is known for various small-rank groups, particularly A_1, A_2, B_2, G_2, A_3, and certain cases of A_4 (see [307, Section 3.12] and the references therein).

But the conjecture at this level of generality is false. Williamson, in a seminal paper [563], proved not only that p must be larger than h, but that no polynomial bound in h will work. There was already a 'problem' with Lusztig's conjecture, because it doesn't give a guess as to what the decomposition of the Weyl module into simple modules is for small primes, $p < h$. Since the so-called 'unstable range', that is, the primes for which Lusztig's conjecture does not hold, grows at least super-polynomially with the Coxeter number, what is needed now is to understand the unstable range, because it is no longer a small-prime problem, but now a many-prime problem.

Having considered the simple modules, how about the p-blocks of $\mathbf{G}$ and $\mathbf{G}^F$? (A p-block of $\mathbf{G}$ can be thought of as a connected component of the graph with vertices all simple modules, and an edge between two if there is non-trivial extension between them.) How do they compare? The linkage principle (Theorem 9.1.5) is the main result in this direction, and proves that the affine Weyl group defines the p-blocks. It shows that the orbits of the affine Weyl group under the dot action are unions of p-blocks of $\mathbf{G}$. Normally it is a single p-block [167] (see also [333, II.7.2(2)]).

We write the linkage principle down again in the first part of the next theorem, since it is so fundamental to this result.

Theorem 9.1.8 *Let λ and μ be dominant weights.*

(i) *If* $\mathrm{Ext}^1_{\mathbf{G}}(L(\lambda), L(\mu)) \neq 0$ *then* $\lambda = \mu \cdot w$ *for some* $w \in W_p$.
(ii) *If* $\mathrm{Ext}^1_{\mathbf{G}}(L(\lambda), L(\mu)) \neq 0$ *then either* $\lambda < \mu$ *or* $\mu < \lambda$. *In particular,* $L(\lambda)$ *cannot extend itself.*
(iii) *Every extension of* $L(\lambda)$ *by* $L(\mu)$ *is a quotient of* $W(\lambda)$. *Therefore*

$$\mathrm{Ext}^1_{\mathbf{G}}(L(\lambda), L(\mu)) \cong \mathrm{Hom}_{\mathbf{G}}(\mathrm{rad}(W(\lambda)), L(\mu)).$$

The last part of this result is [333, II.2.14]. The second part of this follows from the third and the fact that the composition factors of $W(\lambda)$ are modules $L(\mu)$ for $\mu < \lambda$, and one copy of $L(\lambda)$ (from the Jantzen sum formula).

Thus, for each dominant weight λ, the set of $L(\lambda \cdot w)$ for $w \in W_p$ and $\lambda \cdot w$ dominant is a union of blocks.

How does this compare with the situation for the finite group $G = \mathbf{G}^F$? Humphreys in [303], extending work of Dagger [148] to all groups, showed the following theorem.

Theorem 9.1.9 *Let G be a group of Lie type in characteristic p. Every p-block of G has defect group either a Sylow p-subgroup of G or the trivial group.*

But what is the distribution of the $L(\lambda)$ among the blocks? The Steinberg module (i.e., the module $L(\lambda)$ for λ the Steinberg weight) is projective, so lies in a block on its own. Curtis proved [137] that the Steinberg module is the only projective module if $G = G'$, and by Clifford's theorem (Chap. 7) for groups like $\mathrm{PGL}_n(q)$ every projective simple module restricts to the derived subgroup as the Steinberg module. Since the Steinberg module is in a block of defect zero (see Exercise 9.3), it is the reduction modulo p of a module over a field of characteristic 0, so the following definition makes sense.

Definition 9.1.10 The *Steinberg character* is the ordinary character whose p-modular reduction is the character of the Steinberg module.

The Steinberg character satisfies lots of nice properties: see Theorem 9.2.12. For example, the character values $\mathrm{St}(x)$ are easy to describe: if x is p-regular then $\mathrm{St}(x)$ is up to sign the order of a Sylow p-subgroup of $C_G(x)$, and $\mathrm{St}(x) = 0$ otherwise.

What about the blocks of maximal defect? From the remarks at the end of [303], we see that the number of such blocks is equal to the order of $Z(G)$. Furthermore, if M and N are simple kG-modules in the same p-block then a generator z for $Z(G)$ acts on both as a scalar matrix, and in fact on both with the same scalar. (This follows from the definition of a central character in Sect. 2.2.)

Thus, for example, the group $\mathrm{PSL}_n(q)$ for q a power of p has only two p-blocks, one with the Steinberg and one with everything else, quite a different situation to the algebraic group.

If the p-blocks of the finite and algebraic groups are so different, this must be because $\mathrm{Ext}^1(L(\lambda), L(\mu))$ varies significantly depending on whether we have $\mathbf{G}$ or G. Thus we are interested in the restriction map

$$\mathrm{Ext}^1_{\mathbf{G}}(L(\lambda), L(\mu)) \to \mathrm{Ext}^1_{kG}(L(\lambda), L(\mu)),$$

i.e., the map that sends a module for $\mathbf{G}$ that is the extension of $L(\lambda)$ by $L(\mu)$ to the restriction of it to G. This map cannot be surjective, but is it injective?

Theorem 9.1.11 *Let $G = \mathbf{G}^F$ for F a Frobenius endomorphism, and let q be as in Steinberg's restriction theorem, Theorem 9.1.3. Let λ and μ be q-restricted dominant weights. The restriction map*

$$\mathrm{Ext}^1_{\mathbf{G}}(L(\lambda), L(\mu)) \to \mathrm{Ext}^1_{kG}(L(\lambda), L(\mu))$$

is injective.

The original proof follows from more general results of Cline, Parshall, Scott and van der Kallen [119]. But Humphreys noted in [306] that this result actually follows from the fact that the restriction of $W(\lambda)$ to G still has a simple top $L(\lambda)$ if λ is q-restricted, as we see in Exercise 9.5.

There has been lots of work done on the dimensions of Ext^1 between simple modules for finite groups. If $G = \mathrm{SL}_2(q)$ then the dimension of $\mathrm{Ext}^1_{kG}(L(i), L(j))$ has been completely determined for all $0 \leq i, j < q$. This is given in [16, Corollaries 3.9 and 4.5] by Andersen, Jørgensen and Landrock for odd primes: for $p = 2$ it is much easier and was proved by Alperin [3].

Theorem 9.1.12 *Let* $\mathbf{G} = \mathrm{SL}_2$ *in characteristic p, and let* $G = \mathrm{SL}_2(q)$ *for* $q = p^r$. *Let* $0 \leq \lambda, \mu < q$, *and write*

$$\lambda = \sum_{i=0}^{r-1} \lambda_i p^i, \quad \mu = \sum_{i=0}^{r-1} \mu_i p^i,$$

for $0 \leq \lambda_i, \mu_i < p$, *their p-adic expansions.*

(i) *We have that* $\mathrm{Ext}^1_{\mathbf{G}}(L(\lambda), L(\mu))$ *has dimension* 0 *or* 1, *and it has dimension* 1 *if and only if there exists some* j *such that*

- $\lambda_i = \mu_i$ *for* $i \neq j, j+i$, *and there is some* $i < j$ *such that* $\lambda_i \neq p-1$,
- $\lambda_j = p - \mu_j - 2$, *and*
- $\lambda_{j+1} = \mu_{j+1} \pm 1$.

(ii) *Set* $\lambda_r = \lambda_0$ *and* $\mu_r = \mu_0$. *We have that* $\mathrm{Ext}^1_{kG}(L(\lambda), L(\mu))$ *has dimension* 0, 1 *or* 2. *It has dimension* 2 *if and only if* $r = 2$ *and*

$$\{\lambda, \mu\} = \{(p^2 - 3)/2, (p^2 - 2p - 1)/2\},$$

or $q = 9$ *and* $\{\lambda, \mu\} = \{0, 4\}$. *It has dimension* 1 *if and only if there exists some* j *such that*

- $\lambda_i = \mu_i$ *for* $i \neq j, j+i$,
- $\lambda_j = p - \mu_j - 2$, *and*
- $\lambda_{j+1} = \mu_{j+1} \pm 1$.

(It might be instructive to note that for $r = 1$ the group G has a cyclic Sylow p-subgroup, and therefore there are Brauer trees. Example 5.1.5 gives the Brauer trees of $\mathrm{SL}_2(p)$, and as we saw earlier in this section, $L(i)$ is the unique simple module of dimension $i + 1$. One might check that the result above coincides with the dimensions of Ext^1 implied by the structure of the Brauer tree in Example 5.1.5.)

For SL_3, for the algebraic group and $p > 3$, the results appear in the work of Andersen [14], extracted from the Ph.D. thesis of Yehia [573], and Andersen also computes the extensions for $\mathrm{SL}_3(p)$, although the tables are not error-free according to [307, Section 12.11]. In fact, that section goes into detail, describing many situations where extensions (and 1-cohomology) have been computed, and gives a good summary. More or less, it is groups of small rank, or small primes, or small dominant weights. For example, Ext^1 between all pairs of simple modules are determined for $\mathrm{SL}_3(2^n)$ and $\mathrm{SU}_3(2^n)$ in a paper of Sin [524].

The groups G_2 are treated in detail in [307, Chapter 18], and many of the ideas in this section can be seen there, such as the affine Weyl group, the composition factors of the Weyl modules, and some projective modules.

We have of course not given anywhere near a complete overview of the representation theory of these groups in defining characteristic, although at this point it should be possible to skim-read [307] and see what the main topics are.

9.2 Unipotent Classes and Characters

While the defining characteristic case above has some difficult parts, at least the basics are relatively straightforward. With the non-defining characteristic case, even the basic definitions are difficult. We start with the ordinary characters of a group of Lie type $G = G(q)$, and so here k is a field of characteristic 0, and q is a power of a prime r.[2] Specifically, G is the fixed points of a Steinberg endomorphism F of a connected, reductive algebraic group $\mathbf{G}$, and the prime power q is as it appears in Steinberg's restriction theorem, Theorem 9.1.3. However, our results will often apply to simple groups of Lie type, such as $\mathrm{PSL}_n(q)$, which do not fit into this setup.

Recall the 'Jordan decomposition' for elements of torsion groups (i.e., groups all of whose elements have finite order): let $g \in G$ be an element, and let r be a prime. Write $g = g_u g_s$, where g_u is an r-element, g_s is an r'-element, and $g_u g_s = g_s g_u$. (In fact, g_u and g_s must be powers of g, and this expression is unique.) We call g_u the *unipotent part* and g_s the *semisimple part.*

If g is conjugate to h, then we have $g_u g_s = x^{-1} h_u h_s x = h_u^x h_s^x$. Since this Jordan decomposition is unique, we must have $g_u = h_u^x$ and $g_s = h_s^x$. In particular, the semisimple parts of g and h are conjugate. Thus we assume that $g_s = h_s$; then $g_u = h_u^x$ and $g_s = g_s^x$, i.e., $x \in C_G(g_s)$. Thus conjugacy classes in G are parametrized by pairs (s, u), where s is a *semisimple element* (i.e., r'-element) of G up to conjugacy and u is a *unipotent element* (i.e., r-element) of $C_G(s)$ up to conjugacy in $C_G(s)$.

We can attempt to do the same thing with irreducible characters, and parametrize them by pairs (χ_s, χ_u), where χ_s is a 'semisimple character' and χ_u is a 'unipotent character', where it is not clear what either of these means, and also it is not clear for which groups χ_s and χ_u are characters. In fact, we normally don't use χ_s now, and just use a semisimple conjugacy class, but we have to be careful because the class doesn't lie in G itself, but in the *Langlands dual* G^*. Formally, G^* is defined by swapping the roots and coroots of G, but this requires us to define G in terms of a root datum. Informally, G^* has the same type as G (except types B and C are swapped) and the order of $Z(G)$ becomes the order of $G^*/(G^*)'$ (so $\mathrm{PGL}_n(q)$ and $\mathrm{SL}_n(q)$ are dual to one another, and $\mathrm{GL}_n(q)$ is self-dual). Note that there is a Steinberg endomorphism F^* on $\mathbf{G}^*$ that is dual to F on $\mathbf{G}$.

We start with the example of $\mathrm{GL}_n(q)$, where everything is fairly easy to explain.

Example 9.2.1 Let $G = \mathrm{GL}_n(q)$. The unipotent classes of G are parametrized by partitions of n (as they are sums of Jordan blocks). Define a *unipotent character* of G to be a constituent of the permutation character on a Borel subgroup of G, which is just (up to conjugacy) upper triangular matrices. It turns out that these are parametrized by partitions of n as well, and in fact the multiplicities of these

---

[2] You will find in the literature that p is normally the prime over which G is defined, and k has characteristic ℓ. However, in the rest of this book p is the characteristic of k, and I don't think it is a good idea to change notation for this chapter alone.

constituents are independent of q. If $\lambda = (\lambda_1, \ldots, \lambda_r)$ is a partition of n, then the character degree of χ^λ is

$$\chi^\lambda(1) = q^{\sum_{i=1}^r (i-1)\lambda_i} \frac{\prod_{i=1}^n (q^i - 1)}{\prod_{(i,j)\in\lambda} (q^{h_{i,j}} - 1)},$$

where $h_{i,j}$ is the hook length of the box (i, j) of λ, defined in the previous chapter. (Compare this at $q = 1$ to the character degrees of the symmetric group, but remember to remove the factors $(q - 1)$ from the fraction first! Of course, the symmetric group is the Weyl group of G.)

Since in this case $G \cong G^*$, we will be a bit sloppy and pretend that in our pair (s, ϕ) that will label an arbitrary irreducible character, s lies in G rather than G^*. The centralizer of a semisimple element of G is always a Levi subgroup, and is a product of smaller GLs (potentially over fields of order a power of q), and the unipotent characters of a product of groups are just the tensor products of the unipotent characters.

Thus we should obtain a parametrization (s, ϕ) of all characters of $\mathrm{GL}_n(q)$, where s is a semisimple class and ϕ is a unipotent character of $C_G(s)$. (If $s = 1$ then $(1, \phi)$ labels ϕ itself.) The degree of $\chi_{(s,\phi)}$ is

$$|G : C_G(s)|_{r'} \cdot \phi(1),$$

where $m_{r'}$ denotes the r'-part of the integer m.

If $\lambda = (n)$ then χ^λ is the trivial character, and if $\lambda = (1^n)$ then χ^λ is the Steinberg character, of degree $p^{n(n-1)/2}$. (Check this with the formula above.)

A construction of these characters first appears in the work of Green [262]. The first hints of the theory of unipotent characters and Jordan decomposition of characters appear in this work.

Example 9.2.2 We work out the details for $G = \mathrm{GL}_2(q)$, which has order $q(q - 1)^2(q + 1)$. The two partitions of 2 are (2) and (1^2), which yields characters of degrees 1 and q respectively (see the previous example). Any centralizer of a semisimple element s is either G itself or is abelian, of order $(q - 1)^2$ for diagonalizable elements and of order $(q^2 - 1)$ for non-diagonalizable elements, but both tori. There is a unique unipotent character of a torus, of degree 1.

(i) If s is a scalar matrix then $G = C_G(s)$, so $|G : C_G(s)|_{r'} = 1$. We have two characters $(s, 1)$ and (s, St), of degrees 1 and q respectively. There are $q - 1$ such elements s.

(ii) If s is diagonalizable but not a scalar then $|C_G(s)| = (q - 1)^2$, so $|G : C_G(s)|_{r'} = q + 1$. We have a unique character $(s, 1)$, of degree $q + 1$. There are $(q - 1)(q - 2)/2$ such conjugacy classes.

(iii) If s is not diagonalizable then $|C_G(s)| = (q^2 - 1)$, so $|G : C_G(s)|_{r'} = q - 1$. We have a unique character $(s, 1)$, of degree $q - 1$. There are $q(q - 1)/2$ such conjugacy classes.

These match the degrees of the characters of $\mathrm{GL}_2(q)$, as found independently by Jordan [336] and Schur [512] in 1907. One may also check them against the table in [328, Chapter 28].

The group $\mathrm{GL}_n(q)$ is the Platonic ideal for a group of Lie type. All of the other groups fail to live up to its standard in some way or another. One might generally try to define unipotent characters as the constituents of the permutation character on B for B a Borel subgroup, but that doesn't give us all of the unipotent characters. The unipotent characters are no longer parametrized by the characters of the Weyl group (these two facts are connected) like for $\mathrm{GL}_n(q)$, and the numbers of unipotent classes, unipotent characters, and constituents of the permutation module on B, are all different. (For G_2 for example there are four unipotent classes in the algebraic group, or five if $r = 3$, $6 + \gcd(6, r)$ unipotent classes for the finite group, and ten unipotent characters for all primes, six of which lie in the permutation character on B.) The Jordan decomposition of characters itself only works for certain types of Lie type groups, those where the group $\mathbf{G}$ has connected centre (see Sect. 9.4).

We also see here a difference between 'geometric' and 'rational' conjugacy. Two elements of $G = \mathbf{G}^F$ are *rationally conjugate* if they are conjugate in G, and are *geometrically conjugate* if they are conjugate in $\mathbf{G}$. How the geometric conjugacy classes break up into rational classes is important for counting unipotent and semisimple classes of the finite group, and is connected to the connectedness of the centre $Z(\mathbf{G})$ (see Theorem 9.4.4 below). It will also be important when discussing the parametrization of irreducible characters of G in Sect. 9.4.

The rest of this section will discuss unipotent characters only. In order to define them properly we need the concept of Lusztig induction which, while defined later in this section, is too difficult to understand well without some geometry. We will start with Harish-Chandra induction, developed in [274], which is much easier. To be completely formal, we should talk about the algebraic group and a Frobenius endomorphism on it, or a group with a BN-pair; we will take the first approach later, and for now give a finite-group definition in the spirit of groups with a BN-pair, which is a lot simpler but does not fit into the formal framework used by many standard textbooks, which often go via algebraic groups.

(We use the notation $R\mathrm{Irr}(G)$ to mean all R-linear combinations of elements of $\mathrm{Irr}(G)$. Thus $\mathbb{Z}_{\geq 0}\mathrm{Irr}(G)$ is the set of characters of G, and $\mathbb{Z}\mathrm{Irr}(G)$ is the set of *virtual characters*, also known as *generalized characters*.)

Definition 9.2.3 Let L be a Levi subgroup of a group of Lie type G, contained in a parabolic subgroup P of G. *Harish-Chandra induction* from L to G is the map (a functor) $R_L^G : \mathbb{Z}_{\geq 0}\mathrm{Irr}(L) \to \mathbb{Z}_{\geq 0}\mathrm{Irr}(G)$ obtained by inflating a character of L to P, and then inducing to G.

Harish-Chandra restriction ${}^*R_L^G : \mathbb{Z}_{\geq 0}\mathrm{Irr}(G) \to \mathbb{Z}_{\geq 0}\mathrm{Irr}(L)$ is the adjoint to R_L^G, and is the map obtained by restricting a character to P and then taking the sum of those constituents whose restriction to the unipotent radical of P is trivial (this operation is sometimes called *deflation*, as it is the opposite of inflation).

Notice that this means that for $G = \mathrm{GL}_n(q)$, the unipotent characters are precisely the constituents of $R_T^G(1)$, where T is a maximal torus lying in B.

One important point to make is that although we have written R_L^G, we used a parabolic subgroup P containing L in the definition. However, R_L^G is independent of the parabolic subgroup containing L (see for example [161, Proposition 6.1]).

Definition 9.2.4 Let G be a group of Lie type and let T be a maximal torus of G lying in a Borel subgroup B. A *principal-series character* is an irreducible character $\chi \in \mathrm{Irr}(G)$ such that χ is a constituent of $R_T^G(\theta)$ for some $\theta \in \mathrm{Irr}(T)$.

The point is that for $\mathrm{GL}_n(q)$, all unipotent characters are principal series, but that isn't true for the other groups.

Definition 9.2.5 An irreducible character $\chi \in \mathrm{Irr}(G)$ is *cuspidal* if ${}^*R_L^G(\chi) = 0$ for all proper Levi subgroups L of G. This is equivalent to χ not appearing in the image of R_L^G for each Levi subgroup L.

The existence of cuspidal unipotent characters is why understanding unipotent characters is 'hard'; we cannot work inductively using only Harish-Chandra induction.

Example 9.2.6 Let $G = \mathrm{GL}_n(q)$. If $L = (q-1) \cdot \mathrm{GL}_{n-1}(q)$ is a Levi subgroup of G, then R_L^G acts on unipotent characters in the same way as the branching rule for symmetric groups (Theorem 8.1.6): $R_L^G(\chi^\lambda)$ is the sum of all χ^μ for μ obtained from λ by adding a single box to λ.

If L instead were a Levi subgroup $\mathrm{GL}_m(q) \times \mathrm{GL}_{n-m}(q)$ for $1 < m < n$, then we would need to use the Littlewood–Richardson rule, just as for the subgroup $S_m \times S_{n-m}$ inside S_n. We don't have time to discuss this in this book, so see for example [498, Section 4.9] or [327, Section 2.8].

Example 9.2.7 Let $G = \mathrm{GU}_n(q)$. The unipotent characters of G are parametrized just as for $\mathrm{GL}_n(q)$, so by partitions of n. The character degree of χ^λ is the same as that of $\mathrm{GL}_n(q)$ but with q replaced by $-q$ (if this degree is negative, change the sign!). This is a facet of *Ennola duality*, which is broadly speaking the statement that the character table of $\mathrm{GL}_n(q)$ can be turned into the character table of $\mathrm{GU}_n(q)$ by replacing q by $-q$, and changing some roots of unity (as conjectured by Ennola in [189], proved by Hotta and Springer in [301] for large characteristics, and by Kawanaka in general in [341]).

For branching rules now, because G is a twisted group there isn't a Levi subgroup $\mathrm{GU}_{n-1}(q)$ to take Harish-Chandra restriction to, and we have to drop down to a Levi subgroup L that is the product of $\mathrm{GU}_{n-2}(q)$ and a torus $\mathrm{GL}_1(q^2)$ of order (q^2-1). Now $R_L^G(\chi^\lambda)$ is the sum of all χ^μ where μ is obtained from λ by adding a single 2-hook.

But now we have a problem: if λ is a 2-core of size n (i.e., a partition $(m, m-1, m-2, \ldots, 1)$) then ${}^*R_L^G(\chi^\lambda) = 0$. Indeed, this holds for the other proper Levi subgroups as well, and a unipotent character χ^λ is cuspidal if and only if λ is a 2-core.

In order to define a unipotent character, we need to generalize Harish-Chandra induction. We go back to the basics: a principal series unipotent character is a constituent of $R_T^G(1)$ for T a maximal torus lying in a Borel subgroup, but we are missing some unipotent characters doing it this way. What about other maximal tori?

Write $G = \mathbf{G}^F$, where F is a Steinberg endomorphism of the algebraic group $\mathbf{G}$. The completely formal definition of Harish-Chandra induction is as follows: take an F-stable Levi subgroup $\mathbf{L}$ that is contained in an F-stable parabolic subgroup $\mathbf{P}$ of $\mathbf{G}$, and then let $L = \mathbf{L}^F$ and $P = \mathbf{P}^F$, before applying the construction R_L^G.

An F-stable maximal torus $\mathbf{T}$ contained in an F-stable Borel subgroup of $\mathbf{G}$ is called *maximally split*. All maximally split maximal tori are conjugate in G. For the standard Frobenius endomorphism $F = F_q$, $T = \mathbf{T}^F$ is isomorphic to a direct product of copies of $\mathbb{F}_q^\times$ for some q, but for other Frobenius endomorphisms they can look quite different, thus a *split* torus is one isomorphic with direct copies of $\mathbb{F}_q^\times$.

Of course, not all F-stable tori are split, and in fact almost all are not: all maximal tori $\mathbf{T}$ are conjugate in $\mathbf{G}$, but given two of them $\mathbf{T}_1$ and $\mathbf{T}_2$, their fixed points T_1 and T_2 need not be, and are usually not, conjugate in G. They don't even need to have the same order, for example.

Example 9.2.8 Let $G = \mathrm{GL}_n(q)$, and let T be a *Singer cycle*: to define this, note that while the irreducible representations of the cyclic group C_{q^n-1} are definable over $\mathbb{F}_{q^n}$, the sum of an orbit under the Galois group $\mathrm{Gal}(\mathbb{F}_{q^n}/\mathbb{F}_q)$ has dimension n, and is definable over $\mathbb{F}_q$. In other words, there is an embedding

$$C_{q^n-1} \to \mathrm{GL}_n(q),$$

and the image of this is $\mathbf{T}^F$ for a maximal torus $\mathbf{T}$ of GL_n.

Generalizing this example, for $m \mid n$ we may embed $\mathrm{GL}_{n/m}(q^m)$ inside $\mathrm{GL}_n(q)$, then take a standard maximal torus inside the $\mathrm{GL}_{n/m}(q^m)$, to construct an embedding

$$(C_{q^m-1})^{\times(n/m)} \to \mathrm{GL}_n(q).$$

Each of these are of the form $\mathbf{T}^F$ for some maximal torus $\mathbf{T}$ of GL_n.

We cannot define Harish-Chandra induction R_T^G unless T is a maximally split torus. Deligne and Lusztig generalized this to the case where T is no longer maximally split (but is still F-stable). In [157], they constructed a map R_T^G, but it no longer sends characters of T to characters of G, but instead sends characters of T to virtual characters of G, i.e., integral linear combinations of irreducible characters, but where the coefficients can be negative as well as positive.

Again, R_T^G should depend on the choice of a Borel subgroup $\mathbf{B}$ containing $\mathbf{T}$, where $T = \mathbf{T}^F$. However, this time $\mathbf{B}$ is *not* F-stable, so we really do have to move to the algebraic group language to write about it. In [157] though it was proved that R_T^G does not depend on the choice of $\mathbf{B}$, by proving that an analogue of the Mackey formula (Theorem 3.2.14) holds for R_T^G. For R_L^G and R_T^G as defined so far, write $R_{\mathbf{L}\subset\mathbf{P}}^{\mathbf{G}}$ to indicate the choice of parabolic $\mathbf{P}$ containing the F-stable subgroup $\mathbf{L}$.

Definition 9.2.9 Let $\mathbf{L}$ and $\mathbf{M}$ be F-stable Levi subgroups, contained in (possibly not F-stable) parabolic subgroups $\mathbf{P}$ and $\mathbf{Q}$ of $\mathbf{G}$ respectively. We say that *the Mackey formula holds* for $\mathbf{L}$, $\mathbf{M}$, $\mathbf{P}$ and $\mathbf{Q}$ if

$$R_{\mathbf{L}\subset\mathbf{P}}^{\mathbf{G}} \circ {}^*R_{\mathbf{M}\subset\mathbf{Q}}^{\mathbf{G}} = \sum_t (\operatorname{ad} t) \circ {}^*R_{\mathbf{L}^t\cap\mathbf{M}\subset\mathbf{P}^t\cap\mathbf{Q}}^{\mathbf{L}^t} \circ R_{\mathbf{L}^t\cap\mathbf{M}\subset\mathbf{P}^t\cap\mathbf{Q}}^{\mathbf{M}}$$

where the sum runs over all $(\mathbf{L}^F, \mathbf{M}^F)$-double coset representatives t with the property that $\mathbf{L}^t \cap \mathbf{M}$ contains a maximal torus of $\mathbf{G}$, and $\operatorname{ad} t$ denotes the conjugation action by $t \in \mathbf{G}^F$.

Of course, we have not stated that $R_{\mathbf{L}\subset\mathbf{P}}^{\mathbf{G}}$ exists when $\mathbf{L}$ is F-stable but not maximally split (but it does, and it is usually called *Lusztig induction* in this case). By Exercise 9.6, if the Mackey formula holds for a collection of Levi subgroups $\mathbf{L}$ of $\mathbf{G}$, closed under taking subgroups, and parabolics $\mathbf{P}$ containing them, then $R_{\mathbf{L}\subset\mathbf{P}}^{\mathbf{G}}$ does not depend on $\mathbf{P}$ (as long as $\mathbf{P}$ is chosen from the collection for which the Mackey formula holds!) As Deligne and Lusztig proved that the Mackey formula holds whenever $\mathbf{T}$ is an F-stable (but not necessarily maximally split) torus, R_T^G is not dependent on the Borel subgroup containing $T = \mathbf{T}^F$.

Definition 9.2.10 An irreducible character $\chi \in \operatorname{Irr}(G)$ is *unipotent* if χ is a constituent of $R_T^G(1)$ for $\mathbf{T}$ some F-stable maximal torus of $\mathbf{G}$.

This definition is not particularly useful without a definition of R_T^G though. Lusztig induction is actually the cohomology of a variety. Over the years there have been a variety of varieties that have been called Deligne–Lusztig varieties. If we are only interested in unipotent characters then we can simplify things a little [107, Section 7.7] so we will assume that in this explanation. We should note, though, that even the simplified definition involves ℓ-adic cohomology, so we can't exactly do very much more, it is just that the variety is easier to understand.

Write $G = \mathbf{G}^F$. Fix an F-stable maximal torus $\mathbf{T}_0$ inside an F-stable Borel subgroup $\mathbf{B}$, and let w be an element of the Weyl group $N_{\mathbf{G}}(\mathbf{T}_0)/\mathbf{T}_0$. The *Deligne–Lusztig variety* $\mathrm{X}(w)$ is given by

$$\mathrm{X}(w) = \{\mathbf{B}g \in \mathbf{G}/\mathbf{B} \mid (gF)g^{-1} \in \mathbf{B}w\mathbf{B}\}.$$

This set has a right G-action: if $x \in G$ and $\mathbf{B}g \in \mathrm{X}(w)$, we need to check that $\mathbf{B}(gx) \in \mathrm{X}(w)$ as well, i.e., $\big((gx)F\big)(gx)^{-1} \in \mathbf{B}w\mathbf{B}$, but $xF = x$ and so this is

$$\Big((gx)F\Big)x^{-1}g^{-1} = (gF)xx^{-1}g^{-1} = (gF)g^{-1}.$$

What does this have to do with R_T^G? Let $\mathbf{T}$ be another F-stable (but not necessarily maximally split) maximal torus of $\mathbf{G}$. Let $x \in \mathbf{G}$ be such that $\mathbf{T} = \mathbf{T}_0^x$, and note that $(xF)x^{-1}$ normalizes $\mathbf{T}_0$, thus corresponds to some element $w \in W$. We have that

$$R_T^G(1) = \bigoplus_i (-1)^i \mathrm{H}_c^i(\mathrm{X}(w), k)$$

where k is a field $\bar{\mathbb{Q}}_\ell$. (Hence the name ℓ-adic cohomology.)

Notice that if $\mathbf{T}$ is maximally split then we may take $\mathbf{T}_0 = \mathbf{T}$, and the element w is simply the identity. Thus the variety is $\mathrm{X}(1)$, and this variety is structurally much easier to understand than the general case: indeed, it is just the finite set G/B. Thus it is at least plausible that the two definitions of R_T^G coincide in this case.

This hasn't been entirely enlightening, although normally you can just treat unipotent characters as a black box; there are labels, we know the degrees, and they satisfy certain branching rules with respect to Levi subgroups, like we saw with $\mathrm{GL}_n(q)$. The degrees are polynomials in q that only depend on the type of the group, and these polynomials are products of cyclotomic polynomials (and powers of q, which in this theory is an honorary cyclotomic polynomial) and a rational number $1/a$ for a an integer divisible only by 'bad' primes (see the next section), which are 2 for all groups not of type A, 3 for all exceptional types, and 5 for type E_8. Often this is enough information to prove what you want to know about them.

Example 9.2.11 Let G be a symplectic or orthogonal group. The exact isogeny type is not important, since the unipotent characters remain the same for all isogeny types (see, for example, [85]). (Remember that $\mathrm{GO}_{2n}^+(q)$ is not allowed, for example, since the action of this group on the quasisimple subgroup is as a graph automorphism, so it is not a group of type D.)[3]

The parametrization of unipotent characters for all of these groups is by symbols. A *symbol* is an unordered pair (X, Y), where X and Y are both (potentially empty) subsets of $\mathbb{Z}_{\geq 0}$, for example $(\{0, 1\}, \emptyset)$. Fix a symbol $\Lambda = (X, Y)$, and write $X = \{x_1, \dots, x_a\}$ and $Y = \{y_1, \dots, y_b\}$, with $x_i > x_{i+1}$ and $y_i > y_{i+1}$ for all i. The

[3]Most people write $\mathrm{SO}_{2n}^{\pm}(q)$ for the special orthogonal group, i.e., the matrices of determinant 1 that lie in $\mathrm{GO}_{2n}^{\pm}(q)$. Some authors, for example [417], in characteristic 2, write $\mathrm{SO}_{2n}^{\pm}(q)$ for the simple group, so the group I would denote by $\Omega_{2n}^{\pm}(q)$, as is done in [122] and [566]. There are persuasive reasons to do this: the group $\mathrm{GO}_{2n}^+(q)$ induces a graph automorphism on $\Omega_{2n}^+(q)$, so is never the fixed points of a connected group of type D_n. If q is odd then $\mathrm{SO}_{2n}^+(q)$ is a group of type D_n, whereas for q even, $\mathrm{GO}_{2n}^+(q)$ already consists of matrices of determinant 1, so $\mathrm{SO}_{2n}^+(q)$ is not a group of type D_n in characteristic 2. Furthermore, the order of $\mathrm{SO}_{2n}^+(q)$ is given by a polynomial in q, but this polynomial must be doubled in characteristic 2 with the determinant 1 definition. Nevertheless, writing SO for anything other than the *special* transformations is difficult for me, and I will stick with the majority in not trying to redefine a group that has been around for a century. One can be safe and always consider $\mathrm{Spin}_{2n}^{\pm}(q)$ instead, which is always a group of type D_n for all q.

quantity $|X| - |Y|$ (defined up to sign) is the *defect* of Λ, and the quantity

$$\sum_{x \in X} x + \sum_{y \in Y} y - \left\lfloor \frac{(|X| + |Y| - 1)^2}{4} \right\rfloor$$

is the *rank* of Λ. As with β-sets in the previous chapter, introduce an equivalence relation on the set of symbols generated by $(X, Y) \sim (X', Y')$ if

$$X' = \{0\} \cup \{x + 1 \mid x \in X\}, \quad Y' = \{0\} \cup \{y + 1 \mid y \in Y\}.$$

Note that the defect and rank of equivalent symbols are the same.

The unipotent characters of the groups $\mathrm{Spin}_{2n+1}(q)$ and $\mathrm{Sp}_{2n}(q)$, of types B_n and C_n respectively, are both parametrized by symbols of odd defect and rank n. (Since groups of type B and C are the same in characteristic 2, and the unipotent characters and their degrees are independent of q, groups of types B and C have the same parametrization and unipotent—but not all!—character degrees.)

The unipotent characters of the groups $\mathrm{Spin}_{2n}^+(q)$ and $\mathrm{Spin}_{2n}^-(q)$, of types D_n and ${}^2D_n$ respectively, are parametrized by symbols of rank n and defect congruent to 0 modulo 4 and 2 modulo 4 respectively. Each symbol labels a single unipotent character unless $\Lambda = (X, X)$ (which must be for $\mathrm{Spin}_{2n}^+(q)$), in which case it labels two. In this case Λ is called *degenerate*.

We must also give the degree of the character χ_Λ. All of the degrees are similar, but not exactly the same. Write $a = |X|$, $b = |Y|$, n for the rank of Λ, and let $f_\Lambda(q)$ denote the quantity

$$\frac{\left(\prod_{i=1}^{n-1}(q^{2i}-1)\right)\left(\prod_{1 \le i < j \le a}(q^{x_i}-q^{x_j})\right)\left(\prod_{1 \le i < j \le b}(q^{y_i}-q^{y_j})\right)\left(\prod_{i,j}(q^{x_i}+q^{y_j})\right)}{2^{\lfloor (a+b-1)/2 \rfloor} q^{\binom{a+b-2}{2}+\binom{a+b-4}{2}+\cdots}\left(\prod_{i=1}^{a}\prod_{j=1}^{x_i}(q^{2j}-1)\right)\left(\prod_{i=1}^{b}\prod_{j=1}^{y_i}(q^{2j}-1)\right)}.$$

If Λ has odd defect then

$$\chi_\Lambda(1) = (q^{2n} - 1) f_\Lambda(q),$$

if Λ has defect congruent to 0 modulo 4 then

$$\chi_\Lambda(1) = (q^n - 1) f_\Lambda(q),$$

and if Λ has defect congruent to 2 modulo 2 then

$$\chi_\Lambda(1) = (q^n + 1) f_\Lambda(q).$$

The only exception is if Λ is degenerate, in which case the quantity $2^{\lfloor (a+b-1)/2 \rfloor}$ in the denominator is replaced by 2^a. If you are familiar with the orders of the classical groups, you can see why the polynomial factors in the cases differ like this. (See, for example, [107, Section 13.8] for more information.) These results are due to Lusztig [401].

If L is a maximal Levi subgroup of the same type as G, i.e., B_{n-1} inside B_n, C_{n-1} inside C_n, D_{n-1} inside D_n and ${}^2D_{n-1}$ inside ${}^2D_n$, then the branching rule is very similar to the symmetric groups: thinking of each set of the symbol as a β-set for a partition, one may add a single box to either X or Y. Harish-Chandra induction of χ_Λ, a symbol of rank $n-1$, to G, is the sum of all χ_Π, where $\Pi = (X', Y')$ ranges over all symbols where $X' = X$ and Y' is obtained from Y by adding a single box, or vice versa (with two characters if Π is degenerate), and if Λ is degenerate then both characters χ_Λ have the same Harish-Chandra induction.

Thus the cuspidal unipotent characters of G are symbols $(\{0, 1, \dots, r\}, \emptyset)$, and in fact all such symbols are cuspidal (as they don't appear in R_L^G for any other Levi subgroup). The rank of this symbol is easy to compute, and is $r(r+1)/2 - \lfloor r^2/4 \rfloor$. Notice that degenerate symbols other than $(\emptyset, \emptyset)$ cannot be cuspidal, and so G possesses at most one cuspidal unipotent character.

All of the unipotent characters that can be obtained by Harish-Chandra induction from the trivial character for the torus are called principal series characters, as we saw before. If χ is a cuspidal unipotent character for a Levi subgroup L, then all of the unipotent characters that lie in $R_L^G(\chi)$ are defined to be in the same *series*. Since for classical groups χ is unique, we often refer to them as 'L-series characters' rather than 'χ-series characters'.

Why do unipotent characters lie above a single cuspidal for a single Levi subgroup (up to conjugation)? The Mackey formula. Roughly speaking, if a unipotent character χ lies in a Harish-Chandra induction for both L_1 and L_2, then one may find a smaller Levi subgroup L, contained (up to conjugacy) in both of them, and a character ϕ of L such that χ is a constituent of $R_L^G(\phi)$. Thus there is (up to conjugacy) a minimal Levi subgroup such that χ lies in a Harish-Chandra induction from that subgroup (Exercise 9.7).

What is less clear is that there cannot be two cuspidal characters ϕ_1 and ϕ_2 for this Levi subgroup L such that χ is a constituent of both $R_L^G(\phi_i)$. The precise statement is that in this case there is an element $g \in N_G(L)$ such that $\phi_1^g = \phi_2$; this means that they form a single $N_G(L)$-conjugacy class of cuspidal characters. A proof of this, and a formal proof of the previous statement, can be found in [107, Proposition 9.1.5], which applies for all characters, not just unipotent ones. (Compare this definition to that of a source of an indecomposable module from Chap. 3.)

In exceptional groups, the unipotent characters are quite a lot more complicated, and we have to go case by case.

(i) For type G_2 there are ten unipotent characters: six in the principal series and four cuspidal, labelled $G_2[1]$, $G_2[-1]$, $G_2[\theta]$ and $G_2[\theta^2]$.

(ii) For type ${}^3D_4$ there are eight unipotent characters: six in the principal series and two cuspidal, labelled ${}^3D_4[1]$ and ${}^3D_4[-1]$.
(iii) For type F_4 there are 37 unipotent characters: 25 in the principal series, five in the B_2-series (there is a cuspidal unipotent character of B_2, as we can check above), and seven cuspidal, labelled $F_4^{\mathrm{I}}[1]$, $F_4^{\mathrm{II}}[1]$, $F_4[-1]$, $F_4[\theta]$, $F_4[\theta^2]$, $F_4[\mathrm{i}]$ and $F_4[-\mathrm{i}]$.
(iv) For type E_6 there are 30 unipotent characters: 25 in the principal series, three in the D_4-series (there is a cuspidal unipotent character of D_4, as we can check above), and two cuspidal, labelled $E_6[\theta]$ and $E_6[\theta^2]$.
(v) For type ${}^2E_6$ there are also 30 unipotent characters: again, 25 in the principal series; two cuspidal characters, labelled ${}^2E_6[\theta]$ and ${}^2E_6[\theta^2]$; this time a third cuspidal character ${}^2E_6[1]$, and two characters in the ${}^2A_5$-series.
(vi) For type E_7 there are 76 unipotent characters: 60 in the principal series, ten in the D_4-series, two lying above each cuspidal character of E_6, and two cuspidals, labelled $E_7[\mathrm{i}]$ and $E_7[-\mathrm{i}]$.
(vii) For type E_8 there are 166 unipotent characters: 112 in the principal series, 25 in the D_4-series, six lying above each cuspidal character of E_6, two above each cuspidal character of E_7, and thirteen cuspidals, labelled $E_8^{\mathrm{I}}[1]$, $E_8^{\mathrm{II}}[1]$, and $E_8[\alpha]$ for α one of $-1, \theta, \theta^2, -\theta, -\theta^2, \mathrm{i}, -\mathrm{i}$, and ζ^i for $i = 1, 2, 3, 4$.

What are these terms in brackets? They are the *eigenvalues of the Frobenius*; the Frobenius endomorphism acts on the terms of the complex of cohomology of the Deligne–Lusztig variety, and acts as a scalar root of unity times a power of q on the composition factors (the same scalar for each copy of a given module as a composition factor), with the scalar being what is written in square brackets. A cube root of 1 is θ, and a fifth root of 1 is ζ. As suggested, $E_6[\theta]$ is dual to $E_6[\theta^2]$, and in general the dual of a unipotent character is the corresponding character with complex conjugate eigenvalue of Frobenius, so the character is self-dual if and only if its eigenvalue of Frobenius is real.

The tables in [107, Section 13.9] give the full labels and their degrees, together with the Ree and Suzuki groups, which need a bit more setup to work well under this system. For example, the unipotent character degrees are not strictly products of cyclotomic polynomials, because the q involved is a power of $\sqrt{2}$ or $\sqrt{3}$, and over the field $\mathbb{Q}(\sqrt{2})$ the cyclotomic polynomials $\Phi_8(q)$ and $\Phi_{24}(q)$ factorize, and over $\mathbb{Q}(\sqrt{3})$ the polynomial $\Phi_{12}(q)$ factorizes. For some of the degrees one sees these factors of the polynomials, but not the whole polynomial as a factor.

We will also briefly mention how to understand the set of characters in a given Harish-Chandra series, i.e., those characters in the induction $R_L^G(\chi)$ for χ a cuspidal unipotent character of a Levi subgroup L.

The principal series characters are in one-to-one correspondence with the characters of the Weyl group. In fact, the endomorphism ring of the permutation module on the cosets of a Borel subgroup B is a *Hecke algebra*. We cannot talk much about Hecke algebras here (we said a little about them in Sect. 8.3), but they can be associated to a Weyl group, and they are deformations of the group algebra kW, so have the same number of irreducible modules, at least when $k = \mathbb{C}$. It can be shown that the number of distinct constituents of this permutation module is equal to the number of simple modules for the Hecke algebra [138].

We therefore know what to do for other series: take the endomorphism algebra of $R_L^G(\phi)$, where ϕ is a cuspidal unipotent character. This is a Hecke algebra, a deformation of the group algebra $kW_{G/L}$, where $W_{G/L}$ is the Weyl group of the quotient root system of G by that of L (see [107, Section 10.4]). It can be seen inside $\mathbf{G}$ as $(N_{\mathbf{G}}(\mathbf{L})/\mathbf{L})^F$, extending the definition of W as $N_{\mathbf{G}}(\mathbf{T})/\mathbf{T}$ for $\mathbf{T}$ a maximal torus of $\mathbf{G}$. If $\mathbf{L}$ is of type E_6 and $\mathbf{G}$ is of type E_7, the quotient root system E_7/E_6 has rank $7-6=1$, so must be of type A_1. Hence we should expect two irreducible characters in each $E_6[\theta^i]$-series of E_7, agreeing with what we saw above.

For E_6, the quotient root system E_6/D_4 is A_2, so we expect three characters; for E_7, the quotient root system E_7/D_4 is C_3, whose Weyl group has ten characters; the quotient root systems E_8/E_6 and E_8/D_4 are G_2 and F_4 respectively, so we expect six and 25 characters respectively (the number of principal-series characters in those groups) in these series.

If χ lies in a series whose cuspidal character is ϕ, then as polynomials in q we always have that $\phi(1)$ divides $\chi(1)$, and the quotient $\chi(1)/\phi(1)$ can be determined from the Hecke algebra [30], with some methods and calculations in [138]. These *generic degrees* $\chi(1)/\phi(1)$ were explicitly computed for the various Weyl groups in a series of papers (see [29] and the references therein, and a few remaining cases for F_4 were finished in [402]).

We also find that Harish-Chandra induction and restriction for $\mathbf{G}^F$ correspond to standard induction and restriction for the Weyl group of $\mathbf{G}^F$ [302, Theorem 5.9]. Thus it makes perfect sense that Harish-Chandra induction in $\mathrm{GL}_n(q)$ looks like the branching rule for S_n.

The two most famous and important unipotent characters are the trivial character and the Steinberg character. The trivial character is obvious, and the Steinberg character was already defined in Sect. 9.1. We give a few of its properties now; [107, Chapter 6] is devoted to this one character, and all of these properties can be found there.

Theorem 9.2.12 *Let G be a finite group of Lie type. Let I denote an index set of the simple roots of the root system of G, and let P_J denote the parabolic subgroup corresponding to a subset J of I. Write $P_J = U_J L_J$ for a Levi decomposition of P_J.*

(i) *The Steinberg character* St *is equal to the virtual character*

$$\sum_{J\subseteq I}(-1)^{|J|}1_{P_J}\uparrow^G.$$

(In other words, it is the alternating sum of $R_{L_J}^G(1)$ for all Levi subgroups L_J of G.)

(ii) *If g does not lie in any proper parabolic subgroup, then the value of* St *on $g\in G$ is $(-1)^{|I|}$.*

(iii) *If g lies in P_J but not in any parabolic subgroup of smaller rank, then*

$$\mathrm{St}(g) = (-1)^{|J|} \cdot |C_{U_J}(g)|$$

if g lies in L_J, and 0 *if g is not P_J-conjugate to an element of L_J.*

(iv) *Consequently, for any $g \in G$, if g is p-singular then* $\mathrm{St}(g) = 0$, *and* $\mathrm{St}(g) = \pm|C_G(g)|_p$ *if g is p-regular.*

Larsen, Malle and Tiep also proved [385] that St has the largest dimension among all unipotent characters of any finite group G of Lie type.

The final thing to do in this section is to define $R_{\mathbf{L}}^{\mathbf{G}}$ in general, where $\mathbf{L}$ is an F-stable, but non-split, Levi subgroup, and not a torus. It is again the cohomology of a variety, but this time we take the inverse image of the Lang map. Recall that the map $\mathcal{L} : g \mapsto (gF) \cdot g^{-1}$ is called the *Lang map*; it is a standard fact—the *Lang–Steinberg theorem*—that the Lang map is surjective if $\mathbf{G}$ is connected [384, 533].

If $\mathbf{L}$ is an F-stable Levi subgroup lying in a (not necessarily F-stable) parabolic subgroup $\mathbf{P}$, with unipotent radical $\mathbf{U}$, then the variety we want to take ℓ-adic cohomology of is $\mathcal{L}^{-1}(\mathbf{U})$. Because this definition again depends on a parabolic subgroup $\mathbf{P}$, we have to write $R_{\mathbf{L}\subset\mathbf{P}}^{\mathbf{G}}$ for Lusztig induction, at least unless it has been proved that it does not depend on $\mathbf{P}$. If $\mathbf{L}$ is split then this is true, and if $\mathbf{L}$ is a torus then this is true. In general, however, it is not yet known. In 2011, Bonnafé and Michel proved [43] that the Mackey formula holds for all groups $G = \mathbf{G}^F$ for $\mathbf{G}$ simple and F a Frobenius morphism (so excluding ${}^2B_2(q)$, ${}^2G_2(q)$ and ${}^2F_4(q)$) other than ${}^2E_6(2)$, $E_7(2)$ and $E_8(2)$, so with these three potential exceptions we can remove the '$\subset \mathbf{P}$' from $R_{\mathbf{L}\subset\mathbf{P}}^{\mathbf{G}}$. In fact, it is just ${}^2E_6(2)$ that is the problem: the problems with $E_7(2)$ and $E_8(2)$ are just consequences of the induction. Moreover, Taylor proved that the Mackey formula holds for $E_7(2)$ in [537], so it is only the two groups that we ever need to be careful about.

9.3 Unipotent Blocks

Recall that q is a power of a prime r. The first section considered the case where $p = r$, the second where $p = 0$, so this and the next section consider the case where p is neither 0 nor r. Thus the field k over which we take representations has characteristic $p > 0$, and our group of Lie type G is over the field $\mathbb{F}_q$ of characteristic $r \neq p$.

Write d for the multiplicative order of q modulo p, i.e., the smallest d such that $p \mid \Phi_d(q)$, where Φ_d denotes the dth cyclotomic polynomial. If $p = 2$, we change this slightly, and let d denote the order of q modulo 4, so $d \in \{1, 2\}$.

We start with one of the easiest definitions in the book.

Definition 9.3.1 A *unipotent block* is a p-block of a group of Lie type G with a unipotent character in it.

In order to describe the (partly conjectural!) theory of unipotent blocks, we need a quick definition, that of the a-function.

Definition 9.3.2 Let $f = f(q)$ be a polynomial. The *a-function* $a(f)$ is the multiplicity of 0 as a root of f, i.e., the integer such that $f(q)/q^{a(f)}$ is a polynomial for which 0 is not a root. If χ is a unipotent character, define $a(\chi)$ to be $a(f)$, where $f(q) = \chi(1)$ (as a polynomial in q).

The following statements summarize what should hold for the decomposition matrices of unipotent blocks. Some of these are true, some are conjectural, and some do not hold for small primes.

9.3.1 Distribution of Unipotent Characters into Blocks

The overriding theme of the theory of representations of groups of Lie type is that the variables p and q are of less importance than the integer d. Most things should be determined by d, and p and q are just technicalities. For example, the shape of the Brauer trees (Chap. 5) of groups of Lie type depend only on d, not p and q. While p (in fact the order of the defect group) is still encoded, it is only in terms of the exceptionality of the tree, not in its isomorphism type.

We will describe how to distribute the unipotent characters among the p-blocks for groups of Lie type. We can be fairly careless about which version of the group we have (so $\mathrm{SL}_n(q)$ or $\mathrm{PGL}_n(q)$, for example) by [85, Theorem 12], which says that unless p divides the order of the centre of the simply connected version of the group (so $p \mid (q-1)$ for $\mathrm{GL}_n(q)$, $p \mid (q+1)$ for $\mathrm{GU}_n(q)$, $p = 2$ for the other classical groups, $p = 3$ for E_6 and $p = 2$ for E_7) all of the unipotent blocks are the same for the different isogeny types, and in fact there are Morita equivalences between them, even isomorphisms of algebras.

Let's start with the classical groups. For $\mathrm{GL}_n(q)$ and $\mathrm{GU}_n(q)$ we recall from Examples 9.2.1 and 9.2.7 that the unipotent characters are parametrized by partitions of n. The distribution of unipotent characters amongst the blocks was found by Fong and Srinivasan [228].

For $G = \mathrm{GL}_n(q)$, there is an analogue of the Nakayama conjecture: two unipotent characters lie in the same p-block if and only if they have the same d-core. For unitary groups the same statement holds, except that d is replaced by e. Here, $e = d$ if $4 \mid d$, $e = 2d$ if d is odd, and $e = d/2$ otherwise. (Thus e is the order of $-q$ modulo p, where d is the order of q modulo p.) These results were proved only for p odd in [228]: Broué gave a new proof of the results of [228], including the case $p = 2$, in [66]. For $p = 2$ it behaves slightly differently for $\mathrm{GL}_n(q)$ and $\mathrm{GU}_n(q)$, and all unipotent characters lie in the same block [66, p. 174].

For symplectic and orthogonal groups (types B_n, C_n, D_n and ${}^2D_n$), Fong and Srinivasan [230] again gave a partition of the unipotent characters amongst the p-blocks, but now both p and q were required to be odd.

Recall that the unipotent characters of symplectic and orthogonal groups are parametrized by symbols, which are pairs of β-sets. If d is odd then write $e = d$, and if d is even write $e = d/2$. (Thus e is the order of q^2 modulo p, where d is the order of q modulo p.) If $e = d$, then two unipotent characters χ_Λ and χ_Γ lie in the same p-block if and only if Λ and Γ have the same e-core. What this means is that we remove as many e-hooks as possible from Λ to arrive at its e-core; an *e-hook* of $\Lambda = (X, Y)$ is an e-hook of either X or Y. Thus an *e-core* is a symbol whose two β-sets are both e-cores in the partition sense.

If $e = d/2$ on the other hand, then two unipotent characters χ_Λ and χ_Γ lie in the same p-block if and only if Λ and Γ have the same 'e-cocore'. This time we must define an e-cohook. If $\Lambda = (X, Y)$ then removing an *e-cohook* is removing the integer $i \in X$ from X and adding the integer $i - e \notin Y$ to Y, or vice versa. (So it differs from an e-hook by putting $i - e$ back into the other β-set.) The process of removing all possible e-cohooks leaves a well-defined symbol, the *e-cocore*, just as removing e-hooks does (Exercise 9.11).

As we have said though, this only deals with the case where p and q are both odd. If $p = 2$ then there is only one unipotent p-block, by [85, Theorem 13], whereas if q is even then the result is the same as for q odd. This hasn't been proved directly: the distribution of unipotent characters was proved to depend only on d (for p odd of course) by Broué, Malle and Michel [71] for $p > n$, and Cabanes and Enguehard in [86] in general, and then we use the Fong–Srinivasan result above.

Given the dichotomy between the two possible structures, and presaging later results, we give the following definition here.

Definition 9.3.3 Let $G = G(q)$ be a unitary group. The prime p is *linear* if $e = d$ or $e = 2d$ (so d is odd or divisible by 4) and is *unitary* otherwise.

Let $G = G(q)$ be a symplectic or orthogonal group. The prime p is *linear* if $e = d$ (so that d is odd) and is *unitary* otherwise.

The reason for the names 'linear' and 'unitary' is because of the structure of the centralizers of certain semisimple elements.

What about the unipotent character distribution into blocks for exceptional groups? We need a definition to continue: a *good prime* is any prime for groups of type A_n, any odd prime for groups of type B_n, C_n and D_n, any prime greater than 3 for exceptional groups other than E_8, and any prime greater than 5 for E_8. (We met these briefly just before Example 9.2.11.) Of course, if p is not a good prime then it is a *bad prime*. For p an odd good prime (and not $p = 3$ for ${}^3D_4$), it was shown in [71, 86] even for exceptional groups that the distribution of unipotent characters into blocks depends only on d, and given explicitly in the table at the back of [71]. (In this table there are a couple of ambiguities related to non-real eigenvalues of Frobenius.)

This is not true for p bad. We have already seen this for $p = 2$ and classical groups, but it continues for G of type G_2. In [296] the distribution of the ordinary characters amongst the 2-blocks is given, and it differs from the general case of [521]. What happens is that unipotent blocks merge together for bad primes. For

$p = 2, 3$, all unipotent characters lie in the principal block, or are blocks of defect zero (which of course could never merge).

When p is good, all of the unipotent characters have linearly independent reductions modulo p (see the next part on basic sets), but for p bad, linear dependencies between the reductions modulo p of the ordinary characters force unipotent characters that should lie in one block to be pushed into another block, simply because they lie in the span of the unipotent characters from that block.

Enguehard considered the case where p is bad in [187]. The proofs go case by case, with a few general propositions that deal with most situations. There are far fewer unipotent blocks for bad primes, for example for ${}^3D_4$ and $p = 2, 3$, the same statement as for G_2 holds, that all non-principal unipotent blocks have defect zero, but this is not true when the group gets larger. For example, $E_7(q)$ and $E_8(q)$ contain unipotent 2-blocks that are neither the principal block nor of defect zero.

9.3.2 *Basic Sets*

A set I of Brauer characters is a *basic set* if it is linearly independent and the $\mathbb{Z}$-span of I contains all irreducible (and hence all) Brauer characters.

Conjecture 9.3.4 (Basic Set Conjecture) If G is a finite group and p is a prime dividing $|G|$, then the set of reductions modulo p of $\chi \in \mathrm{Irr}(G)$ contains a basic set.

Such a result holds for p-soluble groups by the Fong–Swan theorem (Theorem 7.1.6), but for arbitrary finite groups the conjecture is very open. Geck and Hiss have worked on the basic set problem for groups of Lie type in [243–245]. Of course, for a given block B one may also ask whether the reductions modulo p of a subset of $\mathrm{Irr}(B)$ form a basic set for the Brauer characters of B. For unipotent blocks, the subset of $\mathrm{Irr}(B)$ would be the unipotent characters.

Theorem 9.3.5 (Geck–Hiss [245]) *Let $G = \mathbf{G}^F$ be a group of Lie type and let p be a good prime for G not dividing the order of $(Z(\mathbf{G})/Z(\mathbf{G})^\circ)^F$. The unipotent characters form a basic set for the unipotent blocks of G.*

(See also [88, Theorem 14.4].) As well as requiring p to be good, we include a condition on the centre of G. If G is simple then this is really only an issue for type A, as all primes dividing $Z(G)$ are bad for other types. It says that if you have $\mathrm{SL}_n(q)$ or $\mathrm{SU}_n(q)$ then p cannot divide $|Z(G)|$, but if you have $\mathrm{GL}_n(q)$ or $\mathrm{GU}_n(q)$ (which have connected centre) then this condition disappears.

We'll give a few examples to test the limits of this result now.

Example 9.3.6 Let $G = \mathrm{PGL}_2(q)$ for q odd and $p = 2$. The principal block has two simple modules, as we see from Theorem 6.2.3, and there are two unipotent characters, namely $\chi^{(2)}$ of degree 1 and $\chi^{(1^2)}$ of degree q. They both lie in the principal block, and therefore form a basic set. (Of course, 2 is good for G, so this is to be expected.)

Now restrict to $G' = \mathrm{PSL}_2(q)$ (or equally consider the principal block of $\mathrm{SL}_2(q)$). Now there are three simple modules in the principal block, as we see in Theorem 6.2.2, but there are still two unipotent characters, so they no longer form a basic set, as there are too few of them.

In this case, dividing the order of the centre caused there to be too few unipotent characters, even when the prime was good. However, $p = 2$ behaves differently for $\mathrm{GL}_n(q)$ to odd primes even though it is good, so let us see another example.

Example 9.3.7 Let $G = \mathrm{PSL}_3(4)$. The decomposition matrix for the principal 3-block of G is as follows

	ψ_1	ψ_2	ψ_3	ψ_4	ψ_5
$\chi_1 = 1_1$	1	.	.	.	.
$\chi_2 = 20_1$	1	1	.	.	.
$\chi_3 = 35_1$	1	1	1	.	.
$\chi_4 = 35_2$	1	1	.	1	.
$\chi_5 = 35_3$	1	1	.	.	1
$\chi_6 = 64_1$	.	1	1	1	1

A dot means that the value is 0. As we have said before, dots are used so that the structure of the matrix is clearer. From this table, we can see that ψ_1 has degree 1, ψ_2 has degree 19, and ψ_3, ψ_4 and ψ_5 have degree 15.

On the other hand, the unipotent characters of $\mathrm{SL}_3(4)$ are $\chi^{(3)}$ of degree 1, $\chi^{(2,1)}$ of degree $q(q^2-1)/(q-1) = q(q+1) = 20$, and $\chi^{(1^3)}$ of degree $q^3 = 64$ (the Steinberg character). Thus there are three unipotent characters in the principal block, but five simple modules.

However, the diagonal automorphism of G permutes the three 15-dimensional modules, so in $\mathrm{PGL}_3(4)$ (PGL_3 does have connected centre) there are three unipotent characters, three simple modules, and the unipotent characters indeed form a basic set, in accordance with Theorem 9.3.5.

We again see the same pattern: there are too few unipotent characters for the simple group because some simple modules for the group $\mathrm{PGL}_n(q)$ split on restriction to $\mathrm{PSL}_n(q)$, whereas that does not happen for unipotent characters.

What about if p is bad?

Example 9.3.8 Let $G = \mathrm{Sp}_4(3)$ and let $p = 2$. Since $\mathrm{Sp}_4(3) \cong \mathrm{SU}_4(2)$, we can use results from the first section of this chapter to see that there are seven simple modules in the principal 2-block of G, of dimensions 1, 4, 4, 6, 14, 20 and 20 (and the Steinberg character in a block of defect 0). However, from the previous section we see that there are five principal series characters and one cuspidal character, so six unipotent characters in total. They have degrees

$$1, \quad q(q+1)^2/2, \quad q(q^2+1)/2, \quad q(q^2+1)/2, \quad q(q-1)^2/2, \quad q^4,$$

which for $q = 3$ yield 1, 24, 15, 15, 6 and 81.

However, in the adjoint group of type C_2, which is denoted $\mathrm{PCSp}_4(3)$, the two 4-dimensional, and two 20-dimensional, modules merge, so there are five simple modules. Now there are *more* unipotent characters than simple modules!

So we see the same problem for $\mathrm{PSp}_4(3)$ as for $\mathrm{PSL}_2(q)$, that there are modules that split on restriction to the simple group but the unipotent characters do not. However, for the adjoint group we now find too many unipotent characters: this is because there should be (as in, for good primes) two blocks, the principal block and a block of defect zero. However, they merge for $p = 2$, and while the characters that are meant to be in the principal block are linearly independent (and form a basic set, in fact), the extra one coming from what should be the block of defect zero lies in the span of those from the principal block (which is why the two blocks merge), so the set is no longer linearly independent, and cannot therefore be basic.

If $p = 2$ and G is classical though, then the number of irreducible unipotent Brauer characters is equal to the number of unipotent classes of G, and if r is sufficiently large and G is not twisted then the basic set conjecture was proved by Geck in [244], for connected centre groups. The reason for the caveats is that certain formulae of Lusztig from [406] on generalized Gelfand–Graev characters were only proved in that paper for sufficiently large powers q of sufficiently large primes r. In [535], Taylor removed the prime restrictions, so the basic set conjecture is true for all q for $p = 2$. The recent paper by Chaneb [110] considers the state of affairs for classical groups and $p = 2$, where the precise form of the group is important, and so we do not summarize the situation there, instead referring the reader to [110].

For G exceptional and p bad, explicit decomposition matrices have been computed for small-rank groups, which means that the basic set conjecture can be shown directly.

There are no non-defining bad primes for Suzuki groups; for the small Ree group ${}^2G_2(q)$, we are concerned only with the bad prime $p = 2$. The decomposition matrices are known for small Ree groups and $p = 2$ by Landrock and Michler [383], building on work of Fong [225], where there are five irreducible Brauer characters in the principal block but six unipotent characters, a subset of which form a basic set, just like for $\mathrm{PCSp}_4(3)$ above.

For the bad primes $p = 2, 3$ for $G_2(q)$, Hiss and Shamash computed decomposition numbers with a few unknowns left [295, 296]; the decomposition matrices are lower unitriangular, and therefore there is a basic set consisting of ordinary characters. For both $p = 2$ and $p = 3$, there are seven simple modules in the principal p-block. If $p = 2$ then there are eight unipotent characters in the principal block, and there are nine for $p = 3$. In both cases, a subset of them forms a basic set for that block. The only other unipotent block has defect zero, with the other unipotent characters in it, so in these cases we can also find a basic set consisting of unipotent characters.

An explicit basic set for $p = 3$ and $G = {}^2F_4(q)$ is given in [284, Section 4]; the one chosen by Himstedt in [284] contains a non-unipotent character but there are basic sets consisting entirely of unipotent characters. For ${}^3D_4(q)$ and $p = 2$ (so that q is odd) the proof of the basic set conjecture follows from work of Geck [240], which showed that the decomposition matrices are lower unitriangular.

For the larger exceptional groups and bad primes, the basic set conjecture is not completely known for all primes (although it is known in many cases). For many possible G and p though, a subset of the unipotent characters is known to form a basic set for the unipotent blocks. We may therefore order the rows of the decomposition matrix so as to put the unipotent characters at the top, and we order them by increasing a-function. (Characters with the same a-function can be placed in any order.) Write D_u for the unipotent part of the decomposition matrix, which is a square matrix if the basic set conjecture holds.

9.3.3 Unitriangularity of the Decomposition Matrix

The submatrix D_u is the focus of the rest of the conjectures and statements in this section.

Conjecture 9.3.9 (Unitriangularity Conjecture) Let $\mathbf{G}$ be connected and reductive, and let p be a prime that is good for $\mathbf{G}$ and does not divide the order of $(Z(\mathbf{G})/Z(\mathbf{G})^\circ)^F$, and let B be a unipotent p-block of $\mathbf{G}^F$ with decomposition matrix D. By ordering the unipotent characters by increasing a-function, one may arrange the columns of D_u so that it is lower unitriangular.

This conjecture appears in [246, Conjecture 3.4], along with another conjecture we will talk about soon. It is known for all blocks of $\mathrm{GL}_n(q)$ by work of Dipper [162, 163], based on work of him and James [164] (see also [165] for a proof via the q-Schur algebra).

For the general unitary groups, Geck proved the same result, that the decomposition matrices of all p-blocks are lower unitriangular. The same methods allowed Geck to also prove the lower-unitriangularity property for all p-blocks of ${}^3D_4(q)$ for q odd in [240], as mentioned when we discussed the basic set conjecture.

For the other classical groups, we make the distinction between linear and unitary primes (Definition 9.3.3 above). For unitary primes the result is not known in complete generality, even for p odd, but in many cases it has been proved. However, for linear primes there is a proof, because there is a strong relationship between D and the decomposition matrices of $\mathrm{GL}_n(q)$ for various n, found by Gruber and Hiss [269]. (Their methods also work for the unitary groups.) They proved that the decomposition matrices for these groups are determined by those of $\mathrm{GL}_n(q)$; since the matrices for $\mathrm{GL}_n(q)$ are unitriangular, this in particular proves the same for classical groups at linear primes.

Even for $p = 2$ some results are known, particularly [247], which proves the unitriangularity of the decomposition matrix for the principal 2-block (there is only one unipotent 2-block, as we saw earlier this section) if $\mathbf{G}$ has connected centre. (Again, there are further restrictions on G and q in [247] because of the use of generalized Gelfand–Graev representations, which can be removed by the recent paper of Taylor [535].)

If the block has cyclic defect groups then the decomposition matrix is in general quite sparse, and known (see Sect. 5.3), and it has been checked case by case that the a-function induces a triangular shape.

Thus we move on to exceptional groups. For $G_2(q)$ the full decomposition matrices were almost completely determined for all good primes p dividing $q \pm 1$ (so $d \leq 2$) by work of Hiss in [286] for $p \neq 2, 3$, building on work of Shamash [521, 523], who determined the distribution of characters into blocks. For $d = 3, 6$, the only other relevant d (given that $|G_2(q)|$ is $q^6(q^2 - 1)^2(q^6 - 1)$), the Sylow p-subgroup of $G_2(q)$ is cyclic, and the Brauer trees were given by Shamash [522, 523].

The unipotent blocks of $F_4(q)$ are also lower unitriangular for p and r good (so at least 5) [367, 567]. For E_6 and $d = 3$, for example, Geck and Hiss proved the unitriangularity of the decomposition matrix in [246]. This was also checked for $d = 4$ by Miyachi in [431], using the methods of Geck and Hiss. Again, in these cases both p and r need to be good.

The techniques at this point were Harish-Chandra induction and generalized Gelfand–Graev characters, together with some ad hoc ideas that depended on the situation. At around 2010, the idea of using Deligne–Lusztig theory directly to attack the decomposition number problem appeared on the scene, primarily driven by work of Dudas, and later with Malle, Rouquier and me. Deligne–Lusztig theory gave new projective modules, and meant that previously unknown entries in the decomposition matrix could now be accessed.

For $d = 4$, Dudas and Malle prove the unitriangularity of all of the blocks where the defect group is abelian of rank 2, where p is a good prime. This was more of an illustration of the method rather than the limit of what can be done, and at the time of writing work is currently underway by Brunat, Dudas and Taylor to determine unitriangularity of the decomposition matrices for exceptional groups at good primes.

We should also mention that, while the a-function appears to work as an ordering on the decomposition matrix, the 'correct' function from the geometry is the function $\pi_{\kappa/d}$ that appears in the work of the author [128], and is the function with respect to which there should be a perverse equivalence between the block and its Brauer correspondent (see Sect. 4.4). The reason why several different functions can give unitriangularity is that the a-, A- and $\pi_{\kappa/d}$-functions all give pretty similar orderings, and the entries in the decomposition matrices near the main diagonal are in general quite often equal to 0. Thus perturbing the order slightly can often not affect triangularity.

9.3.4 Reduction Modulo p of Cuspidal Characters

In [242, (6.6)], Geck conjectured that the reduction modulo a good prime p of a cuspidal unipotent character is always irreducible (see also [246, Conjecture 3.4]). This has not been proved in full generality, but almost all cases have been proved.

In [241], Geck confirmed this conjecture for $\mathrm{GU}_n(q)$, and the Gruber–Hiss theorem on decomposition matrices for classical groups at linear primes [269] proves the conjecture for those groups. Geck and Malle proved the result for classical groups for $p = 2$ in [247]. Recently, Dudas and Malle produced a proof in almost complete generality.

Theorem 9.3.10 (Dudas–Malle [177]) *If p and r are good for $\mathbf{G}$, and p does not divide the order of $Z(\mathbf{G})/Z(\mathbf{G})^\circ$, then any cuspidal character of $\mathbf{G}^F$ has irreducible reduction modulo p.*

The restriction on r arises because the methods require the use of generalized Gelfand–Graev characters, which were also mentioned in the results on the basic set conjecture. Since these require r to be good, the same restriction applies here. If Conjecture 9.3.11 below is true then the restriction on r can be removed, simply by replacing q by a power of a different prime.

For $p = 2, 3$ and $G = G_2(q)$, the cuspidal unipotent characters have irreducible reduction modulo p by [295, 296], and if $G = {}^3D_4(q)$ and $p = 2$ then again the reductions are irreducible by work of Himstedt [283]. (We do require $p \neq r$ in this statement! For $\mathrm{PSU}_3(3)$ the cuspidal unipotent character has degree 6, and modulo 3 it is the sum of the natural module and its dual.)

However, we cannot hope for too much. There are no bad primes to check for the Suzuki groups, but for both small and large Ree groups and p bad there are counterexamples, so we cannot allow our groups to be 'very twisted' for bad primes.

For the small Ree groups ${}^2G_2(q)$, there are four cuspidal unipotent characters, but for $p = 2$ only two of these have irreducible reductions. The 2-decomposition matrix is given in [383], and from here we can see that the characters ${}^2G_2^{\mathrm{II}}[\pm \mathrm{i}]$ do not have irreducible reduction modulo 2. For the large Ree groups ${}^2F_4(q)$, the 3-decomposition matrix is not completely known, but enough is known to see that the character ${}^2F_4^{\mathrm{I}}[-1]$, which is cuspidal unipotent, has at least two constituents on reduction modulo 3. (This is character χ_{10} in [284, Table C.5].)

9.3.5 *Known Decomposition Matrices*

Until recently, not all that many decomposition matrices for unipotent blocks were known. If the defect group is cyclic then we described what was happening back in Sect. 5.3, so we will focus entirely on the non-cyclic case.

Conjecture 9.3.11 Let p and p' be two primes, and let q and q' be two powers of primes r and r', chosen so that the multiplicative orders of q and q' modulo p and p' respectively are the same integer d. Let $G_1 = G(q)$ and $G_2 = G(q')$ be two groups of Lie type of the same type (e.g., SL_n). Fix a unipotent character χ of G_i (with the same label in both G_i), and let B_i be the unipotent block of G_i containing χ.

If p and p' are sufficiently large, then the unipotent parts of the decomposition matrices of B_1 and B_2 are equal.

Of course, this conjecture requires us to order the unipotent characters the same way in both B_i, and to order the columns the same way, say so that the decomposition matrices are lower unitriangular. Of course, if we know enough about the decomposition matrix for us to talk about it here, it is highly likely that we already know it is lower triangular with respect to the a-function.

Among the first examples of fully understood decomposition matrices were those of $\mathrm{GL}_n(q)$ for $n \leq 10$, which were determined by James in [326]. From these, the work of Gruber and Hiss [269] mentioned in the unitriangularity section can be used to produce complete decomposition matrices for classical groups at linear primes up to dimension 21, except for a few cases.

For classical groups at unitary primes, things are more complicated. The new work of Dudas and Malle [175] used Deligne–Lusztig varieties to determine the decomposition matrices for $\mathrm{SU}_n(q)$ and $n \leq 9$, and with only a few unknown entries for $n = 10$. Until then, only $\mathrm{SU}_3(q)$ was known. The only caveat is that these decomposition matrices are only valid for $p > n$; for $p \leq n$ there will be differences, because there are issues when $p \mid n$ specifically, and also the Hecke algebras (whose decomposition matrices are submatrices of the decomposition matrices of the groups) are known to behave differently for small primes.

For the other classical groups, little is known except for small rank. The smallest case is $\mathrm{Sp}_4(q)$: White determined almost all of the decomposition matrix of $\mathrm{Sp}_4(q)$ for p odd in [555, 556]. The unipotent part of the principal p-block for $d = 2$ has the following form.

Symbol	Degree	ψ_1	ψ_2	ψ_3	ψ_4	ψ_5
$(\{2\}, \emptyset)$	1	1	.	.	.	.
$(\{0,1,2\}, \emptyset)$	$q(q-1)^2/2$	.	1	.	.	.
$(\{0,1\}, \{2\})$	$q(q^2+1)/2$	1	.	1	.	.
$(\{1,2\}, \{0\})$	$q(q^2+1)/2$	1	.	.	1	.
$(\{0,1,2\}, \{1,2\})$	q^4	1	1	1	α	1

Okuyama and Waki determined this parameter α in [457]: it is 1 if the Sylow p-subgroup is $C_3 \times C_3$, and 2 in all other cases (including $C_{3^a} \times C_{3^a}$ for $a > 1$). This is an example of the small-prime-power decomposition matrices behaving differently to large-prime-power ones: the matrix should stabilize as long as the power p^a dividing the cyclotomic polynomial $\Phi_d(q)$ is sufficiently large. (This is the 'correct' form of Conjecture 9.3.11, but it is slightly more fiddly to state.)

This still leaves $p = 2$: again, White gave the decomposition matrix up to a single parameter [554]. The unipotent part D_u is no longer square, since the unipotent characters do not form a basic set as p is bad, and as mentioned near the start of this section, there is a single unipotent block in this case. There are six unipotent characters and seven simple modules, and D_u is as follows:

Symbol	Degree	ψ_1	ψ_2	ψ_3	ψ_4	ψ_5	ψ_6	ψ_7
$(\{2\}, \emptyset)$	1	1	.	.	.	.	.	.
$(\{0, 1, 2\}, \emptyset)$	$q(q-1)^2/2$	.	1	.	.	.	.	.
$(\{0, 1\}, \{2\})$	$q(q^2+1)/2$	1	1	1	1	.	.	.
$(\{1, 2\}, \{0\})$	$q(q^2+1)/2$	1	.	.	.	1	.	.
$(\{0, 2\}, \{1\})$	$q(q+1)^2/2$	2	.	1	1	1	.	.
$(\{0, 1, 2\}, \{1, 2\})$	q^4	1	$2\alpha+1$	1	1	1	1	1

I don't think the parameter α has been completely determined in general: some quick calculations show that $\alpha = 1$ for $q = 3, 7, 11$ and $\alpha = 0$ for $q = 5, 9$. I think the pattern seems obvious, but of course this is not a proof.

For $\mathrm{Sp}_6(2^a)$ and p odd, White again determined most entries in the decomposition matrix in [557]. Köhler [367] and An–Hiss [13] computed most of the decomposition numbers for $\mathrm{Sp}_6(q)$ (for both q and r odd), and then Himstedt and Noeske finished off the matrix [285], also showing that the unipotent blocks of $\mathrm{SO}_7(q)$ have the same decomposition matrices. (This is a consequence of Conjecture 9.3.11, since $\mathrm{Sp}_{2n}(2^a) = \mathrm{SO}_{2n+1}(2^a)$.) For larger orthogonal groups, the methods of Dudas and Malle, as used in [175], would provide a lot of information, but would require both p and r to be good, thus excluding the case q even. This is the case in [176], where numerous decomposition matrices for those classical groups that appear as Levi subgroups of exceptional groups are determined. Dudas and Malle are currently working on this question. Paolini has almost determined the decomposition matrix for $\mathrm{SO}_8^+(2^a)$ for $d = 2$ (and $p > 3$) in recent work [463].

We move on to exceptional groups. The smallest cases are the Suzuki and small Ree groups: for the Suzuki groups and odd primes p, all Sylow p-subgroups are cyclic, so we ignore these, and for the small Ree groups ${}^2G_2(q)$, only the Sylow 2-subgroup is non-cyclic. The decomposition matrix for this was determined by Landrock and Michler in [383]. The next smallest is $G_2(q)$, where a series of papers by Hiss and Shamash determined the decomposition matrices, including at bad primes, but again up to a couple of parameters [295, 296, 521]. The unknown parameters in the decomposition matrices for p good and $d = 2$ were later found by Dudas in [173].

A similar story holds for ${}^3D_4(q)$, where most of the decomposition numbers were found for $p = 2$ and q odd by Himstedt [283] and for $p > 3$ by Geck [240]. Again, Dudas used his methods to attack the problem for odd p, r in [173], but this time he couldn't obtain the whole matrix.

For F_4 and its twisted version, Köhler [367] and Himstedt [284] respectively gave results, but there are more issues leading to more unknowns in the decomposition matrices. For E_6 and $d = 3, 4$ some results were given in [246] and [431], but there are plenty of unknowns.

The techniques coming from Deligne–Lusztig theory that were not employed when those papers were written have also been brought to bear on the exceptional groups, where Dudas and Malle [176] looked at the decomposition matrices for $d = 4$ and most unipotent blocks of exceptional groups, completely determining the matrices in many cases and with just a few unknowns in the remaining cases.

This is an area in which rapid progress is being made, and the change over the last decade looks set to be repeated over the next.

9.4 General Blocks

Suppose that B is a non-unipotent block of a group of Lie type G. The main aim of much of the theory is to first give some kind of label to B, to understand its ordinary characters, and then to find a unipotent block B' of some subgroup of G such that B and B' are Morita equivalent, but in a strong way that preserves all of the information we want. In particular, the Morita equivalence should preserve defect groups, and send an irreducible character of B to a particular irreducible character of B', so that we can write down the decomposition matrix of B with the character labels given by Jordan decomposition.

This Morita equivalence, in the full generality given here, is still not known. We can give a Morita equivalence, but only to a larger collection of blocks called 'isolated' blocks, rather than just to unipotent blocks, and this is the substance of the Bonnafé–Rouquier theorem (Theorem 9.4.7 below), which has been improved by those two authors and Dat (Theorem 9.4.9).

We start with the Jordan decomposition of characters, mentioned in Sect. 9.2. The best and cleanest theorems occur when the centre of $\mathbf{G}$ is connected, but for now we simply assume that $\mathbf{G}$ is a connected, reductive algebraic group. This material can be found in, for example, [88] and [161]. First, for any irreducible character χ of $G = \mathbf{G}^F$, there is some Deligne–Lusztig character $R_{\mathbf{T}}^{\mathbf{G}}(\theta)$ (where $\mathbf{T}$ is an F-stable maximal torus of $\mathbf{G}$ and $\theta \in \mathrm{Irr}(\mathbf{T}^F)$) such that

$$\langle \chi, R_{\mathbf{T}}^{\mathbf{G}}(\theta) \rangle_{\mathbf{G}^F} \neq 0.$$

This means that if we consider the constituents of all Deligne–Lusztig characters $R_{\mathbf{T}}^{\mathbf{G}}(\theta)$ for all pairs $(\mathbf{T}, \theta)$, where $\mathbf{T}$ is an F-stable maximal torus of $\mathbf{G}$ and $\theta \in \mathrm{Irr}(\mathbf{T}^F)$, then we see all irreducible characters of G. If $\mathbf{T}'$ is another F-stable maximal torus and $\theta' \in \mathrm{Irr}\left((\mathbf{T}')^F\right)$ then the inner product of $R_{\mathbf{T}}^{\mathbf{G}}(\theta)$ and $R_{\mathbf{T}'}^{\mathbf{G}}(\theta')$ is equal to the number of elements of $\mathbf{G}^F$ that conjugate $(\mathbf{T}, \theta)$ to $(\mathbf{T}', \theta')$ divided by $|\mathbf{T}^F|$. This, however, just shows that if $(\mathbf{T}, \theta)$ and $(\mathbf{T}', \theta')$ are not conjugate in $\mathbf{G}^F$ then the inner product of $R_{\mathbf{T}}^{\mathbf{G}}(\theta)$ and $R_{\mathbf{T}'}^{\mathbf{G}}(\theta')$ is zero; it does not say that they have no constituents in common, since these are virtual characters, not actual characters. To give the correct equivalence relation on the set of pairs $(\mathbf{T}, \theta)$, we give a definition.

Definition 9.4.1 Two pairs $(\mathbf{T}, \theta)$ and $(\mathbf{T}', \theta')$ as above are *geometrically conjugate* if there exists some $g \in \mathbf{G}$ such that $\mathbf{T}^g = \mathbf{T}'$ and, if g lies in $\mathbf{G}^{F^n}$ for some $n \in \mathbb{N}$, we have

$$N_{F^n/F} \circ \theta = c_g \circ N_{F^n/F} \circ \theta',$$

where c_g is conjugation by g, and where $N_{F^n/F}$ is the *norm map* $x \mapsto \prod_{i=0}^{n-1} x F^i$.

Example 9.4.2 If $\mathbf{T}$ and $\mathbf{T}'$ are two F-stable maximal tori, then $(\mathbf{T}, 1_{\mathbf{T}})$ and $(\mathbf{T}', 1_{\mathbf{T}'})$ are always geometrically conjugate.

Geometric conjugacy is a good way to split up the irreducible ordinary characters of $G = \mathbf{G}^F$.

Theorem 9.4.3 (Deligne–Lusztig [157]) *If* $(\mathbf{T}, \theta)$ *and* $(\mathbf{T}', \theta')$ *are not geometrically conjugate then the corresponding Deligne–Lusztig characters have no constituents in common. Consequently, the set* $\mathrm{Irr}(\mathbf{G}^F)$ *is the disjoint union of the sets of constituents of* $R_{\mathbf{T}}^{\mathbf{G}}(\theta)$, *as* $(\mathbf{T}, \theta)$ *runs over all pairs in a geometric conjugacy class.*

The set of constituents of $R_{\mathbf{T}}^{\mathbf{G}}(\theta)$ for all $(\mathbf{T}, \theta)$ in a geometric conjugacy class is called a *geometric Lusztig series*. Thus Theorem 9.4.3 states that the geometric Lusztig series form a partition of $\mathrm{Irr}(\mathbf{G}^F)$. A geometric Lusztig series is commonly written $\widetilde{\mathcal{E}}(\mathbf{G}^F, s)$ though (see, for example, [88]), where s is a semisimple element of the Langlands dual $(\mathbf{G}^*)^{F^*}$. This is possible because there is a bijection between the geometric conjugacy classes of pairs $(\mathbf{T}, \theta)$ and F^*-stable conjugacy classes of semisimple elements s of $\mathbf{G}^*$, i.e., geometric conjugacy classes of $(\mathbf{G}^*)^{F^*}$ [161, Proposition 13.12]. (Note that this is *not* the rational class; we will have to worry about geometric and rational conjugacy here.) We have that $\widetilde{\mathcal{E}}(\mathbf{G}^F, 1)$ is the set of unipotent characters.

What about the rational classes? We have to give a finer relation on the set of pairs $(\mathbf{T}, \theta)$ than geometric conjugacy, some sort of 'rational conjugacy'. It is not easy to define, and to do it directly requires the dual group. One vicarious way to define it is via $R_{\mathbf{T}}^{\mathbf{G}}$. Say that $(\mathbf{T}, \theta)$ and $(\mathbf{T}', \theta')$ are equivalent if $R_{\mathbf{T}}^{\mathbf{G}}(\theta)$ and $R_{\mathbf{T}'}^{\mathbf{G}}(\theta')$ share an irreducible constituent (they could still have zero inner product as they are virtual characters), and extend this to an equivalence relation on pairs $(\mathbf{T}, \theta)$. This is *rational conjugacy*. The dual group allows us to give a direct definition of rational conjugacy [88, Definition 8.23]. In this case, it can again be related to semisimple elements, and this time $(\mathbf{T}, \theta)$ and $(\mathbf{T}', \theta')$ are rationally conjugate if and only if the associated semisimple elements of $(\mathbf{G}^*)^{F^*}$ are conjugate in $(\mathbf{G}^*)^{F^*}$. Indeed, the *component group* $C_{\mathbf{G}^*}(s)/C_{\mathbf{G}^*}^{\circ}(s)$ (which is always finite) controls the splitting of the geometric class of s into rational classes. This time we drop the tilde, and write $\mathcal{E}(\mathbf{G}^F, s)$ for the *rational Lusztig series* corresponding to $s \in (\mathbf{G}^*)^{F^*}$. Of course, $\widetilde{\mathcal{E}}(\mathbf{G}^F, s)$ splits as a disjoint union of $\mathcal{E}(\mathbf{G}^F, t)$, for t running over a set of representatives of the rational classes comprising the geometric class of s.

The definition of rational classes looks nicest from a representation-theoretic point of view, and geometric conjugacy looks nicest from a group-theoretic point of view. Thus the best results should come when geometric and rational conjugacy coincide. This happens, i.e., $\widetilde{\mathcal{E}}(\mathbf{G}^F, s) = \mathcal{E}(\mathbf{G}^F, s)$, if and only if $C_{\mathbf{G}^*}(s)^F = (C^{\circ}_{\mathbf{G}^*}(s))^F$, which is the case, for example, if $C_{\mathbf{G}^*}(s)$ is connected. Thus a condition on when centralizers are connected would be nice, and indeed exists.

Theorem 9.4.4 (Springer, Steinberg[4]) *Let $\mathbf{G}$ be a connected, reductive algebraic group and let $s \in \mathbf{G}$ be a semisimple element. If $[\mathbf{G}, \mathbf{G}]$ is simply connected then $C_{\mathbf{G}}(s)$ is connected. Consequently, if $Z(\mathbf{G})$ is connected then $C_{\mathbf{G}^*}(s)$ is connected.*

(See, for example, [107, Theorems 3.5.6 and 4.5.9] for the first statement and the consequence.)

The *Jordan decomposition* gives a parametrization of $\mathcal{E}(\mathbf{G}^F, s)$ in terms of unipotent characters of $C_{\mathbf{G}^*}(s)$. The basic version of the Jordan decomposition is that there is a bijection

$$\mathcal{E}(\mathbf{G}^F, s) \to \mathcal{E}(C_{\mathbf{G}^*}(s)^{F^*}, 1).$$

The set on the right-hand side is simply the unipotent characters of $C_{\mathbf{G}^*}(s)^{F^*}$, which we have so far only defined when $C_{\mathbf{G}^*}(s)$ is connected. This statement was proved in the case when $C_{\mathbf{G}^*}(s)$ is connected by Lusztig in [404, (4.23)], and gives us the following theorem.

Theorem 9.4.5 (Jordan Decomposition) *Let $\mathbf{G}$ be a connected, reductive algebraic group with connected centre, and let F be a Frobenius endomorphism on $\mathbf{G}$. The characters of $\mathbf{G}^F$ are parametrized by pairs (s, χ_s), where s is a semisimple element in $(\mathbf{G}^*)^{F^*}$ up to conjugacy, and χ_s is a unipotent character of $C_{\mathbf{G}^*}(s)^{F^*}$.*

This is merely a labelling of the characters of $\mathbf{G}^F$, but it comes with many good properties. First, one may read off the degree of χ from its associated pair (s, χ_s): it is

$$\chi(1) = |(\mathbf{G}^*)^{F^*} : C_{\mathbf{G}^*}(s)^{F^*}|_{r'} \cdot \chi_s(1),$$

where as earlier in this chapter, $m_{r'}$ is the r'-part of the integer m. Second, if one looks only at characters with semisimple label s, the Deligne–Lusztig character $R_{\mathbf{T}}^{\mathbf{G}}(\theta)$ has constituents with the same multiplicities (up to a global sign) as a corresponding character $R_{\mathbf{T}'}^{C_{\mathbf{G}^*}(s)}(1)$. One would like to obtain a commutative diagram involving Lusztig induction and Jordan decomposition, meaning that taking Jordan decomposition commutes with Lusztig induction from Levi subgroups. While it is likely to be true, such a result in many cases is still conjectural.

---

[4]In [107], Carter attributes this result to Steinberg. For his part, in [532], Steinberg attributes it to Springer, but unpublished, and gives a sketch of a proof. In [529] Springer and Steinberg give a full proof.

If we want to actually use the Jordan decomposition to solve some of the local-global conjectures from Chap. 4, however, we run into two problems. The first is that we usually have the wrong groups: $G = \mathbf{G}^F$ is usually not quasisimple if $\mathbf{G}$ has connected centre. To go from $\mathrm{GL}_n(q)$ to $\mathrm{SL}_n(q)$ requires either Clifford theory on the finite group level, or a Jordan decomposition for groups with disconnected centre. Second, most of the reduction results require some knowledge of the action of automorphisms on the irreducible characters. The action of outer automorphisms of G on *unipotent* characters is known, but nothing above said anything about Jordan decomposition and outer automorphisms of G. We also still haven't distributed the irreducible characters among p-blocks, and even haven't identified which non-unipotent characters belong to unipotent blocks.

Extending Jordan decomposition to groups with disconnected centre was done by Lusztig in [405] (see also the work of Digne and Michel [160]). The result is not as clean as Theorem 9.4.5, however. We first must define a unipotent character for a disconnected group, as this is the one side of the bijection in Jordan decomposition: if $\mathbf{G}$ is disconnected, we first do Deligne–Lusztig induction in $(\mathbf{G}^\circ)^F$, and then induce up to $\mathbf{G}^F$ (see [161, Theorem 13.23]). Thus for an F-stable maximal torus $\mathbf{T}^\circ$ in $\mathbf{G}^\circ$, define $R_{\mathbf{T}^\circ}^{\mathbf{G}}(\theta)$ to be the induction from $(\mathbf{G}^\circ)^F$ to $\mathbf{G}^F$ of $R_{\mathbf{T}^\circ}^{\mathbf{G}^\circ}(\theta)$. With the corresponding definition of $\mathcal{E}(\mathbf{G}^F, 1)$ as above, we obtain a Jordan decomposition for groups with disconnected centre as well.

To distribute the characters into p-blocks, we first define the set $\mathcal{E}_p(\mathbf{G}^F, s)$, which is a union of Lusztig series. If s is a semisimple p'-element of $(\mathbf{G}^*)^{F^*}$, write

$$\mathcal{E}_p(\mathbf{G}^F, s) = \bigcup_{t \in (C_{\mathbf{G}^*}(s)^{F^*})_p} \mathcal{E}(\mathbf{G}^F, st),$$

the union of all Lusztig series $\mathcal{E}(\mathbf{G}^F, st)$, as t ranges over all p-elements of $(\mathbf{G}^*)^{F^*}$ that centralize s.

Theorem 9.4.6 (Broué–Michel [72]) *Let* $\mathbf{G}$ *be connected and reductive. For any semisimple* p'*-element* s *of* $(\mathbf{G}^*)^{F^*}$*, the set* $\mathcal{E}_p(\mathbf{G}^F, s)$ *is a union of* p*-blocks of* $\mathbf{G}^F$.

(See also [88, Theorem 9.12].)[5] This means that to partition $\mathrm{Irr}(G)$ into blocks it suffices to focus on $\mathcal{E}_p(\mathbf{G}^F, s)$ for each p'-element s, so we may label each p-block of G with a semisimple element s. Hiss showed [287] in addition that every Brauer character in a block contained in $\mathcal{E}_p(\mathbf{G}^F, s)$ is already a constituent of a character in $\mathcal{E}(\mathbf{G}^F, s)$.

Notice that, using the Jordan decomposition, if $Z(\mathbf{G})$ is connected then we can see a bijection between $\mathcal{E}_p(\mathbf{G}^F, s)$ and $\mathcal{E}_p(C_{\mathbf{G}^*}(s)^{F^*}, 1)$. Perhaps this yields a relationship between the p-blocks of $\mathbf{G}^F$ lying in $\mathcal{E}_p(\mathbf{G}^F, s)$ and the unipotent p-blocks of $C_{\mathbf{G}^*}(s)$? For most of these blocks, the result has been shown by Bonnafé and Rouquier.

[5] Particularly if you cannot speak French.

Theorem 9.4.7 (Bonnafé–Rouquier [44]) *Let* $\mathbf{G}$ *be a connected, reductive algebraic group with Frobenius endomorphism* F. *Let* s *be a semisimple* p'*-element of* $(\mathbf{G}^*)^{F^*}$, *and suppose that* $C_{\mathbf{G}^*}(s)$ *is contained in an* F*-stable Levi subgroup* $\mathbf{L}^*$ *of* $\mathbf{G}^*$. *The* p*-blocks of* G *in the Lusztig series labelled by* s *are in bijection via Jordan correspondence with the* p*-blocks of* $L = \mathbf{L}^F$ *in the Lusztig series labelled by* s, *and the corresponding blocks are Morita equivalent.*

(See also the write-up in [88].)[6] A relationship between unipotent blocks of centralizers and blocks of the whole group had already been noticed, for example by Fong and Srinivasan in classical groups [228, 230], but the Morita equivalence is incredibly useful in many contexts.

The Jordan decomposition yields a bijection between certain p-blocks of G and certain p-blocks of Levi subgroups L, and induces a Morita equivalence. This leads one to define 'quasi-isolated' semisimple elements as those whose centralizer is *not* contained in a proper Levi subgroup. If even the connected component of the identity of the centralizer is not in a proper Levi subgroup, then we drop the 'quasi-'.

Definition 9.4.8 A semisimple element s in $\mathbf{G}$ is *isolated* if the connected centralizer $C^\circ_{\mathbf{G}}(s)$ is not contained in a proper Levi subgroup of $\mathbf{G}$. The element s is called *quasi-isolated* if the full centralizer $C_{\mathbf{G}}(s)$ is not contained in a Levi subgroup.

An *isolated block* or *quasi-isolated block* is a block whose semisimple label is an isolated element or quasi-isolated element respectively.

Of course, the identity element of $\mathbf{G}$ is isolated, but there are others, for example the two classes of involutions in $E_8(q)$ for q odd. An easier example is $G = \mathrm{PGL}_2(q)$ for q odd, where there are two classes of involutions; one inside $\mathrm{PSL}_2(q)$ and one outside. The only proper Levi subgroups are tori, which are abelian—even cyclic—so if $C_G(s)$ is non-abelian then s must be isolated. One of the classes has centralizer a dihedral group of order $2(q \pm 1)$ (depending on the congruence of q modulo 4) and the other contains a (non-abelian) Sylow 2-subgroup of G. (If $q = 3$ then the one centralizer is Klein four, which is abelian but non-cyclic, so the proof still works.) Thus both classes of involutions are isolated, and even quasi-isolated.

Isolated blocks form a larger class of blocks than just unipotent blocks, and quasi-isolated even more so. Of course, if $\mathbf{G}$ has connected centre then $C_{\mathbf{G}^*}(s)$ is always connected by Theorem 9.4.4, so quasi-isolated and isolated blocks are the same thing.

The Bonnafé–Rouquier theorem gives us a lot of information, but the original result did not prove some important facts, such as that Morita equivalent blocks share a defect group, and also the set of quasi-isolated blocks is somewhat larger than we might like.

The parametrization and distribution of characters into quasi-isolated blocks is beyond the scope of this survey, but we can direct the reader to the relevant papers. The 1999 paper of Cabanes and Enguehard [87] gave a description of all p-blocks

[6] See previous footnote.

whenever $p \geq 5$ is good, so we are only left with bad primes, and $p = 3$ for classical groups. For unipotent blocks, as we mentioned in the previous section, Enguehard determined the blocks for exceptional groups at bad primes in [187], and for symplectic and orthogonal groups there is only one unipotent 2-block. The Bonnafé–Rouquier theorem [44] reduces us to the case of quasi-isolated blocks for bad primes. Enguehard [188] gave the parametrization for quasi-isolated blocks of symplectic and orthogonal groups for $p = 3$ (there are no non-unipotent, quasi-isolated blocks for $p = 2$), although note that one particularly thorny case of certain blocks of $\mathrm{SL}_n(q)$ with defect group a product of groups of order 27 was not accomplished until [349]. The final case is quasi-isolated blocks for exceptional groups at bad primes, which was mostly done in the work of Kessar and Malle in 2013 [349] up to some eigenvalues of Frobenius issues, and a full determination of the p-blocks is currently being investigated by those authors. A uniform labelling of the p-blocks of finite quasisimple groups of Lie type was given by Kessar and Malle [350], although the description is too technical to be given here.

A decade after [44], Bonnafé and Rouquier wrote a sequel paper with Dat [42], which proved that the Morita equivalences in Theorem 9.4.7 preserved defect groups, and also extended the results to include quasi-isolated blocks. A straight Morita equivalence between a quasi-isolated (but not isolated) block of G and a block of a proper Levi subgroup L is not usually possible, because there are cases where they aren't actually Morita equivalent. However, in [42] it is shown that there is a subgroup N of $N_G(L)$ such that the quasi-isolated blocks of G (where the connected centralizer of the semisimple label is contained in L) are Morita equivalent to blocks of N.

However, there is a gap in the proof of the main result of [42], which occurs when $\mathbf{G}$ is simple of type D and the centre $Z(\mathbf{G})$ is non-cyclic. Spotted by Navarro, Ruhstorfer and Späth, this was mostly fixed by Ruhstorfer in [497], but primes p dividing $q^2 - 1$ were omitted. Thus we obtain the following theorem.

Theorem 9.4.9 (Bonnafé–Dat–Rouquier, Ruhstorfer) *Let* $\mathbf{G}$ *be a simple algebraic group. Suppose that* $Z(\mathbf{G})^F$ *is cyclic, or* p *does not divide* $q^2 - 1$ *(necessarily* $\mathbf{G}$ *is of type* D *if* $Z(\mathbf{G})^F$ *is non-cyclic). If* B *is a quasi-isolated block of* $G = \mathbf{G}^F$ *with (quasi-isolated) semisimple label* s*, then there exists a proper Levi subgroup* $L = \mathbf{L}^F$*, whose dual in* $\mathbf{G}^*$ *contains* $C^\circ_{\mathbf{G}^*}(s)$*, a subgroup* N *with* $L \leq N \leq N_G(L)$*, and a block* B' *of* N*, such that* B *and* B' *are Morita equivalent. Furthermore,* B *and* B' *have isomorphic defect groups.*

More or less, we have a fairly good inductive understanding of non-isolated blocks. Isolated p-blocks come from centralizers $C_{\mathbf{G}^*}(s)$ that are maximal-rank subgroups. In general, we know of no good techniques for proving a Morita equivalence between an isolated block and a unipotent block of the centralizer. Some work of Kessar and myself has proved a Morita equivalence for some isolated blocks [133], including almost all blocks with cyclic defect, but a full extension of a Bonnafé–Rouquier-type theorem to all p-blocks, relating them to unipotent blocks of subgroups, is still—at least for now—out of reach.

Exercises

Exercise 9.1 Let k be algebraically closed of characteristic p, and let $\mathbf{G}$ be a simple algebraic group over k. Prove that, if $\lambda = (p-1)\rho$ is the Steinberg weight, then $L(\lambda) = W(\lambda)$.

Exercise 9.2 Let $\mathbf{G}$ be a simple algebraic group with maximal torus $\mathbf{T}$ and root system Φ. Define $s_{\alpha,n}$ acting on the root lattice of Φ via

$$s_{\alpha,n} : v \mapsto v - ((v,\alpha) - n)\alpha^\vee.$$

Show that the group generated by all $s_{\alpha,p}$ for α a simple root is isomorphic to the affine Weyl group as defined in Sect. 9.1 (Hint: $s_{(\alpha,0)}s_{(\alpha,p)}$.)

Exercise 9.3 Use the linkage principle to show that $W(\lambda) = L(\lambda)$, where λ is the Steinberg weight, and that the Steinberg module lies in a block of defect zero.

Exercise 9.4 Let $\mathbf{G}$ be of type A_2, and suppose that $p > 3$. We will fill in the details of Example 9.1.6. For $a \in k^\times$, write

$$s_a = \begin{pmatrix} a & 0 & 0 \\ 0 & a^{-1} & 0 \\ 0 & 0 & 1 \end{pmatrix}, \quad t_a = \begin{pmatrix} 1 & 0 & 0 \\ 0 & a & 0 \\ 0 & 0 & a^{-1} \end{pmatrix}.$$

Let M denote the natural 3-dimensional module for $\mathbf{G}$. Note that $M \otimes M^*$ is the sum of the Lie algebra and a copy of the trivial module. We choose the three positive roots to be

$$s_a \mapsto a^2,\ t_a \mapsto a^{-1},$$
$$s_a \mapsto a,\ t_a \mapsto a,$$
$$s_a \mapsto a^{-1},\ t_a \mapsto a^2.$$

(i) Show that the weight $s_a \mapsto a$, $t_a \mapsto 1$ is dominant, and therefore is the highest weight of M.
(ii) Calculate the images of the dominant weights from Example 9.1.6 under the Weyl group, and then translate by five times an element of the root lattice to show the remaining statements about dominant weights lying in the same orbit under the dot action.

Exercise 9.5 In [570, (2D)] it is shown that the restriction of $W(\lambda)$ to G has a unique maximal submodule whose quotient is $L(\lambda)$. Prove that this implies that the map

$$\mathrm{Ext}^1_{\mathbf{G}}(L(\lambda), L(\mu)) \to \mathrm{Ext}^1_{kG}(L(\lambda), L(\mu))$$

is injective.

Exercise 9.6 This exercise is to prove that if one knows the Mackey formula holds (see Definition 9.2.9) for a given set of parabolics, then Lusztig induction does not depend on the choice of parabolic subgroup from that set.

Let $\mathscr{L}$ be a collection of reductive algebraic groups, closed under taking Levi subgroups. Suppose that the Mackey formula holds for all reductive algebraic groups $\mathbf{G}$ and all Levi subgroups $\mathbf{L}$ and $\mathbf{M}$, where $\mathbf{L}$ and $\mathbf{M}$ lie in $\mathscr{L}$. Prove that if $\mathbf{L}$ is a Levi subgroup of $\mathbf{G}$ in $\mathscr{L}$, and $\mathbf{L}$ is contained in parabolic subgroups $\mathbf{P}$ and $\mathbf{Q}$, then $R^{\mathbf{G}}_{\mathbf{L}\subset\mathbf{P}} = R^{\mathbf{G}}_{\mathbf{L}\subset\mathbf{Q}}$.

Exercise 9.7 Prove that if χ lies in R_L^G and R_M^G for two split Levi subgroups L and M of G, and L and M are minimal with this property, then M and L are conjugate in G.

Exercise 9.8 This exercise develops an alternative definition of the Deligne–Lusztig variety $\mathrm{X}(w)$, to see that it coincides with that given for arbitrary Levi subgroups. In order to do this you will need some basic knowledge of the properties of algebraic groups, as found for example in [417].

(i) Show that the set $\mathcal{B}$ of all Borel subgroups of $\mathbf{G}$ is naturally isomorphic as a $\mathbf{G}$-set to $\mathbf{G}/\mathbf{B}$, via $\mathbf{B}^g \mapsto \mathbf{B}g$.

(ii) Prove that the map from the Weyl group $W \cong N_{\mathbf{G}}(\mathbf{T})/\mathbf{T}$ to $\mathbf{G}$-orbits on $\mathcal{B}\times\mathcal{B}$ taking w to the orbit of w is a bijection.

(iii) Fix a Borel subgroup $\mathbf{B}$. Show that the orbit of w on $\mathcal{B}\times\mathcal{B}$ is isomorphic to

$$\{(\mathbf{B}g_1, \mathbf{B}g_2) \mid g_1 g_2^{-1} \in \mathbf{B}w\mathbf{B}\}.$$

If $(\mathbf{B}g_1, \mathbf{B}g_2)$ lies in the orbit of w, write $\mathbf{B}_1 \xrightarrow{w} \mathbf{B}_2$, where $\mathbf{B}_i = \mathbf{B}^{g_i}$. We say that $\mathbf{B}_1$ is in *relative position* w to $\mathbf{B}_2$.

Show that by transitivity of being in relative position, we may imbue the set of $\mathbf{G}$-orbits on $\mathcal{B}\times\mathcal{B}$ with a multiplication that turns it into a group isomorphic to W.

(iv) Recall that $\mathrm{X}(w)$ is defined by

$$\mathrm{X}(w) = \{\mathbf{B}g \in \mathbf{G}/\mathbf{B} \mid (gF)g^{-1} \in \mathbf{B}w\mathbf{B}\}.$$

Show that this is isomorphic to

$$\{\mathbf{B} \in \mathcal{B} \mid \mathbf{B} \xrightarrow{w} \mathbf{B}\cdot F\}$$

as a $\mathbf{G}^F$-variety, where $\mathbf{B}\cdot F$ is the image of $\mathbf{B}$ under F.

(v) Finally, show that this definition of $\mathrm{X}(w)$ coincides with that of the inverse image of the Lang map for the appropriate unipotent radical corresponding to the F-stable maximal torus $\mathbf{T}$.

Exercise 9.9 Give an alternative formula for the unipotent character degrees for $\mathrm{GL}_n(q)$ that looks more like those for the other classical groups, by giving an analogue of Frobenius's formula for symmetric group character degrees (Exercise 8.1).

Exercise 9.10 Let Λ_1 and Λ_2 be equivalent symbols. Show that the ranks, defects and associated unipotent degrees of Λ_1 and Λ_2 are the same.

Exercise 9.11 Prove that the notion of an e-cocore of a symbol is well defined. (Hint: interleave the two abacuses of the two β-sets X and Y in $\Lambda = (X, Y)$.)

References

1. Jonathan Alperin, *Sylow intersections and fusion*, J. Algebra **6** (1967), 222–241.
2. ———, *The main problem in block theory*, Proc. of the Conference on Finite Groups (Univ. of Utah, Park City, Utah, 1975) (1976), 341–356.
3. ———, *Projective modules for* SL(2, 2^n), J. Pure Appl. Algebra **15** (1979), 219–234.
4. ———, *Local representation theory*, The Santa Cruz Conference on Finite Groups (Univ. California, Santa Cruz, Calif., 1979) (Providence, RI), Proc. Sympos. Pure Math., vol. 37, American Mathematical Society, 1980, pp. 369–375.
5. ———, *The Green correspondence and normal subgroups*, J. Algebra **104** (1986), 74–77.
6. ———, *Local representation theory*, Cambridge Studies in Advanced Mathematics, vol. 11, Cambridge University Press, Cambridge, 1986.
7. ———, *Weights for finite groups*, The Arcata Conference on Representations of Finite Groups (Arcata, Calif., 1986), Proc. Sympos. Pure Math., vol. 47, American Mathematical Society, 1987, pp. 369–379.
8. ———, *A construction of endo-permutation modules*, J. Group Theory **4** (2001), 3–10.
9. Jonathan Alperin, Richard Brauer, and Daniel Gorenstein, *Finite groups with quasi-dihedral and wreathed Sylow* 2*-subgroups*, Trans. Amer. Math. Soc. **151** (1970), 1–261.
10. Jonathan Alperin and Leonard Evens, *Representations, resolutions and Quillen's dimension theorem*, J. Pure Appl. Algebra **22** (1981), 1–9.
11. Jianbei An and Charles Eaton, *Nilpotent blocks of quasisimple groups for odd primes*, J. reine angew. Math. **656** (2011), 131–177.
12. ———, *Nilpotent blocks of quasisimple groups for the prime two*, Algebra Represent. Theory **16** (2013), 1–28.
13. Jianbei An and Gerhard Hiss, *Restricting the Steinberg character in the finite symplectic groups*, J. Group Theory **9** (2006), 251–264.
14. Henning Andersen, *Extensions of simple modules for finite Chevalley groups*, J. Algebra **111** (1987), 388–403.
15. Henning Andersen, Jens Jantzen, and Wolfgang Soergel, *Representations of quantum groups at a p-th root of unity and of semisimple groups in characteristic p: independence of p*, Astérisque **220** (1994), 320pp.
16. Henning Andersen, Jens Jørgensen, and Peter Landrock, *The projective indecomposable modules of* SL(2, p^n), Proc. London Math. Soc. (3) **46** (1983), 38–52.
17. Jaclyn Anderson, *An asymptotic formula for the t-core partition function and a conjecture of Stanton*, J. Number Theory **128** (2008), 2591–2615.
18. George Andrews, Christine Bessenrodt, and Jørn Olsson, *Partition identities and labels for some modular characters*, Trans. Amer. Math. Soc. **344** (1994), 597–615.

D. A. Craven, *Representation Theory of Finite Groups: a Guidebook*, Universitext,
https://doi.org/10.1007/978-3-030-21792-1

19. Louise Archer, *On certain quotients of the Green rings of dihedral* 2*-groups*, J. Pure Appl. Algebra **212** (2008), 1888–1897.
20. Susumu Ariki, *On the decomposition numbers of the Hecke algebra of* $G(m, 1, n)$, J. Math. Kyoto Univ. **36** (1996), 789–808.
21. Michael Aschbacher, Radha Kessar, and Bob Oliver, *Fusion systems in algebra and topology*, London Mathematical Society Lecture Note Series, vol. 391, Cambridge University Press, Cambridge, 2011.
22. Michael Aschbacher and Leonard Scott, *Maximal subgroups of finite groups*, J. Algebra **92** (1985), 44–80.
23. Maurice Auslander and Idun Reiten, *Stable equivalence of dualizing* R*-varieties. V. Artin algebras stably equivalent to hereditary algebras*, Adv. Math. **17** (1975), 167–195.
24. George Avrunin and Leonard Scott, *Quillen stratification for modules*, Invent. Math. **66** (1982), 277–286.
25. Paul Balmer, *Modular representations of finite groups with trivial restriction to Sylow subgroups*, J. Eur. Math. Soc. **15** (2013), 2061–2079.
26. Laurence Barker, *On* p*-soluble groups and the number of simple modules associated with a given Brauer pair*, Quart. J. Math. Oxford (2) **48** (1997), 133–160.
27. Raymundo Bautista, *On algebras of strongly unbounded representation type*, Comment. Math. Helv. **60** (1985), 392–399.
28. V.A. Bašev, *Representations of the group* $Z_2 \times Z_2$ *in a field of characteristic* 2, Dokl. Akad. Nauk. SSSR **141** (1961), 1015–1018, Russian.
29. Clark Benson, *The generic degrees of the irreducible characters of* E_8, Comm. Algebra **7** (1979), 1199–1209.
30. Clark Benson and Charles Curtis, *On the degrees and rationality of certain characters of finite Chevalley groups*, Trans. Amer. Math. Soc. **165** (1972), 251–273.
31. David Benson, *Brauer trees for* $12M_{22}$, J. Algebra **95** (1985), 398–408.
32. ______, *Some remarks on the decomposition numbers for the symmetric groups*, Arcata Conference on Representations of Finite Groups, Proceedings of Symposia in Pure Mathematics, vol. 47, Amer. Math. Soc., 1987, pp. 381–394.
33. ______, *Representations and cohomology, I. Basic representation theory of finite groups and associative algebras*, Cambridge Studies in Advanced Mathematics, vol. 30, Cambridge University Press, Cambridge, 1998.
34. ______, *Representations and cohomology, II. Cohomology of groups and modules*, Cambridge Studies in Advanced Mathematics, vol. 31, Cambridge University Press, Cambridge, 1998.
35. David Benson and Jon Carlson, *Nilpotent elements in the Green ring*, J. Algebra **104** (1986), 329–350.
36. David Benson and Radha Kessar, *Blocks inequivalent to their Frobenius twists*, J. Algebra **315** (2007), 58–599.
37. Thomas Berger and Reinhard Knörr, *On Brauer's height* 0 *conjecture*, Nagoya Math. J. **109** (1988), 109–116.
38. Christine Bessenrodt, Alun Morris, and Jørn Olsson, *Decomposition matrices for spin characters of symmetric groups at characteristic* 3, J. Algebra **164** (1994), 146–172.
39. Christine Bessenrodt and Jørn Olsson, *The* 2*-blocks of the covering groups of the symmetric groups*, Adv. Math. **129** (1997), 261–300.
40. Vitalij Bondarenko, *Representations of dihedral groups over a field of characteristic* 2, Mat. Sbornik **96** (1975), 63–74 (Russian).
41. Vitalij Bondarenko and Yuri Drozd, *The representation type of finite groups*, Zap. Naučn. Sem. Leningrad. Otdel. Mat. Inst. Steklov. (LOMI) **71** (1977), 24–41 (Russian).
42. Cedric Bonnafé, Jean-François Dat, and Raphaël Rouquier, *Derived categories and Deligne–Lusztig varieties II*, Ann. of Math. **185** (2017), 609–670.
43. Cedric Bonnafé and Jean Michel, *Computational proof of the Mackey formula for* $q > 2$, J. Algebra **327** (2011), 506–526.

44. Cedric Bonnafé and Raphaël Rouquier, *Catégories dérivées et variétés de Deligne–Lusztig*, Publ. Math. IHES **57** (2003), 1–57.
45. Serge Bouc, *The Dade group of a p-group*, Invent. Math. **164** (2006), 189–231.
46. Serge Bouc and Jacques Thévenaz, *The group of endo-permutation modules*, Invent. Math. **139** (2000), 275–349.
47. Richard Brauer, *Über die Darstellung von Gruppen in Galoisschen Feldern*, Actualités Sci. Indust. **195** (1935), 15 pages.
48. ———, *Investigations on group characters*, Ann. of Math. **42** (1941), 936–958.
49. ———, *On groups whose order contains a prime number to the first power I*, Amer. J. Math. **64** (1942), 401–420.
50. ———, *On the arithmetic in a group ring*, Proc. Nat. Acad. Sci. U.S.A. **30** (1944), 109–114.
51. ———, *On blocks of characters of groups of finite order, II*, Proc. Nat. Acad. Sci. U.S.A. **41** (1946), 11–19.
52. ———, *On a conjecture by Nakayama*, Trans. Royal Soc. Canada Sect. III **32** (1947), 215–219.
53. ———, *Number theoretical investigations on groups of finite order*, Proceedings of the International Symposium on Algebraic Number Theory (Tokyo and Nikko, 1955), Science Council of Japan, Tokyo, 1956, pp. 55–62.
54. ———, *Zur Darstellungstheorie der Gruppen endlicher Ordnung*, Math. Z. **63** (1956), 406–444.
55. ———, *Representations of finite groups*, Lectures on Modern Mathematics, Vol. I, Wiley, New York, 1963, pp. 133–175.
56. ———, *Some applications of the theory of blocks of characters of finite groups. I*, J. Algebra **1** (1964), 152–167.
57. ———, *Some applications of the theory of blocks of characters of finite groups III*, J. Algebra **3** (1966), 225–255.
58. ———, *On 2-blocks with dihedral defect groups*, Symposia Math. **13** (1974), 366–394.
59. Richard Brauer and Walter Feit, *On the number of irreducible characters of finite groups in a given block*, Proc. Nat. Acad. Sci. U.S.A. **45** (1959), 361–365.
60. Richard Brauer and Cecil Nesbitt, *On the modular characters of groups*, Ann. of Math. **42** (1941), 556–590.
61. Richard Brauer and Michio Suzuki, *On finite groups of even order whose 2-Sylow group is a quaternion group*, Proc. Nat. Acad. Sci. U.S.A. **45** (1959), 1757–1759.
62. John Bray and Robert Wilson, *Examples of 3-dimensional 1-cohomology for absolutely irreducible modules of finite simple groups*, J. Group Theory **11** (2008), 669–673.
63. Sheila Brenner, *Modular representations of p-groups*, J. Algebra **15** (1970), 89–102.
64. Thomas Breuer *et al*, *The Modular Atlas homepage*, http://www.math.rwth-aachen.de/~MOC/, Accessed: 2019-01-08.
65. Michel Brion, *Stable properties of plethysm: On two conjectures of Foulkes*, Manuscripta Math. **80** (1993), 347–371.
66. Michel Broué, *Les l-blocs des groups* $\mathrm{GL}(n,q)$ *et* $\mathrm{U}(n,q^2)$ *et leurs structures locales*, Astérisque **133–134** (1986), 159–188.
67. ———, *Blocs, isométries parfaites, catégories derivées*, C. R. Math. Acad. Sci. Paris **307** (1988), 13–18.
68. ———, *Isométries parfaites, types de blocs, catégories dérivées*, Astérisque **181–182** (1990), 61–92.
69. ———, *Equivalences of blocks of group algebras*, Finite dimensional algebras and related topics (V. Dlab and L. Scott, eds.), Kluwer, 1994, pp. 1–26.
70. Michel Broué and Gunter Malle, *Zyklotomische Heckealgebren*, Astérisque **212** (1993), 119–189.
71. Michel Broué, Gunter Malle, and Jean Michel, *Generic blocks of finite reductive groups*, Astérisque **212** (1993), 7–92.
72. Michel Broué and Jean Michel, *Blocs et séries de Lusztig dans un groupe réductif fini*, J. reine angew. Math. **395** (1989), 56–67.

73. Michel Broué and Lluís Puig, *A Frobenius theorem for blocks*, Invent. Math. **56** (1980), 117–128.
74. Olivier Brunat, *On the inductive McKay condition in the defining characteristic*, Math. Z. **263** (2009), 411–424.
75. ______, *On semisimple classes and semisimple characters in finite reductive groups*, Ann. Inst. Fourier, Grenoble **62** (2012), 1671–1716.
76. Olivier Brunat and Jean-Baptiste Gramain, *Perfect isometries and Murnaghan–Nakayama rules*, Trans. Amer. Math. Soc. **369** (2017), 7657–7718.
77. Olivier Brunat and Frank Himstedt, *On equivariant bijections relative to the defining characteristic*, J. Algebra **334** (2011), 150–174.
78. Jonathan Brundan and Alexander Kleshchev, *Hecke-Clifford superalgebras, crystals of type $A_{2\ell}^{(2)}$ and modular branching rules for $\hat{S}_n$*, Represent. Theory **5** (2001), 317–403.
79. ______, *Projective representations of symmetric groups via Sergeev duality*, Math. Z. **239** (2002), 27–68.
80. ______, *James' regularization theorem for double covers of symmetric groups*, J. Algebra **306** (2006), 128–137.
81. Jonathan Brundan and Jonathan Kujawa, *A new proof of the Mullineux conjecture*, J. Alg. Comb. **18** (2003), 13–39.
82. David Burry and Jon Carlson, *Restrictions of modules to local subgroups*, Proc. Amer. Math. Soc. **84** (1982), 181–184.
83. Marc Cabanes, *Extensions of p-groups and construction of characters*, Comm. Algebra **15** (1987), 1297–1311.
84. ______, *Local structure of the p-blocks of $\tilde{S}_n$*, Math. Z. **198** (1988), 519–543.
85. Marc Cabanes and Michel Enguehard, *Unipotent blocks of finite reductive groups of a given type*, Math. Z. **213** (1993), 479–490.
86. ______, *On unipotent blocks and their ordinary characters*, Invent. Math. **117** (1994), 149–164.
87. ______, *On blocks of finite reductive groups and twisted induction*, Adv. Math. **145** (1999), 188–229.
88. ______, *Representation theory of finite reductive groups*, New Mathematical Monographs, vol. 1, Cambridge University Press, Cambridge, 2004.
89. Marc Cabanes and Britta Späth, *Equivariance and extendibility in finite reductive groups with connected center*, Math. Z. **275** (2013), 689–713.
90. ______, *On the inductive Alperin–McKay condition for simple groups of type* A, J. Algebra **442** (2015), 104–123.
91. ______, *Equivariant character correspondences and inductive McKay condition for type A*, J. reine angew. Math. **728** (2017), 153–194.
92. ______, *Inductive McKay condition for finite simple groups of type C*, Represent. Theory **21** (2017), 61–81.
93. ______, *Descent equalities and the inductive McKay condition for types B and E*, preprint, 2019.
94. Jon Carlson, *The dimensions of periodic modules over modular group algebras*, Illinois J. Math. **23** (1979), 295–306.
95. ______, *The structure of periodic modules over modular group algebras*, J. Pure Appl. Algebra **22** (1981), 43–56.
96. ______, *The varieties and the cohomology ring of a module*, J. Algebra **85** (1983), 104–143.
97. Jon Carlson, David Hemmer, and Nadia Mazza, *The group of endotrivial modules for the symmetric and alternating groups*, Proc. Edin. Math. Soc. **53** (2010), 83–95.
98. Jon Carlson, Nadia Mazza, and Daniel Nakano, *Endotrivial modules for finite groups of Lie type*, J. reine angew. Math. **595** (2006), 93–119.
99. ______, *Endotrivial modules for the symmetric and alternating groups*, Proc. Edin. Math. Soc. **52** (2009), 45–66.

100. ______, *Endotrivial modules for the general linear group in a nondefining characteristic*, Math. Z. **278** (2014), 901–925.
101. ______, *Endotrivial modules for the finite groups of Lie type A in nondefining characteristic*, Math. Z. **282** (2016), 1–24.
102. Jon Carlson, Nadia Mazza, and Jacques Thévenaz, *Endotrivial modules over groups with quaternion or semi-dihedral Sylow* 2*-subgroup*, J. Eur. Math. Soc. **15** (2013), 157–177.
103. ______, *Torsion-free endotrivial modules*, J. Algebra **398** (2014), 413–433.
104. Jon Carlson and Jacques Thévenaz, *Torsion endo-trivial modules*, Algebr. Represent. Theory **3** (2000), 303–355.
105. ______, *The classification of endo-trivial modules*, Invent. Math. **158** (2004), 389–411.
106. ______, *The classification of torsion endo-trivial modules*, Ann. of Math. **162** (2005), 823–883.
107. Roger Carter, *Finite groups of Lie type*, John Wiley & Sons, New York, 1985.
108. Roger Carter and George Lusztig, *On the modular representations of the general linear and symmetric groups*, Math. Z. **136** (1974), 193–242.
109. Roger Carter and M.T.J. Payne, *On homomorphism between Weyl modules and Specht modules*, Math. Proc. Camb. Phil. Soc. **87** (1980), 419–425.
110. Reda Chaneb, *Basic sets for unipotent blocks of finite reductive groups in bad characteristic*, preprint, 2018.
111. Man-Wai Cheung, Christian Ikenmeyer, and Sevak Mkrtchyan, *Symmetrizing tableaux and the 5th case of the Foulkes conjecture*, J. Symbolic Comput. **80** (2017), 833–843.
112. Leo Chouinard, *Projectivity and relative projectivity over group rings*, J. Pure. Appl. Algebra **7** (1976), 287–302.
113. Joseph Chuang and Radha Kessar, *Symmetric groups, wreath products, Morita equivalences, and Broué's abelian defect group conjecture*, Bull. London Math. Soc. **34** (2002), 174–185.
114. Joseph Chuang, Hyohe Miyachi, and Kai Meng Tan, *Parallelotope tilings and q-decomposition numbers*, Adv. Math. **321** (2017), 80–159.
115. Joseph Chuang and Raphaël Rouquier, *Derived equivalences for symmetric groups and* $\mathfrak{sl}_2$*-categorification*, Ann. of Math. **167** (2008), 245–298.
116. ______, *Perverse equivalences*, preprint, 2017.
117. Gerald Cliff, *On centers of* 2*-blocks of Suzuki groups*, J. Algebra **226** (2000), 74–90.
118. Alfred Clifford, *Representations induced in an invariant subgroup*, Ann. of Math. **38** (1937), 533–550.
119. Edward Cline, Brian Parshall, Leonard Scott, and Wilberd van der Kallen, *Rational and generic cohomology*, Invent. Math. **39** (1977), 143–163.
120. Samuel Conlon, *Twisted group algebras and their representations*, J. Austral. Math. Soc. **4** (1964), 152–173.
121. ______, *Certain representation algebras*, J. Austral. Math. Soc. **5** (1965), 83–99.
122. John Conway, Robert Curtis, Simon Norton, Richard Parker, and Robert Wilson, *Atlas of finite groups*, Oxford University Press, Eynsham, 1985.
123. Gene Cooperman, Gerhard Hiss, Klaus Lux, and Jürgen Müller, *The Brauer tree of the principal* 19*-block of the sporadic simple Thompson group*, Experiment. Math. **1** (1997), 293–300.
124. David A. Craven, *The number of t-cores of size n*, preprint, 2006.
125. ______, *Simple modules for groups with abelian Sylow* 2*-subgroups are algebraic*, J. Algebra **321** (2009), 1473–1479.
126. ______, *Algebraic modules and the Auslander–Reiten quiver*, J. Pure Appl. Algebra **215** (2011), 221–231.
127. ______, *The theory of fusion systems: An algebraic approach*, Cambridge Studies in Advanced Mathematics, vol. 131, Cambridge University Press, Cambridge, 2011.
128. ______, *Perverse equivalences and Broué's conjecture II: The cyclic case*, preprint, 2012.
129. ______, *Relating simple modules in weight* 2 *blocks of symmetric groups*, Quart. J. Math. Oxford **63** (2013), 861–872.

130. ______, *Trivial-source endotrivial modules for sporadic and alternating groups*, preprint, 2018.
131. David A. Craven, Olivier Dudas, and Raphaël Rouquier, *Brauer trees of unipotent blocks*, J. Eur. Math. Soc., to appear.
132. David A. Craven, Charles Eaton, Radha Kessar, and Markus Linckelmann, *The structure of blocks with Klein four defect group*, Math. Z. **268** (2011), 441–476.
133. David A. Craven and Radha Kessar, *Brauer trees of isolated blocks of finite groups of Lie type*, in preparation.
134. David A. Craven and Raphaël Rouquier, *Perverse equivalences and Broué's conjecture*, Adv. Math. **248** (2013), 1–58.
135. William Crawley-Boevey, *On tame algebras and bocses*, Proc. London Math. Soc. (3) **56** (1988), 451–483.
136. ______, *Functorial filtrations III: semidihedral algebras*, J. London Math. Soc. (2) **40** (1989), 31–39.
137. Charles Curtis, *The Steinberg character of a finite group with a (B, N)-pair*, J. Algebra **4** (1966), 433–441.
138. Charles Curtis, Nagayoshi Iwahori, and Robert Kilmoyer, *Hecke algebras and characters of parabolic type of finite groups with (B, N)-pairs*, Publ. Math. IHES **43** (1974), 81–116.
139. Everett Dade, *Blocks with cyclic defect groups*, Ann. of Math. **84** (1966), 20–48.
140. ______, *Degrees of modular irreducible representations of p-solvable groups*, Math. Z. **104** (1968), 141–143.
141. ______, *Compounding Clifford's theory*, Ann. of Math. **91** (1970), 236–290.
142. ______, *Isomorphisms of Clifford extensions*, Ann. of Math. **92** (1970), 375–433.
143. ______, *Block extensions*, Illinois J. Math. **17** (1973), 198–272.
144. ______, *Endo-permutation modules over p-groups. I.*, Ann. of Math. **107** (1978), 459–494.
145. ______, *Endo-permutation modules over p-groups. II.*, Ann. of Math. **108** (1978), 317–346.
146. ______, *Extending irreducible modules*, J. Algebra **78** (1982), 357–371.
147. ______, *The Green correspondents of simple group modules*, J. Algebra **78** (1982), 357–371.
148. Stuart Dagger, *On the blocks of the Chevalley groups*, J. London Math. Soc. (2) **3** (1971), 21–29.
149. Susanne Danz, *On vertices of exterior powers of the natural simple module for the symmetric group in odd characteristic*, Arch. Math. (Basel) **89** (2007), 485–496.
150. ______, *Vertices of low-dimensional simple modules for symmetric groups*, Comm. Algebra **36** (2008), 4521–4539.
151. Susanne Danz and Eugenio Giannelli, *Vertices of simple modules of symmetric groups labelled by hook partitions*, J. Group Theory **18** (2015), 313–334.
152. Susanne Danz and Burkhard Külshammer, *Vertices of small order for simple modules of finite symmetric groups*, Algebra Colloq. **17** (2010), 75–86.
153. Susanne Danz, Burkhard Külshammer, and René Zimmermann, *On vertices of simple modules for symmetric groups of small degrees*, J. Algebra **320** (2008), 680–707.
154. Susanne Danz and Jürgen Müller, *Source algebras of blocks, sources of simple modules, and a conjecture of Feit*, J. Algebra **353** (2012), 187–211.
155. Luan Dehuai and Brian Wybourne, *The alternating group: branching rules, products and plethysms for spin representations*, J. Phys. A: Math. Gen. **14** (1981), 1835–1848.
156. ______, *The symmetric group: branching rules, products and plethysms for spin representations*, J. Phys. A: Math. Gen. **14** (1981), 327–348.
157. Pierre Deligne and George Lusztig, *Representations of reductive groups over finite fields*, Ann. of Math. **103** (1976), 103–161.
158. David Denoncin, *Inductive AM condition for the alternating groups in characteristic* 2, J. Algebra **404** (2014), 1–17.
159. Suzie Dent and Johannes Siemons, *On a conjecture of Foulkes*, J. Algebra **226** (2000), 236–249.
160. François Digne and Jean Michel, *On Lusztig's parametrization of characters of finite groups of Lie type*, Astérisque **181–182** (1990), 113–156.

161. ______, *Representations of finite groups of Lie type*, Cambridge University Press, Cambridge, 1991.
162. Richard Dipper, *On the decomposition matrices of the finite general linear groups I*, Trans. Amer. Math. Soc. **290** (1985), 315–344.
163. ______, *On the decomposition matrices of the finite general linear groups II*, Trans. Amer. Math. Soc. **292** (1985), 123–133.
164. Richard Dipper and Gordon James, *Representations of Hecke algebras of general linear groups*, Proc. London Math. Soc. (3) **52** (1986), 20–52.
165. ______, *The q-Schur algebra*, Proc. London Math. Soc. (3) **59** (1989), 23–50.
166. Craig Dodge and Matthew Fayers, *Some new decomposable Specht modules*, J. Algebra **357** (2012), 235–262.
167. Stephen Donkin, *The blocks of a semisimple algebraic group*, J. Algebra **67** (1980), 36–53.
168. ______, *A note on decomposition numbers for general linear and symmetric groups*, Math. Proc. Camb. Phil. Soc. **97** (1985), 57–62.
169. Stephen Donkin and Haralampos Geranios, *Decompositions of some Specht modules, Part I*, preprint, 2018.
170. Peter Donovan, *Dihedral defect groups*, J. Algebra **56** (1979), 184–206.
171. Yuri Drozd, *Tame and wild matrix problems*, Representations and quadratic forms, Akad. Nauk Ukrain. SSR, Inst. Mat., Kiev, 1979, pp. 39–74 (Russian).
172. Olivier Dudas, *Coxeter orbits and Brauer trees*, Adv. Math. **229** (2012), 3398–3435.
173. ______, *A note on the decomposition numbers for groups of Lie type in small rank*, J. Algebra **388** (2013), 364–373.
174. ______, *Coxeter orbits and Brauer trees II*, Int. Math. Res. Not. **15** (2014), 4100–4123.
175. Olivier Dudas and Gunter Malle, *Decomposition matrices for low-rank unitary groups*, Proc. Lond. Math. Soc. (3) **110** (2015), 1517–1557.
176. ______, *Decomposition matrices for exceptional groups at $d = 4$*, J. Pure Appl. Algebra **220** (2016), 1096–1121.
177. ______, *Modular irreducibility of cuspidal unipotent characters*, Invent. Math. **211** (2017), 579–589.
178. Olivier Dudas and Raphaël Rouquier, *Coxeter orbits and Brauer trees III*, J. Amer. Math. Soc. **27** (2014), 1117–1145.
179. Olivier Dudas, Michela Varagnolo, and Eric Vasserot, *Categorical actions on unipotent representations of finite classical groups*, Categorification and higher representation theory, Contemp. Math., vol. 683, Amer. Math. Soc., 2017, pp. 41–104.
180. Olaf Düvel, *On Donovan's conjecture*, J. Algebra **272** (2004), 1–26.
181. Charles Eaton, Florian Eisele, and Michael Livesey, *Donovan's conjecture, blocks with abelian defect groups and discrete valuation rings*, Math. Z., to appear, 2019.
182. Charles Eaton and Michael Livesey, *Donovan's conjecture and blocks with abelian defect groups*, Proc. Amer. Math. Soc. **147** (2019), 963–970.
183. Florian Eisele, *Group rings over the p-adic integers*, Dissertation, RWTH Aachen, 2012.
184. ______, *p-adic lifting problems and derived equivalences*, J. Algebra **356** (2012), 90–114.
185. ______, *Blocks with a generalized quaternion defect group and three simple modules over a 2-adic ring*, J. Algebra **456** (2016), 294–322.
186. ______, *The Picard group of an order and Külshammer reduction*, preprint, 2018.
187. Michel Enguehard, *Sur les l-blocs unipotents des groupes réductifs finis quand l est mauvais*, J. Algebra **230** (2000), 334–377.
188. ______, *Vers une décomposition de Jordan des blocs des groupes réductifs finis*, J. Algebra **319** (2008), 1035–1115.
189. Veikko Ennola, *On the characters of the finite unitary groups*, Ann. Acad. Sci. Fenn. Ser. A No. **323** (1963), 35.
190. Karin Erdmann, *Blocks and simple modules with cyclic vertices*, Bull. London Math. Soc. **9** (1977), 216–218.
191. ______, *Principal blocks of groups with dihedral Sylow 2-subgroups*, Comm. Algebra **5** (1977), 665–694.

192. ______, *Blocks whose defect groups are Klein four groups*, J. Algebra **59** (1979), 452–465.
193. ______, *On 2-blocks with semidihedral defect groups*, Trans. Amer. Math. Soc. **256** (1979), 267–287.
194. ______, *Blocks whose defect groups are Klein four groups: A correction*, J. Algebra **76** (1982), 505–518.
195. ______, *Algebras and dihedral defect groups*, Proc. London Math. Soc. **54** (1987), 88–114.
196. ______, *Algebras and quaternion defect groups I*, Math. Ann. **281** (1988), 545–560.
197. ______, *Algebras and quaternion defect groups II*, Math. Ann. **281** (1988), 561–582.
198. ______, *Algebras and semidihedral defect groups I*, Proc. London Math. Soc. **57** (1988), 109–150.
199. ______, *Algebras and semidihedral defect groups II*, Proc. London Math. Soc. **60** (1990), 123–165.
200. ______, *Blocks of tame representation type and related algebras*, Lecture Notes in Math., vol. 1428, Springer-Verlag, Berlin, 1990.
201. Karin Erdmann and Gerhard Michler, *Blocks with dihedral defect groups in solvable groups*, Math. Z. **154** (1977), 143–151.
202. Anton Evseev, *RoCK blocks, wreath products and KLR algebras*, Math. Ann. **369** (2017), 1383–1433.
203. Anton Evseev and Alexander Kleshchev, *Blocks of symmetric groups, semicuspidal KLR algebras and zigzag Schur–Weyl duality*, Ann. of Math. **188** (2018), 453–512.
204. Anton Evseev, Rowena Paget, and Mark Wildon, *Character deflations and a generalization of the Murnaghan–Nakayama rule*, J. Group Theory **17** (2014), 1035–1070.
205. Hanafi Farahat, Wolfgang Müller, and Michael Peel, *The modular characters of the symmetric groups*, J. Algebra **40** (1976), 354–363.
206. Niamh Farrell and Radha Kessar, *Rationality of blocks of quasi-simple finite groups*, preprint, 2018.
207. Matthew Fayers, *On the structure of Specht modules*, J. London Math. Soc. (2) **67** (2003), 85–102.
208. ______, *Reducible Specht modules*, J. Algebra **280** (2004), 500–504.
209. ______, *Irreducible Specht modules for Hecke algebras of type A*, Adv. Math. **193** (2005), 438–452.
210. ______, *Decomposition numbers for weight three blocks of symmetric groups and Iwahori–Hecke algebras*, J. Algebra **317** (2007), 593–633.
211. ______, *James's conjecture holds for weight four blocks of Iwahori–Hecke algebras*, J. Algebra **317** (2007), 593–633.
212. Matthew Fayers and Sinéad Lyle, *Row and column removal theorems for homomorphisms between Specht modules*, J. Pure Appl. Algebra **185** (2003), 147–164.
213. Matthew Fayers and Stuart Martin, *Homomorphisms between Specht modules*, Math. Z. **248** (2004), 395–421.
214. Matthew Fayers and Kai Meng Tan, *Adjustment matrices for weight three blocks of Iwahori–Hecke algebras*, J. Algebra **306** (2006), 76–103.
215. Walter Feit, *Irreducible modules of p-solvable groups*, The Santa Cruz Conference on Finite Groups (Univ. California, Santa Cruz, Calif., 1979), Proc. Sympos. Pure Math., vol. 37, American Mathematical Society, Providence, RI, 1980, pp. 405–412.
216. ______, *Some consequences of the classification of finite simple groups*, The Santa Cruz Conference on Finite Groups (Univ. California, Santa Cruz, Calif., 1979), Proc. Sympos. Pure Math., vol. 37, American Mathematical Society, Providence, RI, 1980, pp. 175–181.
217. ______, *The representation theory of finite groups*, North–Holland, Amsterdam–New York, 1982.
218. ______, *Possible Brauer trees*, Illinois J. Math. **28** (1984), 43–56.
219. Walter Feit and John Thompson, *Solvability of groups of odd order*, Pacific J. Math. **13** (1963), 775–1029.
220. Zhicheng Feng, Conghui Li, Yanjun Liu, Gunter Malle, and Jiping Zhang, *Robinson's conjecture on heights of characters*, Compos. Math. **155** (2019), 1098–1117.

221. Zhicheng Feng, Zhenye Li, and Jiping Zhang, *On the inductive blockwise Alperin weight condition for classical groups*, preprint, 2019.
222. Peter Fiebig, *An upper bound on the exceptional characteristics for Lusztig's character formula*, J. reine angew. Math. **673** (2012), 1–31.
223. Paul Fong, *Some properties of characters of finite solvable groups*, Bull. Amer. Math. Soc. **66** (1960), 116–117.
224. ______, *On the characters of p-solvable groups*, Trans. Amer. Math. Soc. **98** (1961), 263–284.
225. ______, *On the decomposition numbers of J_1 and $R(q)$*, Sympos. Math. Rome, vol. 13, Academic Press, 1974, pp. 415–422.
226. Paul Fong and Morton Harris, *On perfect isometries and isotypies in finite groups*, Invent. Math. **114** (1993), 139–191.
227. Paul Fong and Bhama Srinivasan, *Blocks with cyclic defect groups in* $\mathrm{GL}(n, q)$, Bull. Amer. Math. Soc. **3** (1980), 1041–1044.
228. ______, *The blocks of finite general linear and unitary groups*, Invent. Math. **69** (1982), 109–153.
229. ______, *Brauer trees in* $\mathrm{GL}(n, q)$, Math. Z. **187** (1984), 81–88.
230. ______, *The blocks of finite classical groups*, J. Reine Angew. Math. **396** (1989), 122–191.
231. ______, *Brauer trees in classical groups*, J. Algebra **131** (1990), 179–225.
232. Ben Ford, *Irreducible restrictions of representations of the symmetric groups*, Bull. London Math. Soc. **27** (1995), 453–459.
233. Ben Ford and Alexander Kleshchev, *A proof of the Mullineux conjecture*, Math. Z. **226** (1997), 267–308.
234. Herbert Foulkes, *Concomitants of the quintic and sextic up to degree four in the coefficients of the ground form*, J. London Math. Soc. **25** (1950), 205–209.
235. James Frame, Gilbert de Beauregard Robinson, and Robert Thrall, *The hook graphs of the symmetric group*, Canad. J. Math. **6** (1954), 316–324.
236. Hans Freudenthal, *Zur Berechnung der Charaktere der halbeinfachen Lieschen Gruppen. I*, Indag. Math. **16** (1954), 369–376.
237. Pierre Gabriel, *Indecomposable representations. - II*, Symposia Mathematica XI, Instituto Nazionale di Alta Matematica Roma, 1973, pp. 81–104.
238. Pierre Gabriel and Christine Riedtmann, *Group representations without groups*, Comment. Math. Helvetici **54** (1979), 240–287.
239. Patrick Gallagher, *Group characters and normal Hall subgroups*, Nagoya Math. J. **21** (1962), 223–230.
240. Meinolf Geck, *Generalized Gel'fand-Graev characters for Steinberg's triality groups and their applications*, Comm. Algebra **19** (1991), 3249–3269.
241. ______, *On the decomposition numbers of the finite unitary groups in non-defining characteristic*, Math. Z. **207** (1991), 83–89.
242. ______, *Brauer trees of Hecke algebras*, Comm. Algebra **20** (1992), 2937–2973.
243. ______, *Basic sets of Brauer characters of finite groups of Lie type II*, J. London Math. Soc. **47** (1993), 255–268.
244. ______, *Basic sets of Brauer characters of finite groups of Lie type, III*, Manuscripta Math. **85** (1994), 195–216.
245. Meinolf Geck and Gerhard Hiss, *Basic sets of Brauer characters of finite groups of Lie type*, J. Reine Angew. Math. **418** (1991), 173–188.
246. ______, *Modular representations of finite groups of Lie type in non-defining characteristic*, Finite reductive groups (Luminy, 1994), Progr. Math., vol. 141, Birkhäuser Boston, Boston, MA, 1997, pp. 195–249.
247. Meinolf Geck and Gunter Malle, *Cuspidal unipotent characters and cuspidal Brauer characters*, J. London Math. Soc. (2) **53** (1996), 63–78.
248. Eugenio Giannelli, *A lower bound on the vertices of Specht modules for symmetric groups*, Arch. Math. (Basel) **103** (2014), 1–9.
249. James Glaisher, *A theorem in partitions*, Messenger of Math. **12** (1883), 158–170.

250. David Gluck, *On the $k(GV)$ problem*, J. Algebra **89** (1984), 46–55.
251. David Gluck and Kay Magaard, *The extraspeical case of the $k(GV)$ problem*, Trans. Amer. Math. Soc. **354** (2002), 287–333.
252. ______, *The $k(GV)$ conjecture for modules in characteristic* 31, J. Algebra **250** (2002), 252–270.
253. David Gluck, Kay Magaard, Udo Riese, and Peter Schmid, *The solution of the $k(GV)$-problem*, J. Algebra **279** (2004), 694–719.
254. David Gluck and Thomas Wolf, *Brauer's height conjecture for p-solvable groups*, Trans. Amer. Math. Soc. **282** (1984), 137–152.
255. ______, *Defect groups and character heights in blocks of solvable groups II*, J. Algebra **87** (1984), 222–246.
256. Dominic Goodwin, *Regular orbits of linear groups with an application to the $k(GV)$-problem, 1*, J. Algebra **227** (2000), 395–432.
257. ______, *Regular orbits of linear groups with an application to the $k(GV)$-problem, 2*, J. Algebra **227** (2000), 433–473.
258. Daniel Gorenstein and John Walter, *The characterization of finite groups with dihedral Sylow 2-subgroups I,II,III*, J. Algebra **2** (1965), 85–151, 218–270, 334–393.
259. Roderick Gow and Wolfgang Willems, *Quadratic geometries, projective modules, and idempotents*, J. Algebra **160** (1993), 257–272.
260. Jean-Baptiste Gramain, *On a conjecture of G. Malle and G. Navarro on nilpotent blocks*, Electronic J. Comb. **18** (2011), P217.
261. Andrew Granville and Ken Ono, *Defect zero p-blocks for finite simple groups*, Trans. Amer. Math. Soc. **148** (1996), 331–347.
262. James Green, *The characters of the finite general linear groups*, Trans. Amer. Math. Soc. **80** (1955), 402–447.
263. ______, *On the indecomposable representations of a finite group*, Math. Z. **70** (1959), 430–445.
264. ______, *Blocks of modular representations*, Math. Z. **79** (1962), 100–115.
265. ______, *A transfer theorem for modular representations*, J. Algebra **1** (1964), 73–84.
266. ______, *Some remarks on defect groups*, Math. Z. **107** (1968), 133–150.
267. ______, *Walking around the Brauer tree*, J. Austral. Math. Soc. **17** (1974), 197–213.
268. Jesper Grodal, *Endotrivial modules for finite groups via homotopy theory*, preprint, 2016.
269. Jochen Gruber and Gerhard Hiss, *Decomposition numbers of finite classical groups for linear primes*, J. reine angew. Math. **485** (1997), 55–91.
270. Robert Guralnick, *The dimension of the first cohomology group*, Representation theory, II (Ottawa, Ont., 1984), Lecture Notes in Math., vol. 1178, Springer, Berlin, 1986, pp. 94–97.
271. Robert Guralnick and Corneliu Hoffman, *The first cohomology group and generation of simple groups*, Groups and geometries (Siena, 1996), Trends Math., Birkhäuser, Basel, 1998, pp. 81–89.
272. Robert Guralnick and Pham Huu Tiep, *First cohomology groups of Chevalley groups in cross characteristic*, Ann. of Math. **174** (2011), 543–559.
273. ______, *Sectional rank and cohomology*, J. Algebra, to appear, 2019.
274. Harish-Chandra, *Eisenstein series over finite fields*, Functional analysis and related fields (Proc. Conf. M. Stone, Univ. Chicago, Chicago, Ill., 1968), Springer-Verlag, 1970, pp. 76–88.
275. Morton Harris and Reinhard Knörr, *Brauer correspondence for covering blocks of finite groups*, Comm. Algebra **13** (1985), 1213–1218.
276. Morton Harris and Markus Linckelmann, *Splendid derived equivalences for blocks of finite p-solvable groups*, J. London Math. Soc. **62** (2000), 85–96.
277. Alex Heller, *The loop-space functor in homological algebra*, Proc. Amer. Math. Soc. **12** (1961), 640–643.
278. Alex Heller and Irving Reiner, *Indecomposable representations*, Illinois J. Math. **5** (1961), 314–323.
279. Hans-Werner Henn and Stewart Priddy, *p-nilpotent, classifying space indecomposability, and other properties of almost all finite groups*, Comment. Math. Helv. **69** (1994), 335–350.

280. Martin Hertweck, *A counterexample to the isomorphism problem for integral group rings*, Ann. of Math. **154** (2001), 115–138.
281. Donald Higman, *Modules with a group of operators*, Duke Math. J. **21** (1954), 369–376.
282. ———, *Relative cohomology*, Canad. J. Math. **9** (1957), 19–34.
283. Frank Himstedt, *On the 2-decomposition numbers of Steinberg's triality groups* ${}^3D_4(q)$, *q odd*, J. Algebra **309** (2007), 569–593.
284. ———, *On the decomposition numbers of the Ree groups* ${}^2F_4(q^2)$ *in non-defining characteristic*, J. Algebra **325** (2011), 365–403.
285. Frank Himstedt and Felix Noeske, *Decomposition numbers for* $\mathrm{SO}_7(q)$ *and* $\mathrm{Sp}_6(q)$, J. Algebra **413** (2014), 15–40.
286. Gerhard Hiss, *On the decomposition numbers of* $G_2(q)$, J. Algebra **120** (1989), 339–360.
287. ———, *Regular and semisimple blocks of finite reductive groups*, J. London Math. Soc. **41** (1990), 63–68.
288. ———, *The Brauer trees of the Ree groups*, Comm. Algebra **19** (1991), 871–888.
289. ———, *Morita equivalences between blocks of finite Chevalley groups*, Proceedings of 'Representation Theory of Finite and Algebraic Groups' (N. Kawanaka, G. Michler, and K. Uno, eds.), Osaka University, 2000, pp. 128–136.
290. Gerhard Hiss and Radha Kessar, *Scopes reduction and Morita equivalence classes of blocks in finite classical groups*, J. Algebra **230** (2000), 378–423.
291. ———, *Scopes reduction and Morita equivalence classes of blocks in finite classical groups II*, J. Algebra **283** (2005), 522–563.
292. Gerhard Hiss and Frank Lübeck, *The Brauer trees of the exceptional Chevalley groups of types* F_4 *and* ${}^2E_6$, Arch. Math. (Basel) **70** (1998), 16–21.
293. Gerhard Hiss, Frank Lübeck, and Gunter Malle, *The Brauer trees of the exceptional Chevalley groups of type* E_6, Manuscripta Math. **87** (1995), 131–144.
294. Gerhard Hiss and Klaus Lux, *Brauer trees of sporadic groups*, Clarendon Press, Oxford, 1989.
295. Gerhard Hiss and Josephine Shamash, *3-blocks and 3-modular characters of* $G_2(q)$, J. Algebra **131** (1990), 371–387.
296. ———, *2-blocks and 2-modular characters of the Chevalley groups* $G_2(q)$, Math. Comp. **59** (1992), 645–672.
297. Gerhard Hochschild, *Relative homological algebra*, Trans. Amer. Math. Soc. **82** (1956), 246–269.
298. Peter Hoffman and John Humphreys, *Projective representations of the symmetric group*, Oxford Mathematical Monographs, Clarendon Press, Oxford, 1992.
299. Miles Holloway, *Derived equivalences for group algebras*, Ph.D. thesis, University of Bristol, 2001.
300. Thorsten Holm, Radha Kessar, and Markus Linckelmann, *Blocks with a quaternion defect group over a 2-adic ring: the case* $\tilde{A}_4$, Glasgow Math. J. **49** (2007), 29–43.
301. Ryoshi Hotta and Tonny Springer, *A specialization theorem for certain Weyl group representations and an application to the Green polynomials of unitary groups*, Invent. Math. **41** (1977), 113–127.
302. Robert Howlett and Gus Lehrer, *Representations of generic algebras and finite groups of Lie type*, Trans. Amer. Math. Soc. **280** (1983), 753–779.
303. James Humphreys, *Defect groups for finite groups of Lie type*, Math. Z. **119** (1971), 149–152.
304. ———, *Modular representations of classical Lie algebras and semisimple groups*, J. Algebra **19** (1971), 51–79.
305. ———, *Introduction to Lie algebras and representation theory*, Graduate Texts in Mathematics, vol. 9, Springer-Verlag, New York, 1972.
306. ———, *Non-zero* Ext^1 *for Chevalley groups (via algebraic groups)*, J. London Math. Soc. (2) **31** (1985), 463–467.
307. ———, *Modular representations of finite groups of Lie type*, London Mathematical Society Lecture Note Series, vol. 326, Cambridge University Press, Cambridge, 1994.
308. John Humphreys, *The projective characters of the Mathieu group* M_{22}, J. Algebra **76** (1982), 1–24.

309. ———, *Blocks of projective representations of the symmetric groups*, J. London Math. Soc. (2) **33** (1986), 441–452.
310. Shin-ichi Kato, *On the Kazhdan–Lusztig polynomials for affine Weyl groups*, Adv. Math. **55** (1985), 103–130.
311. I. Martin Isaacs, *Invariant and extendible group characters*, Illinois J. Math. **14** (1970), 70–75.
312. ———, *Characters of solvable and symplectic groups*, Amer. J. Math. **95** (1973), 594–635.
313. ———, *Character theory of finite groups*, Dover, 1994.
314. I. Martin Isaacs, Gunter Malle, and Gabriel Navarro, *A reduction theorem for the McKay conjecture*, Invent. Math. **170** (2007), 33–101.
315. I. Martin Isaacs and Gabriel Navarro, *New refinements of the McKay conjecture for arbitrary finite groups*, Ann. Math. **156** (2002), 333–344.
316. Noboru Itô, *On the degrees of irreducible representations of a finite group*, Nagoya Math. J. **3** (1951), 5–6.
317. Gordon James, *The irreducible representations of the symmetric groups*, Bull. London Math. Soc. **8** (1976), 229–232.
318. ———, *On the decomposition matrices of the symmetric groups, I*, J. Algebra **43** (1976), 42–44.
319. ———, *On the decomposition matrices of the symmetric groups, II*, J. Algebra **43** (1976), 45–54.
320. ———, *Representations of the symmetric groups over the field of order* 2, J. Algebra **38** (1976), 280–308.
321. ———, *A characteristic-free approach to the representation theory of* $\mathfrak{S}_n$, J. Algebra **46** (1977), 430–450.
322. ———, *Some counterexamples in the theory of Specht modules*, J. Algebra **46** (1977), 457–461.
323. ———, *On a conjecture of Carter concerning irreducible Specht modules*, Math. Proc. Camb. Phil. Soc. **83** (1978), 11–17.
324. ———, *The representation theory of the symmetric groups*, Lecture Notes in Mathematics, vol. 682, Springer-Verlag, Berlin, 1978.
325. ———, *On the decomposition matrices of the symmetric groups, III*, J. Algebra **71** (1981), 115–122.
326. ———, *The decomposition matrices of* $\mathrm{GL}_n(q)$ *for* $n \leq 10$, Proc. London Math. Soc. (3) **60** (1990), 225–265.
327. Gordon James and Adalbert Kerber, *The representation theory of the symmetric group*, Encyclopedia of Mathematics and its Applications, Addison-Wesley Publishing Co., Reading, MA, 1981.
328. Gordon James and Martin Liebeck, *Representations and characters of groups*, Cambridge University Press, New York, 2001.
329. Gordon James and Andrew Mathas, *The irreducible Specht modules in characteristic* 2, Bull. London Math. Soc. **31** (1999), 457–462.
330. Gordon James and Gwendolen Murphy, *The determinant of the Gram matrix for a Specht module*, J. Algebra **59** (1979), 222–235.
331. Gordon James and Adrian Williams, *Decomposition numbers for symmetric groups by induction*, J. Algebra **228** (2000), 119–142.
332. Jens Jantzen, *Moduln mit einem höchsten gewicht*, Lecture Notes in Mathematics, vol. 750, Springer-Verlag, Berlin, 1979.
333. ———, *Representations of algebraic groups*, American Mathematical Society, Providence, RI, 2003.
334. Jens Jantzen and Gary Seitz, *On the representation theory of the symmetric groups*, Proc. London Math. Soc. **65** (1992), 475–504.
335. Gerald Janusz, *Indecomposable modules for finite groups*, Ann. of Math. **89** (1969), 209–241.
336. Herbert Jordan, *Group-characters of various types of linear groups*, Amer. J. Math. **29** (1907), 387–405.

337. Thomas Jost, *Morita equivalence for blocks of finite general linear groups*, Manuscripta Math. **91** (1996), 121–144.
338. Vicor Kac and Boris Weisfeiler, *Coadjoint action of a semi-simple algebraic group and the center of the enveloping algebra in characteristic p*, Indag. Math. **38** (1976), 136–151.
339. Gregory Karpilovsky, *Clifford theory for group representations*, Mathematics Studies, vol. 156, North-Holland, Amsterdam, 1989.
340. Masaki Kashiwara, *Crystal bases of modified quantized enveloping algebra*, Duke Math. J. **73** (1994), 383–413.
341. Noriaki Kawanaka, *Generalized Gelfand-Graev representations and Ennola duality*, Algebraic groups and related topics (Kyoto/Nagoya, 1983), Adv. Stud. Pure Math., vol. 6, North-Holland, Amsterdam, 1985, pp. 175–206.
342. Adalbert Kerber and Michael Peel, *On the decomposition numbers of symmetric and alternating groups*, Mitt. Math. Sem. Univ. Giessen **91** (1971), 45–81.
343. Radha Kessar, *Blocks and source algebras for the double covers of the symmetric and alternating groups*, J. Algebra **186** (1996), 872–933.
344. ———, *Source algebra equivalences for blocks of finite general linear groups over a fixed field*, Manuscripta Math. **104** (2001), 145–162.
345. ———, *Scopes reduction for blocks of finite alternating groups*, Quart. J. Math. Oxford **53** (2002), 442–454.
346. ———, *A remark on Donovan's conjecture*, Arch. Math. (Basel) **82** (2004), 391–394.
347. ———, *On duality inducing automorphisms and sources of simple modules in classical groups*, J. Group Theory **12** (2009), 331–349.
348. Radha Kessar and Markus Linckelmann, *On perfect isometries for tame blocks*, Bull. London Math. Soc. **34** (2002), 46–54.
349. Radha Kessar and Gunter Malle, *Quasi-isolated blocks and Brauer's height zero conjecture*, Ann. of Math. **178** (2013), 321–384.
350. ———, *Lusztig induction and ℓ-blocks of finite reductive groups*, Pac. J. Math. **279** (2015), 269–298.
351. ———, *Brauer's height zero conjecture for quasi-simple groups*, J. Algebra **475** (2017), 43–60.
352. Radha Kessar and Mary Schaps, *Crossover Morita equivalences for blocks of the covering groups of the symmetric and alternating groups*, J. Group Theory **9** (2006), 715–730.
353. Byungchan Kim and Jeremy Rouse, *Explicit bounds for the number of p-core partitions*, Trans. Amer. Math. Soc. **366** (2014), 875–902.
354. Ian Kiming, *A note on a theorem of A. Granville and K. Ono*, J. Number Theory **60** (1996), 97–102.
355. Alexander Kleshchev, *On restrictions of irreducible modular representations of semisimple algebraic groups and symmetric groups to some natural subgroups, I*, Proc. London Math. Soc. **69** (1994), 515–540.
356. ———, *Branching rules for modular representations of symmetric groups, i*, J. Algebra **178** (1995), 493–511.
357. ———, *Branching rules for modular representations of symmetric groups, ii*, J. reine Angew. Math. **459** (1995), 163–212.
358. ———, *Branching rules for modular representations of symmetric groups iii*, J. London Math. Soc. **54** (1996), 25–38.
359. ———, *On decomposition numbers and branching coefficients for symmetric and special linear groups*, Proc. London Math. Soc. **75** (1997), 497–558.
360. ———, *Branching rules for modular representations of symmetric groups, iv*, J. Algebra **201** (1998), 547–572.
361. ———, *Branching rules for symmetric groups and applications*, Algebraic Groups and Their Representations (Roger Carter and Jan Saxl, eds.), NATO ASI Series C, vol. 517, Kluwer Academic, Dordrecht-Boston-London, 1998, pp. 103–130.
362. Alexander Kleshchev and Vladimir Schchigolev, *Modular branching rules for projective representations of symmetric groups and lowering operators for the supergroup $Q(n)$*, Mem. Amer. Math. Soc. **220** (2012), no. 1034, xvii+123pp.

363. Reinhard Knörr, *Morita-injektive Moduln über kommutativen Dedekind-Ringen*, Dissertation, Eberhard-Karls-Universität Tübingen, 1973.
364. ———, *Blocks, vertices and normal subgroups*, Math. Z. **148** (1976), 53–60.
365. ———, *On the vertices of irreducible modules*, Ann. Math. **110** (1979), 487–499.
366. Reinhard Knörr and Geoffrey Robinson, *Some remarks on a conjecture of Alperin*, J. London Math. Soc. **39** (1989), 48–60.
367. Christoph Köhler, *Unipotente Charaktere und Zerlegungszahlen der endlichen Chevalleygruppen vom Typ F_4*, Dissertation, RWTH Aachen, 2006.
368. Christoph Köhler and Herbert Pahlings, *Regular orbits and the $k(GV)$-problem*, Groups and Computation III (Columbus, OH, 1999) (Berlin), de Gruyter, 2001, pp. 209–228.
369. Shigeo Koshitani and Naoko Kunugi, *Broué's conjecture holds for principal 3-blocks with elementary abelian defect group of order 9*, J. Algebra **248** (2002), 575–604.
370. Shigeo Koshitani, Naoko Kunugi, and Katsushi Waki, *Broué's conjecture for non-principal 3-blocks of finite groups*, J. Pure Appl. Algebra **173** (2002), 177–211.
371. Shigeo Koshitani and Caroline Lassueur, *Splendid Morita equivalences for principal 2-blocks with dihedral defect groups*, J. Algebra, to appear, 2019.
372. Shigeo Koshitani and Britta Späth, *The inductive Alperin-McKay and blockwise Alperin weight conditions for blocks with cyclic defect groups and odd primes*, J. Group Theory **19** (2016), 777–813.
373. ———, *The inductive Alperin-McKay condition for 2-blocks with cyclic defect groups*, Arch. Math. (Basel) **106** (2016), 107–116.
374. Bertram Kostant, *A formula for the multiplicity of a weight*, Trans. Amer. Math. Soc. **93** (1959), 53–73.
375. Stanislav Kruglyak, *Representations of the group (p, p) over a field of characteristic p*, Dokl. Akad. Nauk SSR **153** (1963), 1253–1256 (Russian).
376. Burkhard Külshammer, *Donovan's conjecture, crossed products and algebraic group actions*, Israel J. Math. **92** (1995), 295–306.
377. Burkhard Külshammer and Lluís Puig, *Extensions of nilpotent blocks*, Invent. Math. **102** (1990), 17–71.
378. Naoko Kunugi, *Morita equivalent 3-blocks of the 3-dimensional projective special linear groups*, Proc. London Math. Soc. **80** (2000), 575–589.
379. Herbert Kupisch, *Projektive Moduln endlicher Gruppen mit zyklischer p-Sylow-Gruppe*, J. Algebra **10** (1968), 1–7.
380. ———, *Unzerlegbare Moduln endlicher Gruppen mit zyklischer p-Sylow-Gruppe*, Math. Z. **108** (1969), 77–104.
381. Peter Landrock, *A counterexample to a conjecture on the Cartan invariants of a group algebra*, Bull. London Math. Soc. **5** (1973), 223–224.
382. ———, *The non-principal 2-blocks of sporadic simple groups*, Comm. Algebra **6** (1978), 1865–1891.
383. Peter Landrock and Gerhard Michler, *Principal 2-blocks of the simple groups of Ree type*, Trans. Amer. Math. Soc. **260** (1980), 83–111.
384. Serge Lang, *Algebraic groups over finite fields*, Amer. J. Math. **78** (1956), 555–563.
385. Michael Larsen, Gunter Malle, and Pham Huu Tiep, *The largest irreducible representations of simple groups*, Proc. Lond. Math. Soc. (3) **106** (2013), 65–96.
386. Alain Lascoux, Bernard Leclerc, and Jean-Yves Thibon, *Hecke algebras at roots of unity and crystal bases of quantum affine algebras*, Commun. Math. Phys. **181** (1996), 205–263.
387. Caroline Lassueur, Gunter Malle, and Elisabeth Schulte, *Simple endotrivial modules for quasi-simple groups*, J. reine angew. Math. **712** (2016), 141–174.
388. Walter Ledermann, *Introduction to group characters*, second ed., Cambridge University Press, Cambridge, 1987.
389. Conghui Li and Jiping Zhang, *The inductive blockwise Alperin weight condition for $\mathrm{PSL}_n(q)$ and $\mathrm{PSU}_n(q)$ with cyclic outer automorphism groups*, J. Algebra **495** (2018), 130–149.
390. Markus Linckelmann, *Derived equivalence for cyclic blocks over a P-adic ring*, Math. Z. **207** (1991), 293–304.

391. ______, *The isomorphism problem for cyclic blocks and their source algebras*, Invent. Math. **125** (1996), 265–283.
392. ______, *On derived equivalences and local structure of blocks of finite groups*, Turk. J. Math. **22** (1998), 93–107.
393. ______, *On splendid derived and stable equivalences between blocks of finite groups*, J. Algebra **242** (2001), 819–843.
394. ______, *The block theory of finite group algebras (2 vols)*, London Mathematical Society Student Texts, vol. 91–92, Cambridge University Press, Cambridge, 2018.
395. Markus Linckelmann and Lluís Puig, *Structure des p'-extensions des blocs nilpotents*, C. R. Acad. Sci. Paris Sér. I Math. **304** (1987), 181–184.
396. Dudley Littlewood and Archibald Richardson, *Group characters and algebra*, Philos. Trans. Roy. Soc. London, Ser. A **233** (1934), 99–142.
397. Michael Livesey, *Broué's perfect isometry conjecture holds for the double covers of the symmetric and alternating groups*, Algebra Represent. Theory **19** (2016), 783–826.
398. Joseph Loubert, *Homomorphisms from an arbitrary Specht module to one corresponding to a hook*, J. Algebra **485** (2017), 97–117.
399. Frank Lübeck, *Computation of Kazhdan–Lusztig polynomials and some applications to finite groups*, preprint, 2016.
400. Pablo Luka, *Small self-centralizing subgroups in defect groups of finite classical groups*, Dissertation, Technische Universität Kaiserslautern, 2017.
401. George Lusztig, *Irreducible representations of finite classical groups*, Invent. Math. **43** (1977), 125–175.
402. ______, *Unipotent representations of a finite Chevalley group of type E_8*, Quart. J. Math. Oxford **30** (1979), 315–338.
403. ______, *Some problems in the representation theory of finite Chevalley groups*, The Santa Cruz Conference on Finite Groups (Univ. California, Santa Cruz, Calif., 1979) (Providence, RI), Proc. Sympos. Pure Math., vol. 37, American Mathematical Society, 1980, pp. 313–317.
404. ______, *Characters of reductive groups over a finite field*, Annals of Mathematics Studies, vol. 107, Princeton University Press, Princeton, NJ, 1984.
405. ______, *On the representations of reductive groups with disconnected centre*, Astérisque **168** (1988), 157–166.
406. ______, *A unipotent support for irreducible representations*, Adv. Math. **94** (1992), 139–179.
407. Sinéad Lyle, *Some reducible Specht modules*, J. Algebra **269** (2003), 536–543.
408. Lukas Maas, *Modular spin characters of symmetric groups*, Dissertation, Essen, 2011.
409. Saunders Mac Lane, *Categories for the working mathematician*, Springer-Verlag, 1971.
410. George Mackey, *On induced representations of groups*, Amer. J. Math. **73** (1951), 576–592.
411. Gunter Malle, *Height* 0 *characters of finite groups of Lie type*, Represent. Theory **11** (2007), 192–220.
412. ______, *The inductive McKay condition for simple groups not of Lie type*, Comm. Algebra **36** (2008), 455–463.
413. ______, *On the inductive Alperin–McKay and Alperin weight conjecture for groups with abelian Sylow subgroups*, J. Algebra **397** (2014), 190–208.
414. ______, *On a minimal counterexample to Brauer's $k(B)$-conjecture*, Israel J. Math. **228** (2018), 527–556.
415. Gunter Malle and Gabriel Navarro, *Blocks with equal height zero degrees*, Trans. Amer. Math. Soc. **363** (2011), 6647–6669.
416. Gunter Malle and Britta Späth, *Characters of odd degree*, Ann. of Math. **184** (2016), 869–908.
417. Gunter Malle and Donna Testerman, *Linear algebraic groups and finite groups of Lie type*, Cambridge University Press, 2011.
418. Andrei Marcus, *On equivalences between blocks of group algebras: reduction to the simple components*, J. Algebra **184** (1996), 372–396.
419. ______, *Broué's abelian defect group conjecture for alternating groups*, Proc. Amer. Math. Soc. **132** (2003), 7–14.

420. Stuart Martin, *On the ordinary quiver of the principal block of certain symmetric groups*, Quart. J. Math. Oxford **40** (1989), 209–223.
421. ______, *Projective indecomposable modules for symmetric groups I*, Quart. J. Math. Oxford **44** (1993), 87–99.
422. Ursula Martin, *Almost all p-groups have automorphism group a p-group*, Bull. Amer. Math. Soc. **15** (1986), 78–82.
423. Johannes Maslowski, *Equivariant character bijections in groups of Lie type*, Dissertation, Technische Universität Kaiserslautern, 2010.
424. Andrew Mathas, *Iwahori–Hecke algebras and Schur algebras of the symmetric group*, University Lecture Series, vol. 15, American Mathematical Society, Providence, RI, 1999.
425. Nadia Mazza and Jacques Thévenaz, *Endotrivial modules in the cyclic case*, Arch. Math. (Basel) **89** (2007), 497–503.
426. John McKay, *A new invariant for simple groups*, Notices Amer. Math. Soc. **18** (1971), 397.
427. ______, *Irreducible representations of odd degree*, J. Algebra **20** (1972), 416–418.
428. Tom McKay, *On plethysm conjectures of Stanley and Foulkes*, J. Algebra **319** (2008), 2050–2071.
429. Gerhard Michler, *Brauer's conjectures and the classification of finite simple groups*, Representation Theory II: Groups and Orders. Proceedings, Ottawa, 1984 (Vlastimil Dlab, Pierre Gabriel, and Gerhard Michler, eds.), Springer-Verlag, 1986, pp. 129–142.
430. ______, *A finite simple group of Lie type has p-blocks with different defects, $p \neq 2$*, J. Algebra **104** (1986), 220–230.
431. Hyohe Miyachi, *Rouquier blocks in Chevally groups of type E*, Adv. Math. **217** (2008), 2841–2871.
432. Kiiti Morita, *Duality for modules and its applications to the theory of rings with minimum condition*, Sci. Rep. Tokyo Kyoiku Daigaku **6** (1958), 83–142.
433. Alun Morris, *The spin representation of the symmetric group*, Proc. London Math. Soc. (3) **12** (1962), 55–76.
434. ______, *The spin representation of the symmetric group*, Canad. J. Math. **17** (1965), 543–549.
435. Alun Morris and Abdul Yaseen, *Some combinatorial results involving shifted Young diagrams*, Math. Proc. Camb. Phil. Soc. **99** (1986), 23–31.
436. ______, *Decomposition matrices for spin characters of symmetric groups*, Proc. Royal Soc. Edin. **108A** (1988), 145–164.
437. Brian Mortimer, *The modular permutation representations of the known doubly transitive groups*, Proc. London Math. Soc. (3) **41** (1980), 1–20.
438. Jürgen Müller, *The 2-modular decomposition matrices of the symmetric groups S_{15}, S_{16}, and S_{17}*, Comm. Algebra **28** (2000), 4997–5005.
439. ______, *Brauer trees for the Schur cover of the symmetric group*, J. Algebra **266** (2003), 427–445.
440. Jürgen Müller, Max Neunhöffer, Frank Röhr, and Robert Wilson, *Completing the Brauer trees for the sporadic simple Lyons group*, LMS J. Comput. Math. **5** (2002), 18–33.
441. Jürgen Müller and René Zimmermann, *Green vertices and sources of simple modules of the symmetric group labelled by hook partitions*, Arch. Math. (Basel) **89** (2007), 97–108.
442. Glen Mullineux, *Bijections of p-regular partitions and p-modular irreducibles of the symmetric groups*, J. London Math. Soc. **20** (1979), 60–66.
443. ______, *A bijection on the set of 3-regular partitions*, J. Comb. Theory, Series A **28** (1980), 115–124.
444. Gwendolen Murphy, *On decomposability of some Specht modules for symmetric groups*, J. Algebra **66** (1980), 156–168.
445. Michael Naehrig, *Die Brauerbäume des Monsters M in Charakteristik* 29, Diplomarbeit, RWTH Aachen, Germany, 2002.
446. Tadasi Nakayama, *On some modular properties of irreducible representations of symmetric groups, II*, Jap. J. Math. **17** (1941), 411–423.
447. Gabriel Navarro, *Characters and blocks of finite groups*, London Mathematical Society Lecture Note Series, vol. 250, Cambridge University Press, Cambridge, 1998.

448. ———, *The McKay conjecture and Galois automorphisms*, Ann. Math. **160** (2004), 1129–1140.
449. Gabriel Navarro and Britta Späth, *On Brauer's height zero conjecture*, J. Eur. Math. Soc. **16** (2014), 695–747.
450. Gabriel Navarro, Britta Späth, and Carolina Vallejo, *A reduction theorem for the Galois-McKay conjecture*, preprint, 2019.
451. Gabriel Navarro and Pham Huu Tiep, *A reduction theorem for the Alperin weight conjecture*, Invent. Math. **184** (2011), 529–565.
452. ———, *Brauer's height zero conjecture for the 2-blocks of maximal defect*, J. reine angew. Math. **669** (2013), 225–247.
453. ———, *Characters of relative p'-degree with respect to a normal subgroup*, Ann. of Math. **178** (2013), 1135–1171.
454. Maxim Nazarov, *Young's orthogonal form of irreducible projective representations of the symmetric group*, J. London Math. Soc. (2) **42** (1990), 437–451.
455. Tetsuro Okuyama and Masayuri Wajima, *Irreducible characters of p-solvable groups*, Proc. Japan Acad. Ser A **55** (1979), 309–312.
456. ———, *Character correspondence and p-blocks of p-solvable groups*, Osaka J. Math. **17** (1980), 801–806.
457. Tetsuro Okuyama and Katsushi Waki, *Decomposition numbers of* $\mathrm{Sp}(4, q)$, J. Algebra **199** (1998), 544–555.
458. Jørn Olsson, *On 2-blocks with quaternion and quasidihedral defect groups*, J. Algebra **36** (1975), 212–241.
459. ———, *Frobenius symbols for partitions and degrees of spin characters*, Math. Scand. **61** (1987), 223–247.
460. ———, *On the p-blocks of symmetric and alternating groups and their covering groups*, J. Algebra **128** (1990), 188–213.
461. Markus Ottensmann, *Vervollständigung der Brauerbäume von* $3.ON$ *in Charakteristik* 11, 19 *und* 31 *mit Methoden der Kondensation*, Diplomarbeit, RWTH Aachen, Germany, 2000.
462. Rowena Paget and Mark Wildon, *Generalized Foulkes modules and maximal and minimal constituents of plethysms of Schur functions*, Proc. Lond. Math. Soc. **118** (2019), 1153–1187.
463. Alessandro Paolini, *On the decomposition numbers for* $\mathrm{SO}_8^+(2^f)$, J. Pure Appl. Algebra **222** (2018), 3982–4003.
464. Michael Peel, *Hook representations of the symmetric groups*, Glasgow Math. J. **12** (1971), 136–149.
465. ———, *Specht modules and symmetric groups*, J. Algebra **36** (1975), 88–97.
466. Alexander Premet, *Weights of infinitesimally irreducible representations of Chevalley groups over a field of prime characteristic*, Mat. Sb. **133** (1987), 167–183 (Russian), translated Math. USSR-Sb. 61 (1988), 167–183.
467. Lluís Puig, *Pointed groups and construction of characters*, Math. Z. **176** (1981), 265–292.
468. ———, *Nilpotent blocks and their source algebras*, Invent. Math. **93** (1988), 77–116.
469. ———, *Notes sur les algèbres de Dade*, unpublished manuscript, 1988.
470. ———, *Affirmative answer to a question of Feit*, J. Algebra **131** (1990), 513–526.
471. ———, *On Joanna Scopes' criterion of equivalence for blocks of symmetric groups*, Algebra Colloq. **1** (1994), 25–55.
472. Daniel Quillen, *The spectrum of an equivariant cohomology ring: I*, Ann. of Math. **94** (1971), 549–572.
473. ———, *The spectrum of an equivariant cohomology ring: II*, Ann. of Math. **94** (1971), 573–602.
474. William Reynolds, *Blocks and normal subgroups of finite groups*, Nagoya Math. J. **22** (1988), 15–32.
475. Matthew Richards, *Some decomposition numbers for Hecke algebras of general linear groups*, Math. Proc. Camb. Phil. Soc. **119** (1996), 383–402.
476. Jeremy Rickard, *Derived categories and stable equivalence*, J. Pure Appl. Alg. **61** (1989), 303–317.

477. ______, *Morita theory for derived categories*, J. London Math. Soc. **39** (1989), 436–456.
478. ______, *Splendid equivalences: derived categories and permutation modules*, Proc. London Math. Soc. (3) **72** (1996), 331–358.
479. ______, *Some recent developments in modular representation theory*, Algebras and Modules I (Reiten, Smalø, and Solberg, eds.), Canad. Math. Soc. Conference Proceedings, vol. 23, 1998, pp. 157–178.
480. ______, *Equivalences of derived categories for symmetric algebras*, J. Algebra **257** (2002), 460–481.
481. Udo Riese, *The quasisimple case of the $k(GV)$ conjecture*, J. Algebra **235** (2001), 45–65.
482. ______, *On the extraspecial case of the $k(GV)$-conjecture*, Arch. Math. (Basel) **78** (2002), 177–183.
483. Udo Riese and Peter Schmid, *Real vectors for linear groups and the $k(GV)$-problem*, J. Algebra **267** (2003), 725–755.
484. Claus Michael Ringel, *The indecomposable representations of the dihedral 2-groups*, Math. Ann. **214** (1975), 19–34.
485. Noelia Rizo, *p-blocks relative to a character of a normal subgroup*, J. Algebra **514** (2018), 254–272.
486. Geoffrey Robinson, *Local structure, vertices, and Alperin's Conjecture*, J. London Math. Soc. **72** (1996), 312–330.
487. ______, *Further reductions for the $k(GV)$-problem*, J. Algebra **195** (1997), 141–150.
488. ______, *A note on perfect isometries*, J. Algebra **226** (2000), 71–73.
489. Geoffrey Robinson and John Thompson, *On Brauer's $k(B)$ problem*, J. Algebra **184** (1996), 1143–1160.
490. Gilbert de Beauregard Robinson, *On a conjecture by Nakayama*, Trans. Royal Soc. Canada Sect. III **41** (1947), 20–25.
491. ______, *Representation theory of the symmetric group*, Edinburgh University Press, Edinburgh, 1961.
492. Klaus Roggenkamp and Leonard Scott, *Isomorphisms of p-adic group rings*, Ann. of Math. **126** (1987), 593–647.
493. Frank Röhr, *Die Brauer-Charaktere der sporadisch einfachen Rudvalis-Gruppe in den Charakteristiken* 13 *und* 29, Diplomarbeit, RWTH Aachen, Germany, 2000.
494. Andrei Roiter, *Unboundedness of the dimension of the indecomposable representations of an algebra which has infinitely many indecomposable representations*, Izv. Akad. Nauk SSSR **32** (1968), 1275–1282.
495. Bruce Rothschild, *Degrees of irreducible modular characters of blocks with cyclic defect groups*, Bull. Amer. Math. Soc. **73** (1967), 102–104.
496. Raphaël Rouquier, *Derived equivalences and finite dimensional algebras*, International Congress of Mathematicians. Vol. II, Eur. Math. Soc., Zürich, 2006, pp. 191–221.
497. Lucas Ruhstorfer, *On the Bonnafé–Dat–Rouquier Morita equivalence*, preprint, 2018.
498. Bruce Sagan, *The symmetric group*, second ed., Graduate Texts in Mathematics, vol. 203, Springer-Verlag, New York, 2001.
499. Adam Salminen, *On the sources of simple modules in nilpotent blocks*, J. Algebra **319** (2008), 4559–4574.
500. ______, *Endo-permutation modules arising from the action of a p-group on a defect zero block*, J. Group Theory **12** (2009), 201–207.
501. Benjamin Sambale, *Blocks of finite groups and their invariants*, Lecture Notes in Mathematics, vol. 2127, Springer-Verlag, Berlin, 2014.
502. ______, *On the Brauer-Feit bound for abelian defect groups*, Math. Z. **276** (2014), 785–797.
503. ______, *Cartan matrices and Brauer's $k(B)$-conjecture IV*, J. Math. Soc. Japan **69** (2017), 735–754.
504. ______, *On blocks with abelian defect groups of small rank*, Results Math. **71** (2017), 411–422.
505. Amanda Schaeffer Fry, *Odd-degree characters and self-normalizing Sylow subgroups: A reduction to simple groups*, Comm. Algebra **44** (2016), 1882–1904.

506. ______, *Galois automorphisms on Harish-Chandra series and Navarro's self-normalizing Sylow 2-subgroup conjecture*, Trans. Amer. Math. Soc. **372** (2019), 457–483.
507. Amanda Schaeffer Fry and Jay Taylor, *On self-normalising Sylow 2-subgroups in type A*, J. Lie Theory **28** (2018), 139–168.
508. Klaus-Dieter Schaper, *Charakterformeln für Weyl-Moduln und Specht-Moduln in Primcharakteristik*, Diplomarbeit, Universität Bonn, 1981.
509. Peter Schmid, *Clifford theory of simple modules*, J. Algebra **119** (1988), 185–212.
510. Gerhard Schneider, *Die 2-modularen Darstellungen dere Mathieu-Gruppe M_{12}*, Dissertation, Universität Duisberg-Essen, 1981.
511. Elisabeth Schulte, *The inductive blockwise Alperin weight condition for $G_2(q)$ and ${}^3D_4(q)$*, J. Algebra **466** (2016), 314–369.
512. Issai Schur, *Untersuchungen über die Darstellung der endlichen Gruppen durch gebrochene lineare Substitutionen*, J. Reine Angew. Math. **132** (1907), 85–137.
513. ______, *Über die Darstellung der symmetrischen und der alternierenden Gruppen durch gebrochene lineare Substitutionen*, J. Reine Angew. Math. **139** (1911), 155–250.
514. ______, *Zur additiven Zahlentheorie*, S.-B. Preuss. Akad. Wiss. Phys.-Math. Kl. (1926), 488–495, Reprinted in I. Schur, Gesammelte Abhandlungen, vol.3, Springer, Berlin, 1973, 43–50.
515. Marcel-Paul Schützenberger, *La correspondence de Robinson*, Combinatoire et Répresentation du Groupe Symétrique, Lecture notes in Math., vol. 579, Springer-Verlag, New York, 1977, pp. 59–135.
516. Joanna Scopes, *Cartan matrices and Morita equivalence for blocks of the symmetric groups*, J. Algebra **142** (1991), 441–455.
517. ______, *Symmetric group blocks of defect two*, Quart. J. Math. Oxford **46** (1995), 201–234.
518. Leonard Scott, *Modular permutation representations*, Trans. Amer. Math. Soc. **175** (1973), 101–121.
519. ______, *Some new examples in 1-cohomology*, J. Algebra **260** (2003), 416–425.
520. Alexander Sergeev, *The Howe duality and the projective representations of symmetric groups*, Represent. Theory **3** (1999), 416–434.
521. Josephine Shamash, *Blocks and Brauer trees in the groups $G_2(q)$ for primes dividing $q \pm 1$*, Comm. Algebra **17** (1989), 1901–1949.
522. ______, *Brauer trees for blocks of cyclic defect in the groups $G_2(q)$ for primes dividing $q^2 \pm q + 1$*, J. Algebra **123** (1989), 378–396.
523. ______, *Blocks and Brauer trees for the groups $G_2(2^k)$, $G_2(3^k)$*, Comm. Algebra **20** (1992), 1375–1387.
524. Peter Sin, *Extensions of simple modules for $\mathrm{SL}_3(2^n)$ and $\mathrm{SU}_3(2^n)$*, Proc. London Math. Soc. **65** (1992), 265–296.
525. Britta Späth, *Inductive McKay condition in defining characteristic*, Bull. London Math. Soc. **44** (2012), 426–438.
526. ______, *A reduction theorem for the Alperin–McKay conjecture*, J. reine angew. Math. **680** (2013), 153–189.
527. ______, *A reduction theorem for the blockwise Alperin weight conjecture*, J. Group Theory **16** (2013), 159–220.
528. Wilhelm Specht, *Die irreduziblen Darstellungen der symmetrischen Gruppe*, Math. Z. **39** (1935), 696–711.
529. Tonny Springer and Robert Steinberg, *Conjugacy classes*, Seminar on Algebraic Groups and Related Finite Groups, Lecture Notes in Mathematics, no. 131, Springer, 1970, pp. 167–266.
530. Radu Stancu, *Control of fusion in fusion systems*, J. Algebra Appl. **5** (2006), 817–837.
531. Dennis Stanton, *Open positivity conjectures for integer partitions*, Trends Math. **2** (1999), 19–25.
532. Robert Steinberg, *Representations of algebraic groups*, Nagoya Math. J. **22** (1963), 33–56.
533. ______, *Endomorphisms of linear algebraic groups*, Mem. Amer. Math. Soc. (1968), no. 80, 108pp.
534. Richard Swan, *The Grothendieck ring of a finite group*, Topology **2** (1963), 85–110.

535. Jay Taylor, *Generalized Gelfand–Graev representations in small characteristics*, Nagoya Math. J. **224** (2016), 93–167.
536. ______, *Action of automorphisms on irreducible characters of symplectic groups*, J. Algebra **505** (2018), 211–246.
537. ______, *On the Mackey formula for connected centre groups*, J. Group Theory **21** (2018), 439–448.
538. Jacques Thévenaz, *Extensions of group representations from a normal subgroup*, Comm. Algebra **11** (1983), 391–425.
539. ______, *G-algebras and modular representation theory*, Oxford Mathematical Monographs, Oxford, 1995.
540. Glânffrwd Thomas, *Baxter algebras and Schur funçtions*, Ph.D. thesis, University College of Wales, 1974.
541. ______, *On Schensted's construction and the multiplication of Schur-functions*, Adv. in Math. **30** (1978), 8–32.
542. John Thompson, *Defect groups are Sylow intersections*, Math. Z. **100** (1967), 146.
543. ______, *Vertices and sources*, J. Algebra **6** (1967), 1–6.
544. Robert Thrall, *On symmetrized Kronecker powers and the structure of the free Lie ring*, Amer. J. Math. **64** (1942), 371–388.
545. Kar To Law, *Results on the decomposition matrices for the symmetric groups*, Ph.D. thesis, University of Cambridge, 1983.
546. Shunsuke Tsuchioka and Masaki Watanabe, *Schur partition theorems via perfect crystal*, preprint, 2016.
547. Will Turner, *Rock blocks*, Mem. Amer. Math. Soc. **202** (2009), no. 947, vii+102pp.
548. Alexandre Turull, *Strengthening the McKay conjecture to include local fields and local Schur indices*, J. Algebra **319** (2008), 4853–4868.
549. Harold Ward, *The analysis of representations induced from a normal subgroup*, Michigan Math. J. **15** (1968), 417–428.
550. Charles Weibel, *An introduction to homological algebra*, Cambridge Studies in Advanced Mathematics, vol. 38, Cambridge University Press, Cambridge, 1994.
551. Hermann Weyl, *Theorie der Darstellung kontinuierlicher halb-einfacher Gruppen durch lineare Transformationen. I*, Math. Z. **23** (1925), 271–309.
552. ______, *Theorie der Darstellung kontinuierlicher halb-einfacher Gruppen durch lineare Transformationen. II*, Math. Z. **24** (1926), 328–376.
553. ______, *Theorie der Darstellung kontinuierlicher halb-einfacher Gruppen durch lineare Transformationen. III*, Math. Z. **24** (1926), 376–395.
554. Donald White, *The 2-decomposition numbers of* $\mathrm{Sp}(4, q)$, *q odd*, J. Algebra **131** (1990), 703–725.
555. ______, *Decomposition numbers of* $\mathrm{Sp}(4, q)$ *for primes dividing* $q \pm 1$, J. Algebra **132** (1990), 488–500.
556. ______, *Decomposition numbers of $Sp_4(2^a)$ in odd characteristics*, J. Algebra **177** (1995), 264–276.
557. ______, *Decomposition numbers of unipotent blocks of $Sp_6(2^a)$ in odd characteristics*, J. Algebra **227** (2000), 172–194.
558. Mark Wildon, *Two theorems on the vertices of Specht modules*, Arch. Math. (Basel) **81** (2003), 505–511.
559. ______, *Character values and decomposition matrices of symmetric groups*, J. Algebra **319** (2008), 3382–3397.
560. ______, *Vertices of Specht modules and blocks of the symmetric group*, J. Algebra **323** (2010), 2243–2256.
561. Wolfgang Willems, *Bermerkungen zur modularen Darstellungstheorie. 3. Induzierte und eingeschränkte Moduln*, Arch. Math. (Basel) **26** (1976), 497–503.
562. Adrian Williams, *Symmetric group decomposition numbers for some three-part partitions*, Comm. Algebra **34** (2006), 1599–1613.

563. Geordie Williamson, *Schubert calculus and torsion explosion*, J. Amer. Math. Soc. **30** (2017), 1023–1046.
564. Robert Wilson, *The Brauer tree for J_3 in characteristic* 17, J. Symbolic Comput. **15** (1993), 325–330.
565. ———, *The McKay conjecture is true for the sporadic simple groups*, J. Algebra **207** (1998), 294–305.
566. ———, *The finite simple groups*, Graduate Texts in Mathematics, vol. 251, Springer-Verlag, London, 2009.
567. Elmar Wings, *Über die nilpotenten Charaktere der Chevalley-Gruppen vom typ F_4 in guter Charakteristik*, Dissertation, RWTH Aachen, 1995.
568. Thomas Wolf, *Characters of p'-degree in solvable groups*, Pacific J. Math. **74** (1978), 267–271.
569. ———, *Defect groups and character heights in blocks of solvable groups*, J. Algebra **72** (1981), 1097–1111.
570. Warren Wong, *Irreducible modular representations of finite Chevllay groups*, J. Algebra **20** (1972), 355–367.
571. Maozhi Xu, *On Mullineux' conjecture in the representation theory of symmetric groups*, Comm. Algebra **25** (1997), 1797–1803.
572. Abdul Yaseen, *Modular spin representatins of the symmetric group*, Ph.D. thesis, University of Wales, Aberystwyth, 1987.
573. Samy Yehia, *Extensions of simple modules for the universal Chevalley groups and its parabolic subgroups*, Ph.D. thesis, University of Warwick, 1982.
574. Jiping Zhang, *Vertices of irreducible representations and a question of L. Puig*, Algebra Colloq. **1** (1994), 139–148.

Index of Names

D. A. Craven, *Representation Theory of Finite Groups: a Guidebook*, Universitext,
https://doi.org/10.1007/978-3-030-21792-1

Index

D. A. Craven, *Representation Theory of Finite Groups: a Guidebook*, Universitext,
https://doi.org/10.1007/978-3-030-21792-1

Printed in the United States
By Bookmasters